Membrane-Based Desalination

An Integrated Approach (MEDINA)

EU report Series

Membrane-Based Desalination

An Integrated Approach (MEDINA)

Editors
Prof. Enrico Drioli
Dr. Eng. Alessandra Criscuoli
Dr. Eng. Francesca Macedonio

IWA
Publishing
London • New York

Published by **IWA Publishing**
Alliance House
12 Caxton Street
London SW1H 0QS, UK
Telephone: +44 (0)20 7654 5500
Fax: +44 (0)20 7654 5555
Email: publications@iwap.co.uk
Web: www.iwapublishing.com

First published 2011

Typeset by MPS Limited, a Macmillan Company, Chennai, India
Printed by Lightning Source

British Library Cataloguing in Publication Data
A CIP catalogue record for this book is available from the British Library

Library of Congress Cataloging-in-Publication Data
A catalog record for this book is available from the Library of Congress

ISBN13: 9781843393214
ISBN: 1843393212

Contents

Acknowledgement ... xi
List of contributors ... xiii

Introduction ... 1

Chapter 1
Water quality assessment tools ... 3
1.1 Introduction ... 4
1.2 Water quality characterization tools ... 5
1.2.1 Liquid chromatography – organic carbon detection ... 5
1.2.2 Fluorescence excitation emission matrix ... 7
1.2.3 Enumeration, identification and diversity of marine microorganisms including bacteria and phytoplankton ... 10
1.3 Development of particulate fouling indices ... 14
1.3.1 SDI and MFI ... 14
1.3.2 Modified fouling index ultrafiltration at constant flux ... 16
1.3.3 Crossflow sampler modified fouling index ultrafiltration ... 21
1.4 Assessment of biofouling potential of salt and brackish water for RO desalination plants – BDOC/LC-OCD ... 23
1.4.1 Method ... 23
1.4.2 Sampling ... 25
1.4.3 Test procedure ... 25
1.4.4 Results ... 25
1.5 Determination of the concentrations of AOC and ATP in seawater ... 26
1.5.1 Introduction ... 26
1.5.2 Materials and methods ... 27
References ... 31

Chapter 2
Evaluation and comparison of seawater and brackish water pre-treatment ... 33
2.1 Removal of organic matter and microbial characterization in pretreatment MBR system for RO seawater desalination ... 34

2.1.1 Introduction ... 34
2.1.2 Results ... 35
2.2 Submerged ultrafiltration for reducing nom in SWRO desalination ... 39
2.2.1 Introduction ... 39
2.2.2 Experimental activity ... 39
2.2.3 Results and discussions ... 42
2.2.4 Conclusions ... 44
2.3 Comparison of granular media filtration and low-pressure membrane filtration for seawater pretreatment ... 44
2.3.1 Introduction ... 44
2.3.2 Pilot-scale experimental set up ... 44
2.3.3 Water quality parameters and fouling indices ... 46
2.3.4 Results and interpretation ... 46
2.3.5 Conclusions ... 50
2.4 Comparison of different pretreatment methods for RO desalination ... 51
2.4.1 Biofilter as pretreatment to membrane based desalination: evaluation in terms of fouling index ... 51
2.4.2 Fibre media filtration as a pretretment for seawater ... 53
2.4.3 Submerged microfiltration coupled with physico-chemical processes as pre-treatment to sea water desalination ... 56
2.5 Submerged hollow fiber system as pre-treatment for the sea water reverse osmosis: Effect of the operation conditions and the characterizations of the permeate quality ... 58
2.5.1 Introduction ... 58
2.5.2 Method and materials ... 58
2.5.3 Data analysis ... 59
2.5.4 Results and discussions ... 59
2.5.5 Conclusions ... 63
2.6 Ultrafiltration-based hybrid processes for pre-treatment to seawater reverse osmosis desalination ... 63
2.6.1 Introduction ... 63
References ... 69

Chapter 3
Development of tools for RO fouling characterization and understanding ... 71
3.1 Membrane autopsies ... 71
3.1.1 Introduction ... 71
3.1.2 Presentation of studied sites ... 72
3.1.3 Membrane sampling protocol and analytical tools ... 72
3.1.4 Results ... 73
3.2 Specific organic compounds analyses ... 78
3.2.1 ^{13}C-NMR ... 78
3.2.2 Flash pyrolysis – GC/MS analysis ... 79
3.2.3 Thermochemolysis TMAH analysis: fatty acids analysis ... 81
3.3 Quantitative biomass parameters ... 82
3.4 Molecular analysis ... 84
3.4.1 Adaptation of molecular tools to analyze microbial community structure: homogeneity and reproducibility ... 84
3.4.2 Bacterial diversity at different desalination plants ... 85
3.4.3 Evolution of bacterial communities in SWRO membranes from one full-scale desalination plant (Site D) according to module usage time ... 85
3.4.4 Conclusions ... 86
3.5 General conclusion on autopsy tools relevancy ... 87
3.5.1 Tools for microscopic observation ... 87
3.5.2 Inorganic matter characterization ... 88
3.5.3 Organic matter characterization ... 88
3.5.4 Microbial characterization ... 89
References ... 90

Chapter 4
Development of cleaning strategies for RO membranes ... 93
4.1 Cleaning of spiral-wound membranes ... 93
4.1.1 Introduction ... 93
4.1.2 Study objectives ... 94
4.1.3 Summary of results ... 94
4.1.4 General discussion ... 96
4.1.5 Conclusions and recommendations ... 97
4.2 Development of a laboratory method for testing membrane cleaning procedures ... 97
4.2.1 Introduction ... 97
4.2.2 Principle of the test ... 98
4.2.3 Production of biofilm samples ... 98
4.2.4 Biofilm samples and biomass concentrations ... 98
4.2.5 Cleaning test procedures ... 99
4.2.6 Cleaning efficiency for biofilms on polymer tubing ... 99
4.2.7 Validation tests ... 101
4.2.8 Discussion and conclusions ... 102
4.3 Effects of chemicals on membrane permeability and foulants in laboratory tests ... 102
4.3.1 Introduction ... 102
4.3.2 Methods and materials ... 103
4.3.3 Analysis of inorganic compounds ... 104
4.3.4 Results and discussion ... 106
4.3.5 Conclusions ... 110
4.4 Effects of cleaning on membrane performance in a pilot plant ... 111
4.4.1 Introduction ... 111
4.4.2 Pilot plant, membranes and test procedures ... 111
4.4.3 Results ... 113
4.4.4 Discussion and conclusions ... 115
References ... 115

Chapter 5
Process strategies for mitigation of impact of concentrates on the environment ... 117
5.1 Introduction ... 118
5.2 WT 5.1: Reduction of brine volume ... 118
5.2.1 BWRO concentrate disposal by deep well injection: design criteria for BWRO plants and field test results ... 118
5.2.2 Vacum membrane distillation ... 124
5.3 WT 5.2: Recovery of dissolved salts as crystalline product ... 132
5.3.1 Membrane crystallization ... 133
5.3.2 Wind-Aided Intensified eVaporation ... 140
5.4 WT 5.3: Economic evaluation ... 143
5.4.1 Description and results of the economical evaluation ... 143
5.5 Conclusions ... 148
References ... 148

Chapter 6
Innovative technologies to reduce energy consumption in seawater desalination facilities ... 151
6.1 Introduction ... 151
6.2 Preparation and characterization of membranes ... 152
6.3 Membrane distillation and solar energy ... 156
6.3.1 Development and validation of a meteorogical model ... 157
6.3.2 Experiments at lab-scale on configurations coupling solar collector and VMD ... 158
6.3.3 Design of a semi-industrial pilot plant ... 158
6.4 Study and development of solar systems coupled with membrane distillation ... 159
References ... 162

Chapter 7
Optimization and modelling of seawater and brackish water reverse osmosis desalination processes 163
7.1 Introduction 163
7.2 Optimization of NF membranes used in the pre-treatment of membrane based desalination 164
7.2.1 Membrane preparation 164
7.2.2 Membrane characterization 164
7.2.3 Development of NF membranes with low fouling properties 168
7.2.4 Conclusions and outlook 169
7.3 Use of membrane contactors for controlling the water gas composition 169
7.3.1 Introduction 169
7.3.2 Membrane characterization 170
7.3.3 Results and discussion 170
7.3.4 Conclusions and outlook 173
7.4 Comprehensive modelling of the RO desalination process and optimizing hydraulics in spiral wound elements 175
7.4.1 Introduction 175
7.4.2 Materials and methods 176
7.4.3 CFD modelling 178
7.4.4 Results and discussion 179
7.5 Summary 183
References 184

Chapter 8
Integrated system configuration 187
8.1 Introduction 187
8.2 WT 8.1: Critical state of the art of the desalination technologies 188
8.3 WT 8.2 and WT 8.3: Study and optimization of different integrated systems 189
8.3.1 Integrated membrane processes: systems' design and analysis 191
8.3.2 Integrated membrane processes: systems' modelling 194
8.4 WT 8.4: Economic evaluation of the integrated membrane systems 195
8.5 WT 8.5: Quantitative indicators 197
8.6 Conclusions 198
References 198

Chapter 9
Environmental impact assessment (EIA) and life cycle analysis (LCA) of membrane-based desalination plants 201
9.1 Introduction 201
9.2 Environmental impacts 202
9.2.1 Concentrate disposal 202
9.2.2 Pretreatment and cleaning chemicals 203
9.2.3 Energy use 204
9.2.4 Other environmental concerns 204
9.3 Environmental Impact Assessment (EIA) 205
9.3.1 EIA process 205
9.3.2 Environmental monitoring 206
9.4 Best Available Techniques (BAT) 207
9.5 Decision support system 208
9.5.1 Case study 208
9.6 Energy and exergy analysis of integrated membrane systems 213
9.6.1 Integrated flow-sheets 213
9.6.2 Results and discussion 213

9.7 Conclusions ... 215
9.8 Acknowledgements ... 216
References ... 216

Conclusion ... 217

Appendix I ... 221

Appendix II ... 225

Acknowledgement

All Medina participants wish to acknowledge and to thank the European Commission for approving and supporting the project, in particular Dr. Avelino González-González, Dr. Leonore Klemchen and Dr. Michel Schouppe serving as Project Officers within unit I.3 of MEDINA project in Brussels in the last 3–4 years.

A special acknowledgement to Dr. Andrea Tilche, Head of the Unit "Environmental Technologies and Pollution Prevention", for his interest in this project.

List of contributors

Amy G. L.
UNESCO-IHE, Institute for Water Education, Westvest 7, 2611AX Delft, The Netherlands
King Abdullah University of Science and Technology, KAUST, Water Desalination and Reuse Center, Al-Jazri Bldg Office 4231, Thuwal 23955-6900, Saudi Arabia

Barbe C.
Anjou-Recherche, Veolia Environment, Chemin de la Digue, 78600 Maisons-Laffitte, France

Ben-David E.
Zuckerberg Institute for Water Research (ZIWR), J. Blaustein Institutes for Desert Research, Ben- Gurion University of the Negev, POB 653, Beer-Sheva, Israel, 84105

Brisson P.
Anjou-Recherche, Veolia Environment, Chemin de la Digue, 78600 Maisons-Laffitte, France

Cabassud C.
Université de Toulouse; INSA, UPS, INP; LISBP, 135 Avenue de Rangueil, F-31077 Toulouse, France INRA, UMR792 Ingénierie des Systèmes Biologiques et des Procédés, F-31400 Toulouse, France CNRS, UMR5504, F-31400 Toulouse, France

Chen V.
UNESCO centre for membrane Science and Technology, University of New South Wales, Sydney, NSW, 2052, Australia

Chinu K. J.
University of Technology, Sydney (UTS), P.O. Box 123 Broadway, NSW 2007, Australia

Cornelissen E. R.
KWR Watercycle Research Institute. P.O. Box 1072, 3430 BB Nieuwegein, The Netherlands

Criscuoli A.
Institute on Membrane Technology ITM-CNR c/o University of Calabria, Via P. Bucci CUBO 17/C, 87030 Arcavacata di Rende (CS), Italy

Croué J.-P.
CNRS UMR 6008, Laboratoire de Chimie et Microbiologie de l'Eau (LCME), Université de Poitiers, 40 Avenue du Recteur Pineau, 86022 Poitiers, France
Water Desalination and Reuse Center, King Abdullah University of Science and Technology, Thuwal, Kingdom of Saudi Arabia

Curcio E.
Department of Chemical Engineering and Materials, University of Calabria, Via P. Bucci CUBO 45/A, 87030 Arcavacata di Rende (CS), Italy

Di Profio G.
Department of Chemical Engineering and Materials, University of Calabria, Via P. Bucci CUBO 45/A, 87030 Arcavacata di Rende (CS), Italy
Institute on Membrane Technology ITM-CNR c/o University of Calabria, Via P. Bucci CUBO 17/C, 87030 Arcavacata di Rende (CS), Italy

Drioli E.
Department of Chemical Engineering and Materials, University of Calabria, Via P. Bucci CUBO 45/A, 87030 Arcavacata di Rende (CS), Italy
Institute on Membrane Technology ITM-CNR c/o University of Calabria, Via P. Bucci CUBO 17/C, 87030 Arcavacata di Rende (CS), Italy
Hanyang University, WCU Energy Engineering Department, 17 Haengdang-dong, Seongdong-gu, Seoul 133-791 S. Korea

Fane A. G.
UNESCO centre for membrane Science and Technology, University of New South Wales, Sydney, NSW, 2052, Australia

Flemming H.-C.
IWW Water Centre, Moritzstrasse 26, 45476 Mülheim an der Ruhr, Germany

Gabsi S.
University of Gabès, ENIG, Route de Médenine, 6029 Gabès (Tunisia)

Gaeta S.
GVS, via Roma 50, 40069 Zola Predosa (BO), Italy

Gilron J.
Zuckerberg Institute for Water Research (ZIWR), J. Blaustein Institutes for Desert Research, Ben- Gurion University of the Negev, POB 653, Beer-Sheva, Israel, 84105

Grobe S.
IWW Water Centre, Moritzstrasse 26, 45476 Mülheim an der Ruhr, Germany

Groenendijk M.
Water supply company Brabant Water, 's-Hertogenbosch, the Netherlands

Herulah B.
UNESCO centre for membrane Science and Technology, University of New South Wales, Sydney, NSW, 2052, Australia

Herzberg M.
Zuckerberg Institute for Water Research (ZIWR), J. Blaustein Institutes for Desert Research, Ben- Gurion University of the Negev, POB 653, Beer-Sheva, Israel, 84105

Hijnen W. A. M.
KWR Watercycle Research Institute. P.O. Box 1072, 3430 BB Nieuwegein, The Netherlands

Huber S.
DOC-Labor Dr. Huber, Eisenbahnstr. 6, 76229 Karlsruhe, Germany

Ji X.
Department of Chemical Engineering and Materials, University of Calabria, Via P. Bucci CUBO 45/A, 87030 Arcavacata di Rende (CS), Italy

Johir A. H.
University of Technology, Sydney (UTS), P.O. Box 123 Broadway, NSW 2007, Australia

Kandasamy J.
University of Technology, Sydney, P.O. Box 123 Broadway, NSW 2007, Australia

Keller M.
University of Duisburg-Essen (Germany)

Kennedy M. D.
UNESCO-IHE, Institute for Water Education, Westvest 7, 2611AX Delft, The Netherlands

Kushmaro A.
Department of Biotechnology Engineering, Ben- Gurion University of the Negev, POB 653, Beer-Sheva, Israel, 84105

Laborie S.
Université de Toulouse; INSA, UPS, INP; LISBP, 135 Avenue de Rangueil, F-31077 Toulouse, France INRA, UMR792 Ingénierie des Systèmes Biologiques et des Procédés, F-31400 Toulouse, France CNRS, UMR5504, F-31400 Toulouse, France

Laroche J. F.
Anjou-Recherche, Veolia Environment, Chemin de la Digue, 78600 Maisons-Laffitte, France

Lattemann S.
Institute for Chemistry and Biology of the Marine Environment (ICBM) at the Carl von Ossietzky University Oldenburg, Germany.
King Abdullah University of Science and Technology, KAUST, Water Desalination and Reuse Center, Saudi Arabia.

Lebaron P.
CNRS, UMR 7621, Laboratoire d'Océanographie Microbienne (LOMIC) , Observatoire Océanologique, F-66650 Banyuls/Mer, France
CNRS, UMS 2348, Observatoire Océanologique, F-66650 Banyuls/Mer, France
Université Pierre et Marie Curie 06 (UPMC), UMR 7621, LOMIC, Laboratoire Océanologique, F-66650 Banyuls/Mer, France
Université Pierre et Marie Curie 06 (UPMC), UMS 2348, Observatoire Océanologique, F-66650 Banyuls/Mer, France

Lee J. J.
University of Technology, Sydney (UTS), P.O. Box 123 Broadway, NSW 2007, Australia

Leparc J.
Anjou-Recherche, Veolia Environment, Chemin de la Digue, 78600 Maisons-Laffitte, France

Linder C.
Zuckerberg Institute for Water Research (ZIWR), J. Blaustein Institutes for Desert Research, Ben- Gurion University of the Negev, POB 653, Beer-Sheva, Israel, 84105

Macedonio F.
Department of Chemical Engineering and Materials, University of Calabria, Via P. Bucci CUBO 45/A, 87030 Arcavacata di Rende (CS), Italy
Institute on Membrane Technology ITM-CNR c/o University of Calabria, Via P. Bucci CUBO 17/C, 87030 Arcavacata di Rende (CS), Italy

Machinal C.
Anjou-Recherche, Veolia Environment, Chemin de la Digue, 78600 Maisons-Laffitte, France

Manes C-L de O.
CNRS, UMR 7621, Laboratoire d'Océanographie Microbienne (LOMIC), Observatoire Océanologique, F-66650 Banyuls/Mer, France
Université Pierre et Marie Curie 06 (UPMC), UMR 7621, LOMIC, Laboratoire Océanologique, F-66650 Banyuls/ Mer, France

Méricq J.-P.
Université de Toulouse; INSA, UPS, INP; LISBP, 135 Avenue de Rangueil, F-31077 Toulouse, France INRA, UMR792 Ingénierie des Systèmes Biologiques et des Procédés, F-31400 Toulouse, France CNRS, UMR5504, F-31400 Toulouse, France

Messalem R.
Zuckerberg Institute for Water Research (ZIWR), J. Blaustein Institutes for Desert Research, Ben- Gurion University of the Negev, POB 653, Beer-Sheva, Israel, 84105

Mondamert L.
CNRS UMR 6008, Laboratoire de Chimie et Microbiologie de l'Eau (LCME), Université de Poitiers, 40 Avenue du Recteur Pineau, 86022 Poitiers, France

Oren Y.
Zuckerberg Institute for Water Research (ZIWR), J. Blaustein Institutes for Desert Research, Ben- Gurion University of the Negev, POB 653, Beer-Sheva, Israel, 84105

Panglish S.
IWW Water Centre, Moritzstrasse 26, 45476 Mülheim an der Ruhr, Germany

Petrowski K.
IWW Water Centre, Moritzstrasse 26, 45476 Mülheim an der Ruhr, Germany

Pütz D.
IWW Water Centre, Moritzstrasse 26, 45476 Mülheim an der Ruhr, Germany

Raat K. J.
KWR Watercycle Research Institute. P.O. Box 1072, 3430 BB Nieuwegein, The Netherlands

Rapenne S.
Anjou-Recherche, Veolia Environment, Chemin de la Digue, 78600 Maisons-Laffitte, France

Remize P.-J.
Anjou-Recherche, Veolia Environment, Chemin de la Digue, 78600 Maisons-Laffitte, France

Robert C.
Anjou-Recherche, Veolia Environment, Chemin de la Digue, 78600 Maisons-Laffitte, France

Salinas Rodriguez S. G.
UNESCO-IHE, Institute for Water Education, Westvest 7, 2611AX Delft, The Netherlands

Schaule G.
IWW Water Centre, Moritzstrasse 26, 45476 Mülheim an der Ruhr, Germany

Schippers J. C.
UNESCO-IHE, Institute for Water Education, Westvest 7, 2611AX Delft, The Netherlands

Schrotter J.-C.
Anjou-Recherche, Veolia Environment, Chemin de la Digue, 78600 Maisons-Laffitte, France

Shon H. K.
University of Technology, Sydney (UTS), P.O. Box 123 Broadway, NSW 2007, Australia

Sim L. N.
UNESCO centre for membrane Science and Technology, University of New South Wales, Sydney, NSW, 2052, Australia

Strathmann M.
IWW Water Centre, Moritzstrasse 26, 45476 Mülheim an der Ruhr, Germany

Stuyfzand P. J.
KWR Watercycle Research Institute. P.O. Box 1072, 3430 BB Nieuwegein, The Netherlands
VU University, Amsterdam, The Netherlands

Tansakul C.
Université de Toulouse; INSA, UPS, INP; LISBP, 135 Avenue de Rangueil, F-31077 Toulouse, France INRA, UMR792 Ingénierie des Systèmes Biologiques et des Procédés, F-31400 Toulouse, France CNRS, UMR5504, F-31400 Toulouse, France

van de Wetering S.
Water supply company Brabant Water, 's-Hertogenbosch, the Netherlands

van der Kooij D.
KWR Watercycle Research Institute. P.O. Box 1072, 3430 BB Nieuwegein, The Netherlands

Veenendaal H.
KWR Watercycle Research Institute. P.O. Box 1072, 3430 BB Nieuwegein, The Netherlands

Vigneswaran S.
University of Technology, Sydney (UTS), P.O. Box 123 Broadway, NSW 2007, Australia

Weihs-Fimbres G.
UNESCO centre for membrane Science and Technology, University of New South Wales, Sydney, NSW, 2052, Australia

Wiley D.
UNESCO centre for membrane Science and Technology, University of New South Wales, Sydney, NSW, 2052, Australia

West N. J.
CNRS, UMS 2348, Observatoire Océanologique, F-66650 Banyuls/Mer, France
Université Pierre et Marie Curie 06 (UPMC), UMS 2348, Observatoire Océanologique, F-66650 Banyuls/Mer, France

Ye Y.
UNESCO centre for membrane Science and Technology, University of New South Wales, Sydney, NSW, 2052, Australia

Introduction

Prof. Enrico Drioli and Dr. Eng. Francesca Macedonio

Membrane-Based Desalination: An Integrated Approach (acronym ***MEDINA***) was a project financed by the European Commission within the scope of its 6th Framework Program.

The project formally started on 15th October 2006 and ended on 14th January 2010.

The main aim of the project was to improve the overall performance of membrane-based water desalination processes by applying an innovative approach based on the integration of different membrane operations in the reverse osmosis (RO) pre-treatment and post-treatment stages accordingly to the philosophy of Process Intensification.

In order to achieve its aim, project Consortium collected 13 research teams of proven and complementary expertise in seawater and brackish water desalination (Table 1): 7 between universities and research centres from 4 Member States; two Australian, one Israeli and a Tunisian research centres; 2 industrial companies. In addition to the core research partners, MEDINA project included the participation of several desalination plants and utilities and, thereby, the research project team benefited from direct contact with full-scale plant experience. This real-world field of experience allowed the researchers to test, fine tune and try to transfer directly their findings and applications directly to the practitioners.

Objective of MEDINA project was to address the technological issues and vacant questions that are not, or only partly, answered to date, in sight of improving and optimising membrane desalination as well as the integration of its technology into existing infrastructures and water management practices. The implementation of an integrated approach to the design of advanced membrane desalination systems represents an attractive opportunity because of the synergic effects that can be reached, the simplicity of these units, and the possibility of advanced levels of automation and remote control. In the pre-treatment steps, the integration of different tools (such as water quality characterisation, membrane cleaning strategies, selection of the most appropriate pre-treatment processes) leads to the minimisation of membrane replacement needs thereby reducing the operating costs. In the RO post-treatment stages, the presence of Membrane Contactors (MC) and/or Membrane Distillation (MD) and/or Membrane Crystallizer (MCr) and/or Wind Intensified Enhanced Evaporation (WAIV) working on the brine streams, offers the possibility to produce more fresh water thus increasing water recovery factor of current desalination plants, reducing brine disposal problem and approaching the concept of "*zero-liquid-discharge*", "*total raw materials utilization*" and "*low energy consumption*".

The performed research activities covered all the aspects related to membrane desalination: from the characterization of water sources to the development of appropriate pre-treatment methods; from the investigation of alternative designs and engineering approaches able to decrease the environmental impacts to the adoption of renewable energy sources for driving some unit operations; from the development of innovative post-treatment strategies to the optimization and integration of the different units.

Table 1 List of MEDINA Participants

Participant name	Country	Role*
University of Calabria	Italy	CO
Anjou Recherche – Veolia Water	France	CR
UNESCO – IHE	The Netherlands	CR
KIWA	The Netherlands	CR
Universität Duisburg-Essen-IWW	Germany	CR
Ben Gurion University	Israel	CR
Centre National de la Recherche Scientifique Laboratoire d'Océanographie Biologique Laboratoire de Chimie de l'Eau et de l'Environnement	France	CR
INSA Toulouse	France	CR
GVS S.P.A.	Italy	CR
University of Technology, Sidney	Australia	CR
University of New South Wales	Australia	CR
Carl von Ossietzky University of Oldenburg, Institute for Chemistry and Biology of the Marine Environment	Germany	CR
Ecole Nationale d'Ingénieurs de GABES/Institut Supérieur des Etudes Technologiques de SFAX	Tunisia	CR

**CO = Coordinator; CR = Contractor.*

The project was subdivided in the following nine distinct work packages (WP), consisting of closely related activities:
WP1 – Water Quality Assessment Tools
WP2 – Evaluation and Comparison of Seawater and Brackish Water Pretreatment Processes
WP3 – Development of Tools for RO Fouling Characterization and Understanding
WP4 – Development of Cleaning Strategies for RO Membranes
WP5 – Process strategies for mitigation of impact of concentrates on the environment
WP6 – Innovative Technologies to Reduce Energy Consumption in Seawater Desalination Facilities
WP7 – Optimization and Modelling of Seawater and Brackish Water Reverse Osmosis Desalination Processes
WP8 – Integrated system configuration
WP9 – Environmental Impact Assessment and Life Cycle Analysis (LCA) of Membrane-Based Desalination Plants.
The most relevant results achieved in the thirty-nine months of the project will be described in the following chapters.

Chapter 1

Water quality assessment tools

G. L. Amy[1], S. G. Salinas Rodriguez[1], M. D. Kennedy[1], J. C. Schippers[1], S. Rapenne[2], P.-J. Remize[2], C. Barbe[2], C.-L. de O. Manes[3,4], N. J. West[5,6], P. Lebaron[3,4,5,6], D. van der Kooij[7], H. Veenendaal[7], G. Schaule[8], K. Petrowski[8], S. Huber[9], L. N. Sim[10], Y. Ye[10], V. Chen[10] and A. G. Fane[10]

[1] *UNESCO-IHE (The Netherlands)*
[2] *Anjou-Recherche, Veolia Environment (France)*
[3] *UPMC, UMR 7621, LOMIC, Laboratoire Océanologique (France)*
[4] *CNRS, UMR 7621, LOMIC, Observatoire Océanologique (France)*
[5] *UPMC, UMS 2348, Observatoire Océanologique (France)*
[6] *CNRS, UMS 2348, Observatoire Océanologique (France)*
[7] *KWR (The Netherlands)*
[8] *IWW (Germany)*
[9] *DOC-Labor Dr. Huber (Germany)*
[10] *University of New South Wales (Australia)*

In this chapter, the most interesting results carried out in Work Package 1 of MEDINA project are summarized. The WP1 leader has been Prof. Gary L. Amy. The contributors of this WP have been as follows:

- UNESCO-IHE (Sergio G. Salinas Rodriguez, Maria D. Kennedy, Jan C. Schippers and Gary L. Amy);
- AR-VW (Sophie Rapenne, Pierre-Jean Remize, Caroline Barbe);
- LOB (C-L de O. Manes, N. J. West, P. Lebaron);
- KWR (Dick van der Kooij, Harm Veenendaal);
- IWW (Gabriela Schaule, Kathrin Petrowski, Stefan Huber);
- UNSW (Lee Nuang Sim, Yun Ye, Vicki Chen, Anthony G. Fane).

WP1 was structured into the following complementary Work Tasks:

- WT 1.1 Water Quality Characterisation Tools;
- WT 1.2 Development of Particulate/Colloidal Fouling Indicators;
- WT 1.3 Development of scaling indicators

 Membrane-Based Desalination: An Integrated Approach (MEDINA). Edited by Enrico Drioli, Alessandra Criscuoli and Francesca Macedonio. ISBN: 9781843393214. Published by IWA Publishing, London, UK.

- WT 1.4 Development of Biofouling Indicators;
- WT 1.5 Validation of Analytical Methods and Fouling Assessment Tool at Full-Scale Desalination Facilities.

In the following pages, some of the research activities carried on the 39 months and some of the final results are reported.

1.1 INTRODUCTION

Fouling represents the major constraint to more cost-effective, and therefore expanded, application of membrane technology in drinking water, particularly for reverse osmosis systems. Indeed, spiral-wound membrane modules cannot be pneumatically backwashed on a regularly basis and only chemical cleaning can restore normal performance after fouling. Fouling can occur in several forms:

- *Particulate fouling:* small particles and colloids (sub-micron particles) not retained by upstream pre-treatment impart resistance and/or increased salt polarisation and reduce flux as cake-layer deposits accumulate onto the membrane surface.
- *Organic fouling:* natural organic matter (NOM) present in the feed water passing through the pre-treatment processes may be adsorbed onto the membrane surface as a gel-layer, reducing the permeability of the membrane and thereby decreasing production. Moreover, the biodegradable organic matter (BOM) retained on the membrane surface can be utilized by microorganisms as nutrients and may contribute to biological growth.
- *Biofouling:* the combined presence of microorganisms and BOM or nutrients in the membrane feed water may lead to the formation and the development of a biofilm. The adsorbed and retained microorganisms multiply and the associated extra-cellular polymeric substances (EPS) can cover the membrane surface in a layer of up to several microns. The established biofilm promotes further fouling through entrapment of organic molecules, colloidal particles, suspended particles and bacteria cells.
- *Scaling (inorganic fouling):* salt precipitation occurs on the surface of the membrane due to localized super saturation conditions.

The occurrence and importance of organic fouling will depend on several parameters among which are the structure of the biological communities of the influent seawater and its potential fouling through the release of extracellular polymeric substances (EPS) and the growth of bacteria. The development of tools such as an early-warning system related to the fouling potential of seawater can be developed by using some key parameters related to the structure of biological communities.

Various analytical protocols and surrogate bench-scale tests have been developed to assess feed water quality in terms of fouling potential with respect to (*i*) particulate/colloidal fouling, (*ii*) natural organic matter (NOM) fouling, and (*iii*) biological fouling (biofouling). Analytical protocols provide a *direct* measure of the amount, and ideally the character, of potential foulants in a feed stream, while surrogate tests provide an *indirect* indication of anticipated fouling trends, often in terms of a fouling index derived from a bench-scale test.

Traditionally, an indirect estimate of particulate fouling potential has been done through the silt density index (SDI) and, more recently, the modified fouling index (MFI) at constant pressure; both rely on the use of a $0.45\,\mu m$ filter to simulate flux decline trends. The SDI is derived from a simple filtrated volume versus time curve, with a SDI value of less than 3 targeted for RO feed water. The MFI, modified from the SDI, has been used to indicate the particulate fouling potential of a feed water. While it is not as widely used, it exhibits more accuracy than the SDI since it is based on a cake filtration mechanism and it is dependent on particle size and particle concentration (Schippers and Verdouw, 1980). In general, smaller particles forming a cake layer result in higher MFI values. As cake layer formation is the dominant mechanism in particulate fouling, the MFI can be used as a basis for modelling flux decline in membrane systems.

Flow cytometry (FCM) may be very useful to provide information on the abundance and structure of both phytoplanktonic and bacterial communities. The high sensitivity of the most recent instruments allows one to count and to characterize the structure of picoplanktonic communities which may partly control its fouling properties. FCM can also provide information on the abundance of active bacterial cells in seawater. The recent development of aquatic instruments provide a new and very interesting opportunity for testing the potential role of this tool as an early warning system for biofouling.

Traditionally, the NOM fouling potential of a feed water has been assessed in terms of dissolved organic carbon (DOC), UV absorbance, and color, however, NOM fouling rates do not appear to correlate with these traditional water quality parameters. A problem is that DOC only indicates the *amount* but not the *character* of the NOM. More recently, specific UV absorbance (SUVA) has been used to indicate the aromatic character of NOM but SUVA is a direct measure of humic substances which are less problematic as foulants compared to non-humic materials. A quality parameter that can describe the level of organic fouling expected for a particular membrane feed water (surface, brackish and seawater) is urgently required. There has been no surrogate test developed for indirect assessment of NOM fouling potential.

The biofouling potential of seawater, after pre-treatment for particle removal, depends on the presence of microbial biomass and compounds promoting the growth of bacteria. The majority of this has been done with representative seawater and, to a lesser extent, brackish waters obtained from participating utilities/end users. For studying the biofouling potential of water, it determined the relationship between the concentration of microbial biomass, analyzed as adenosine triphosphate (ATP) and total direct cell (TDC) counts, as well as the concentration of easily assimilable organic carbon (AOC) on the rate and extent of biofouling using a flat-sheet membrane model system. An important area of the work presented is adapting the AOC protocol, developed for freshwaters, to saline conditions. Furthermore the adaptation of the two methods Biodegradable Dissolved Organic Carbon (BDOC) and LC-OCD to saline matrix have been performed and their combination (BDOC/LC-OCD) have developed a praxis proofed and sensitive tool.

This chapter focuses on the developments on water quality assessment tools to directly and/or indirectly predict the fouling potential of feed waters applied to seawater (SWRO) reverse osmosis membranes. These tools range from new methods to existing methods developed for freshwater feeds and adapted to seawater and brackish water feeds.

1.2 WATER QUALITY CHARACTERIZATION TOOLS

1.2.1 Liquid chromatography – organic carbon detection

Liquid chromatography – organic carbon detection (LC-OCD) (also called size exclusion chromatography, SEC-DOC) can be used to effectively monitor NOM components with a lower SUVA. LC-OCD has been successfully applied to monitoring changes in NOM associated with fresh water treatment (Her *et al.*, 2002). It has also been used to identify problematic NOM components in membrane fouling (Her *et al.*, 2004). LC-OCD separates NOM according size/molecular weight (MW) classes ranging from higher to lower MW: biopolymers (BP), humic substances (HS), bulding blocks (BB), low MW acids (LMA) and neutrals (N). The magnitude of the BP peak has been linked with fouling potential in UF membranes (Amy and Her, 2004).

A typical chromatogram of NOM contained in surface water is illustrated in Figure 1.1. The first fraction identified after approximately 25–45 min is the biopolymer peak with significant response by organic carbon detection (OCD) only. The organic colloids and proteins present in this fraction provide response in OCD and UV detection. The second and third fraction responses in OCD and UVD are attributed to humic substances and building blocks, respectively. Building blocks are a weathering product of humic substances. The fourth response to OCD and UV detection is attributed to low MW organics (acids and neutrals).

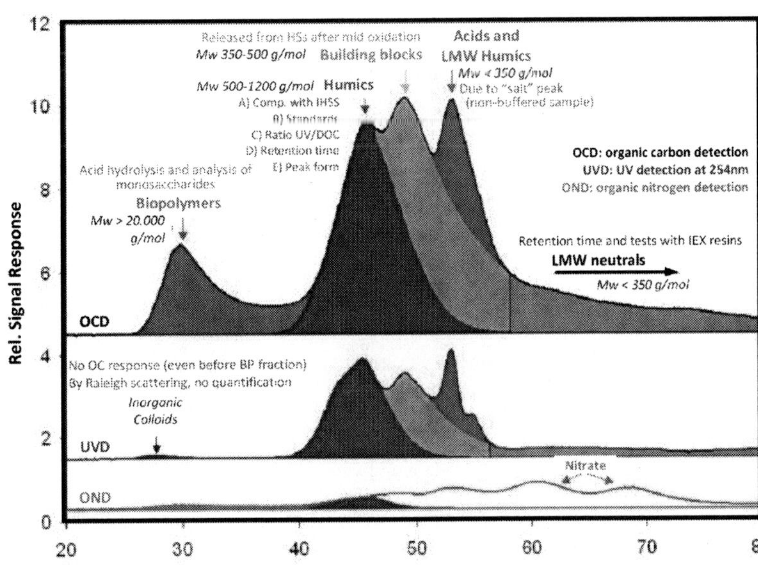

Figure 1.1 Typical NOM chromatogram of a fresh water sample (Huber, 2007)

1.2.1.1 Ionic strength effect

The low DOC content and high salinity of seawater, require the adaptation of the standard LC-OCD method validated for fresh water.

The high ionic strength of seawater lead to changes on OM configuration and modify the interaction of the OM with the mobile phase and size exclusion resin. Some of the observed effects when using the standard method for fresh water in high salinity water are: a shift on the elution time of the organic matter fractions (longer elution time); negative depression on the base line at elution times corresponding to acids. The negative peak is probably linked to a strong UV scavenging of salts below 200 nm eluted at this time.

For optimizing the method for high salinity waters, several variables of the test were studied, such as: ionic strength of the mobile phase, longer elution time (longer column length), oxidant concentrations, dilution of the sample.

A more concentrated buffer did not improve the resolution of the chromatogram and the shift in elution time continued. A longer column did not improve the resolution of the chromatogram and the test increased from 2 hours up to 4 hours. Oxidant addition (0.5 g/L potassium peroxidisulphate) to the acidic phase improve the signal in the organic carbon detector.

The suggested testing conditions for high salinity waters are summarized in Table 1.1.

Table 1.1 LC-OCD conditions for high salinity waters

Parameter	Description
Column	Toyopearl HW 50S (30 µm particle size)
Eluent	20 mM
Eluent flow	0.7 ml/min
Acidic phase	0.5 g/L potassium peroxidisulfate addition
Dilution factor	2
Software integration	Fiffikus

The combination of low organic matter concentration and high ionic strength are difficult to solve in liquid chromatography. The new conditions allow an acceptable separation of the different organic fractions. An accurate quantification of these fractions is possible with a variation coefficient lower than 12%.

1.2.1.2 Applications

Various raw waters were characterized by LC-OCD (Figure 1.2). For the three "seawater-representative" locations the DOC content is on average 1 mg/L where the humic substances represent about 50% of the DOC content. In all cases the fraction with size larger than 20 kDa (Biopolymers) represents about 7%. In the case of estuarine water, the DOC content is around 5 mg/L. Additionally, the LC-OCD results for canal water are included to illustrate main differences.

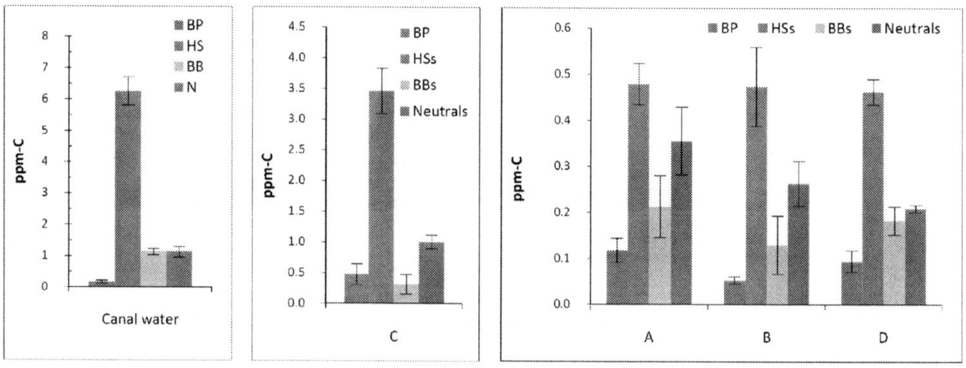

Figure 1.2 LC-OCD results for: Fresh water (left), estuarine water (middle) and Seawater (right). This figure is available in colour at http://www.iwapublishing.com/template.cfm?name=isbn9781843393214

The DOC and SUVA values for the plants of sites A, B, C and D are shown in Table 1.2.

Table 1.2 Raw water DOC (ppm-C) and SUVA values (L/mg.m)

	Site A (seawater)	Site B (seawater)	Site C (estuarine water)	Site D (seawater)
SUVA	1.14 ± 0.59	0.70 ± 0.05	3.03 ± 0.54	0.89 ± 0.22
DOC	1.16 ± 0.16	0.92 ± 0.09	5.26 ± 0.56	0.95 ± 0.03

Figure 1.3 shows the average values for all analyzed samples. Variation represents the maximum and minimum values of the samples. Higher variation is present in humic substances and building blocks in comparison with neutrals and biopolymers.

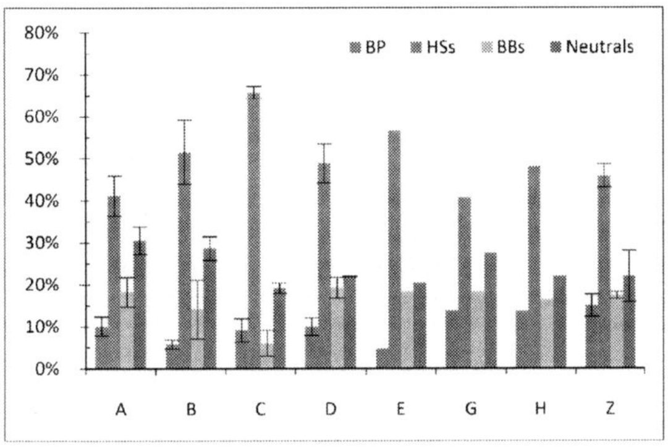

Figure 1.3 Typical organic matter fractions (%)

LC-OCD results show that humic substances are the higher concentration fraction in seawater and estuarine water compared with biopolymers, building blocks and neutrals. Typically, humic substances (0.5–5 kDa in size) represent ~50% of the DOC content while the biopolymer fraction (>20 kDa) is less than 8% of the total DOC. In the case of estuarine water, the humic substances represent ~65% of the total DOC and biopolymers ~10%.

The average LC-OCD results for the sampling campaigns are presented in Figure 1.4.

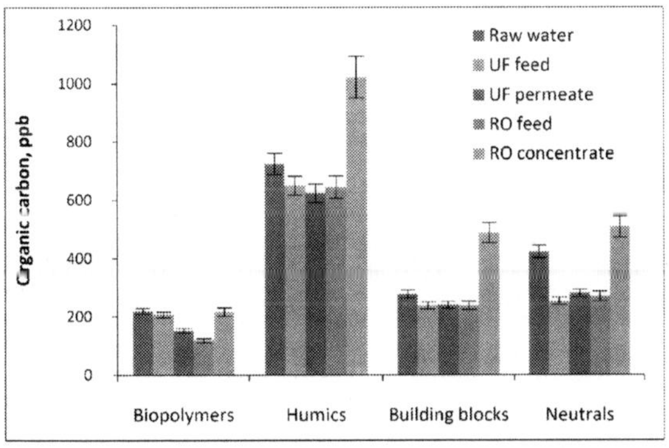

Figure 1.4 Organic matter fractions along the treatment plant. This figure is available in colour at http://www.iwapublishing.com/template.cfm?name=isbn9781843393214

Figure 1.4 shows the organic matter concentration for biopolymers (BPs), humic substances (HSs), building blocks (BBs) and neutrals.

1.2.2 Fluorescence excitation emission matrix

A fluorescence excitation-emission matrix (F-EEM) is developed by scanning over an excitation range of 240 to 450 nm by 10-nm increments and an emission range of 290 to 530 nm by 2-nm increments using a FluoroMax-3 spectrofluorometer (HORIBA Jobin Yvon, Inc., USA). The fluorescence samples are first adjusted to 1 mg/L of DOC and a pH of ~2.8. The result is a three-dimension spectrum in which fluorescence intensity (normalized to Raman units) is represented as a function of excitation and emission wavelengths. In order to further remove the Rayleigh scattering effects, emission measurement data made in the region of the excitation wavelength + 20 nm were deleted, and a set of zeros were inserted in a triangular-shaped region where the emission wavelength is less than excitation wavelength (upper corner of left side of EEM).

Protein-like organic matter, hypothesized to be a principal membrane foulant (polysaccharides are potential foulants as well), exhibits a dominant peak at lower excitation/emission wavelengths while humic/fulvic substances show dominant primary and secondary peaks at higher excitation/emission wavelengths. Table 1.3 shows the typical wavelenghts for various compounds.

Table 1.3 Typical EEM peak values

Description	Fluorescence range	
	Ex	Em
Humic-Like Primary Peak	330–350	420–480
Humic-Like Secondary Peak	250–260	380–480
Amino acid-Like (Tyrosine) Peak	270–280	300–320
Amino acid-Like (Tryptophan) Peak	270–280	320–350
Protein-Like (Albumin) Peak	280	320

Based on an F-EEM, a fluorescence index (FI) can be calculated by the ratio of fluorescence intensity at emission 450 and 500 nm at excitation 370 nm. A higher FI (~1.7–~2.0) reflects organic matter of an autochthonous (microbial) origin while a lower FI (~1.3–~1.4) reflects organic matter of an allochthonous (terrestrial) origin (McKnight *et al.*, 2001).

1.2.2.1 Effect of ionic strength

Ionic strength showed to have an important effect in the fluorescence intensity. As it can be observed in Figure 1.5 the measured intensity increases linearly with the increase of salinity represented as TDS. In brackish water (typically TDS < 10 g/L), an increase of 5 to 8% is observed for amino acids and humic substances while 24% is observed for a big protein like albumin. In saline environment (TDS = 35 g/L) an increase of 13 to 24% is observed for amino acids and humic substances.

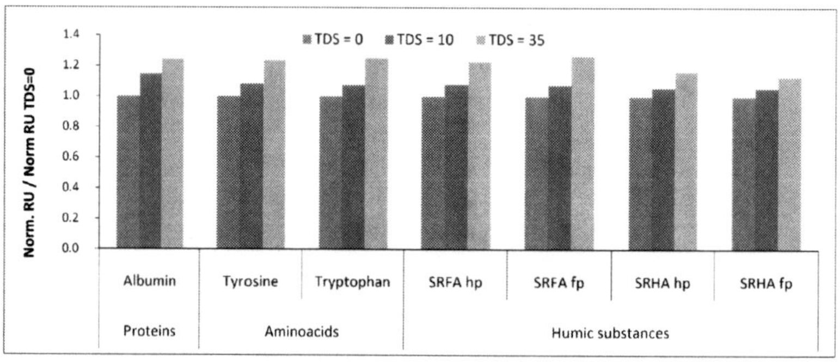

Figure 1.5 Salinity effect on fluorescence EEM measurements. The intensities are expressed as ratios with respect to the intensity of the sample with TDS = 0 g/L.

1.2.2.2 Effect of pH

pH of a set of samples with constant concentration of OM was modified by the addition of hydrochloric acid (HCl) or sodium hydroxide (NaOH), and as a result, a variation of the fluorescence intensity was observed (see Figure 1.6). This variation may be attributed to the degradation of OM in low and high pH environments.

1.2.2.3 Effect of concentration

Analyses of different compound solutions with controlled concentrations show that the observed peak intensities do not always follow a linear trend with the concentration (see Figure 1.7). Albumin shows an acceptable linear trend ($R^2 = 0.9983$). For the other observed compounds, a parabolic or a third order polynomial gave a better fit.

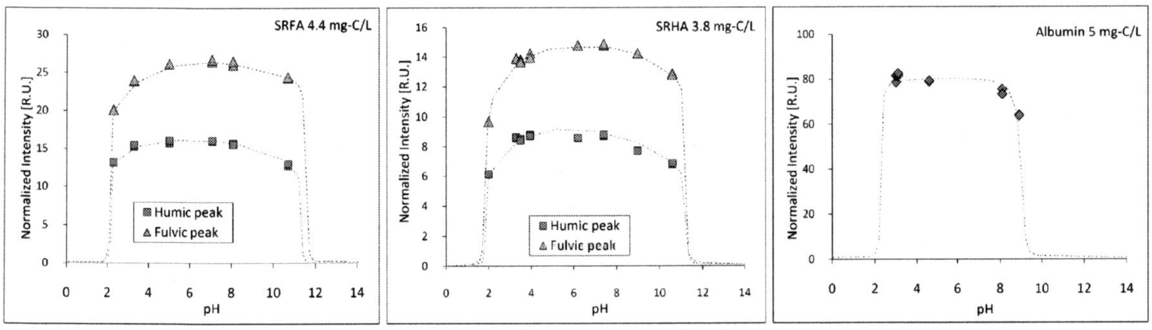

Figure 1.6 Effect of pH on the measured EEM peak intensities. For humic substances, the intensities at both peaks are shown.

For humic substances and as for DSR (deep seawater reference material, which contains humic substances), both peaks, humic and fulvic showed a non linear behaviour. It was observed that the intensities on the fulvic peak followed a trend with higher curvature than the intensities in the humic peak.

The physical bases of this phenomena are not well understood; however, this effect may be related to the absorption of the emitted light by nearby molecules in the sample. This "obstruction" effect increases with the concentration.

The use on PARAFAC (parallel factor analysis) analysis is based upon de assumption of linearity of the response. Thus, linearization of the measurements would be required to obtain accurate results.

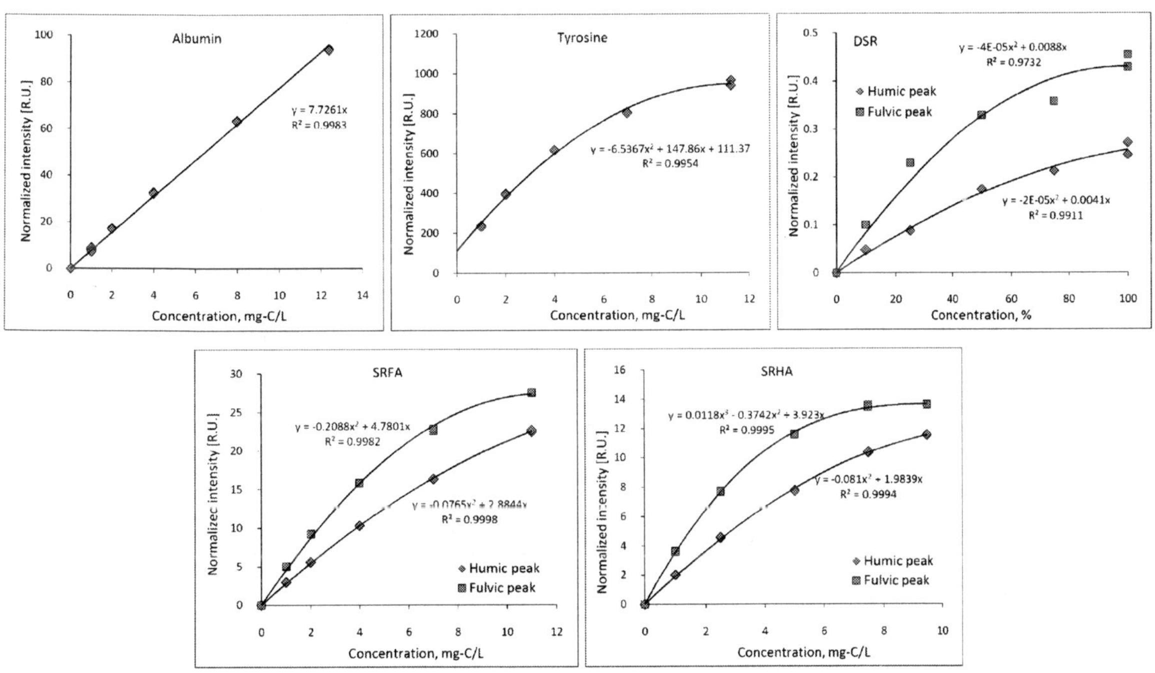

Figure 1.7 Behaviour of the maximum normalized intensity R.U. with the concentration for different OM compounds. For humic substances, the intensities at both peaks are shown

1.2.2.4 Applications

F-EEM is a good tool to characterize protein-like, humic-like compounds and recent advances with PARAFAC (Parallel factor analysis, a multi-way analysis technique) forecast a major application of fluorescence for water treatment applications. Figure 1.8 shows the F-EEM for Mediterranean water and for North Sea water, highlighting the different peak locations of typical organic matter as shown in Table 1.3.

The results from fluorescence spectrometry for a plant treating brackish water are presented in Figure 1.9. In this case, the fluorescence intensities are presented according to Table 1.3. Humic substances (humic, fulvic and marine-humic like materials) are dominating the spectra with higher fluorescence intensities than the amino-acid like materials.

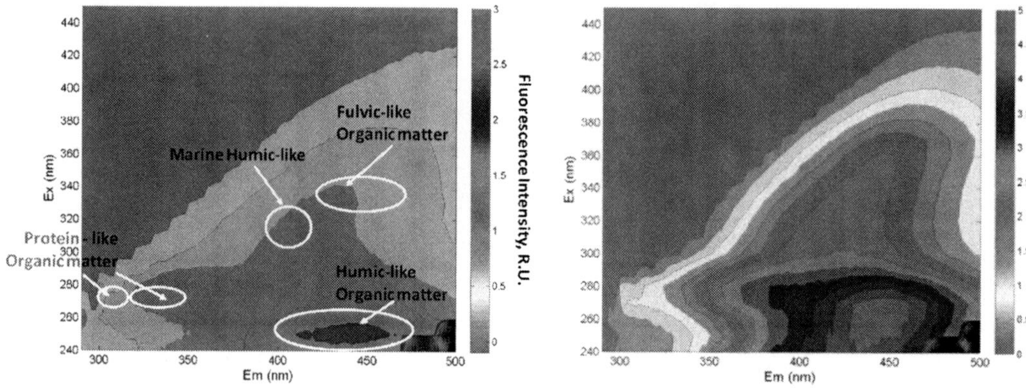

Figure 1.8 Typical F-EEM for Mediterranean water (left) and for North Seawater (right)

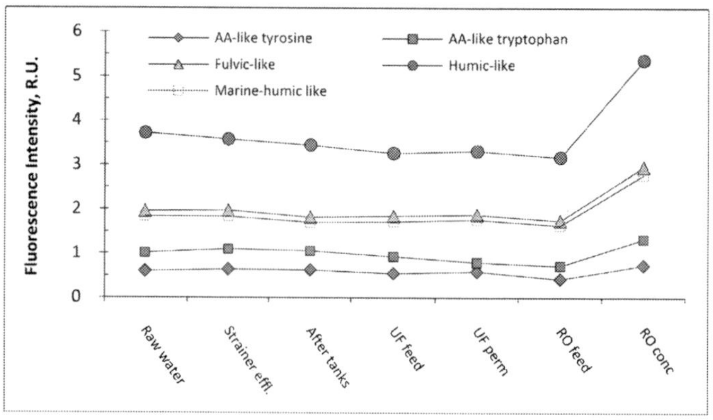

Figure 1.9 Fluorescence intensities (R.U.) along the treatment plant

From Figure 1.9, the AA-like tyrosine peak was removed by 23% and the AA-like tryptophan peak was removed by 22% after the UF units. Coagulation (Strainer effluent – UF feed) removed 15% AA-like material and 6%, 9% and 6% for fulvic-like, humic-like and marine-humic-like material, respectively.

1.2.2.5 Comments

F-EEM is a reliable analytical technique to track changes along the water treatment in a plant as the sensitivity is quite high. It is fast and low price.

1.2.3 Enumeration, identification and diversity of marine microorganisms including bacteria and phytoplankton

Water samples were analysed quantitatively and molecularly for microbial community diversity at different compartments of both pilot (Site A) and full-scale desalination plants (Sites B, D, G and H). Consistent with the site, water samples were taken at different locations prior to and after pre-treatment steps. For Site A: raw water RW, after dual-media filtration DMF and after micro-filtration MF. Whereas for Site B and G: intake raw water RW and from the beach well (representing a slow sand filtration, SSF). For Site D and H samples came from open intake OI water and after dual-media filtration DMF. The number of samples analysed per site including their respective pre-treatments is summarized in Table 1.4. Water samples from open intake and after different pre-treatments were analysed by flow cytometry (FCM) for total bacteria and phytoplankton counts. At molecular level, community structure was analysed by 16S ribosomal gene fingerprinting using the capillary electrophoresis single-stranded conformational polymorphism CE-SSCP technique. An in-depth analysis of bacterial community composition by means of 16S ribosomal RNA gene clone libraries was undertaken for the water samples from Site D at one time point (Nov/2007).

The following experimental framework was conducted for the characterization of microbial communities in seawater:

1. Autotrophic/Heterotrophic cell counts (FCM); for cell abundances
2. Genetic fingerprinting (CE-SSCP); for bacterial community structure and diversity
3. 16S ribosomal RNA gene sequencing and phylogenetic affiliation of clones; for bacterial community composition of the major (dominant) species.

Table 1.4 Denomination of pilot (Site A) and full-scale desalination plants (Sites B, D, G and H) used in this study and the number of samples analysed for each sample type (OI, open intake water; DMF, after dual media filtration; MF, after microfiltration; SSF, after slow sand filtration)

Water origin and pre-treatments	No. samples
Site A (OI, DMF, MF)	14
Site B (OI, SSF)	6
Site D (OI, DMF)	3
Site G (OI, SSF)	1
Site H (OI, DMF)	1

1.2.3.1 Comparison of pre-treatment efficacy in removing autotrophic and heterotrophic microorganisms

It was observed a high variability in total bacteria and phytoplankton counts along time for the studied Sites. In general, *Synechococcus* was the major phytoplankton group with variable abundance levels along the sampling period for Sites A, B and D. However, after pre-treatments phytoplankton counts were not significant. Intake water quality in terms of bacterial cells contents also varied according to the Plant location. Table 1.5 shows that in average, higher bacterial cell numbers entered Sites D and B compared to Sites A, G and H.

Table 1.5 Average bacterial cell concentration in open intake waters from the studied sites

Site	Total bacteria counts (cells/ml)
Site A (OI), n = 14	3.5E + 05
Site B (OI), n = 8	5.2E + 05
Site D (OI), n = 4	5.4E + 05
Site G (OI), n = 1	1.2E + 04
Site H (OI), n = 1	1.6E + 04

Pre-treatments efficiency in removing microorganisms was evaluated in three sites. Table 1.6 displays the average removal percentage of both phytoplankton and bacterial cells. According to our data, MF and SSF were the most efficient pre-treatment removing respectively 96% and 89% of bacterial cells and 99% and 93% of phytoplankton cells.

Table 1.6 Autotrophic (phytoplankton) and heterotrophic (bacteria) cell removal following different pre-treatment procedures on Sites A, B and D

	Site A (DMF/MF)	Site B (SSF)	Site D (DMF)
Total Bacteria cells removal	66/96%	89%	63%
Phytoplankton cells removal (Syn)	71/99%	93%	51%

1.2.3.2 Bacterial diversity: spatial and temporal analysis in Site D

SSCP profiles were obtained for all the samples analysed so far from Sites A, B, D, and G. In general for Sites A, B, G and H: SSCP profile comparison indicated loss of dominance for some groups majorly represented in RW samples as compared to samples after pre-treatments. However, for Site D in sampling period from Nov/2007 and Site H, pre-treatment did not significantly change microbial diversity, just relative abundances (data not shown). We have analyzed the evolution of bacterial community structure in Site D for an 8 months period through genetic fingerprinting (Figure 1.10A).

Statistics analysis of community profiles revealed that bacterial communities are more similar to each other before and after pre-treatment than according to the sampling date suggesting that seasonal variation affects feed water quality (Figure 1.10B).

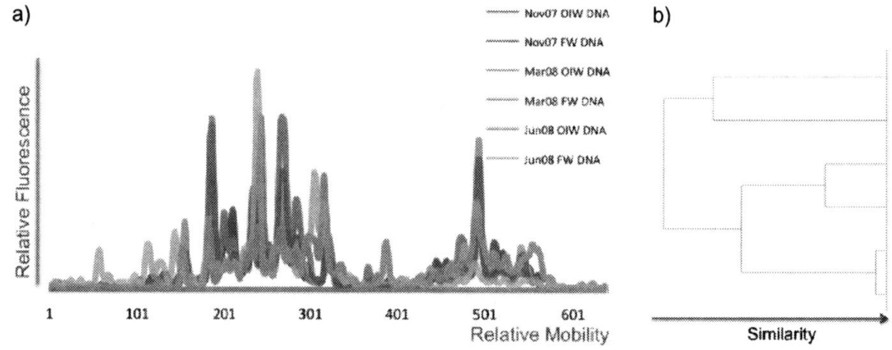

Figure 1.10 Bacterial community structure in a full-scale desalination plant (FSDP), Site D, as revealed by capillary electrophoresis single-stranded conformational polymorphism (CE-SSCP). (A) Superimposed normalised DNA profiles showing different peak profiles for seawater samples at different sampling periods respectively. OI, open intake seawater; F, SWRO membrane feed water after pre-treatment. (B) Similarity of the DNA CE-SSCP profiles from samples taken from different compartments of a FSDP; open intake seawater (OI), seawater reverse osmosis (SWRO) membrane feed water after pre-treatment (F) at different sampling periods revealed in the dendrogram. Bray Curtis similarities were calculated between pairwise comparisons of the sample profiles and the dendrogram was constructed by the average linkage method. This figure is available in colour at http://www.iwapublishing.com/template.cfm?name=isbn9781843393214

1.2.3.3 Taxonomic classification of 16S rRNA gene clones for Site D

Clone libraries were performed for water samples, OI and F, from sampling period 1 (Nov/2007) in order to identify the major bacterial groups present in intake water and also to evaluate the pre-treatment efficacy in removing and/or favouring certain bacterial groups from open intake waters. Figure 1.11 displays the taxonomic classification of the clones from both OI and F libraries. The data shows that *Proteobacteria* phylum representatives dominate both OI and RO feed water (F) samples. *Alphaproteobacteria* (49.2% and 44.7% respectively) and *Gammaproteobacteria* (33.3% and 21.3% respectively) classes were dominant but they differed in their relative abundances (Figure 1.11). Cluster analysis into operational taxonomic units (OTUs) and the phylogenetic affiliation of clones revealed that in the water samples the SAR11 clade (OI1, 22%; F1, 40%), an *Actinobacteria* OTU belonging to the *Acidimicrobinae* (OI1, 7%; F1, 4%), and a *Caldithrix* OTU (SAR406 clade) mostly restricted to the F1 water (5%) were the dominant bacterial groups (Table 1.7).

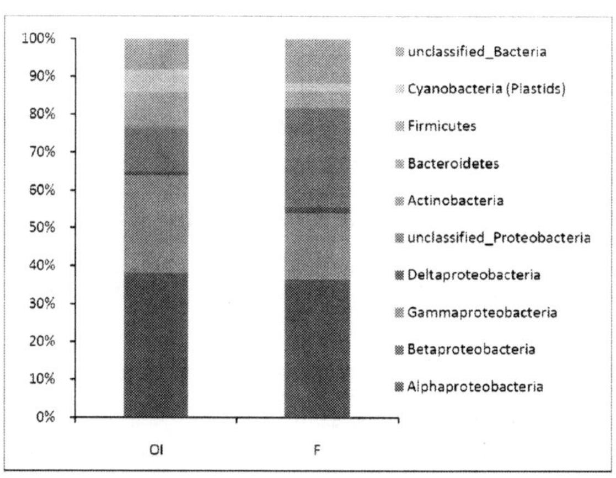

Figure 1.11 Schematic representation of the hierarchical classification and distribution of the 16S ribosomal RNA gene sequences into major bacterial groups from clone libraries obtained from water samples a full-scale desalination plant (FSDP) comprising open intake seawater (OI) and seawater reverse osmosis (SWRO) membrane feed water after pretreatment (F). This figure is available in colour at http://www.iwapublishing.com/template.cfm?name=isbn9781843393214

Table 1.7 Phylogenetic affiliation and closest relatives of representative 16S rRNA gene clones for each operational taxonomic unit (OTU) and their relative abundance in the water clone libraries. OI, open intake seawater; F, seawater reverse osmosis (SWRO) membrane feed water after pretreatment. Unique OTUs from each clone libraries are shaded in grey

Closest relative ID	**Similarity %**	**OTU classification#**	**Abundance %**	
			OI1	**F1**
		Alphaproteobacteria		
AACY020273010	98	Defluvicoccus*	2	
FJ716871	98	Rhodobacteraceae	3	1
EF471647	100	Roseovarius	1	1
EF471704	99	marine group AEGEAN-169*		2
AACY023897748	98	marine group AEGEAN-169*		1
EU799440	100	SAR116*	2	
AY664095	98	SAR116*	1	1
DQ009166	99	SAR11clade*	3	6
EU805335	99	SAR11clade*	2	1
EU237296	100	Rhodospirillales	1	
EP879548	99	SAR11clade*	**22**	**40**
FJ745192	99	Rhodobium_2*	2	
EU804393	99	Rickettsiales*		2
		Deltaproteobacteria		
EF574189	99	SAR324clade*	2	
		Gammaproteobacteria		
AACY020555669	99	SAR86 clade*	3	4
FJ745006	99	SAR86 clade*	3	3
EF574960	99	SAR86 clade*	3	2
FJ745145	100	SAR86 clade*		3
EF575172	99	SAR86 clade*	1	1
		Actinobacteria		
EP609371	99	Acidimicrobineae*	7	4
		Bacteroidetes		
EF572094	100	Flavobacteriaceae		3
AF406540	100	Flexibacteraceae	3	3
		Deferribacteres		
AACY023373955	99	SAR406 clade*	1	5
Singletons%			16	7

 # OTU classification according to RDP-10 database.
** OTU classification according to silva 100 database.*

1.2.3.4 Comments

Water samples, open intake and after different pre-treatments, coming from the different studied Sites were analysed both quantitatively and molecularly for microbial community diversity. Samples were analysed by flow cytometry (FCM) for total bacteria and phytoplankton counts and at the molecular level, community structure was analysed by 16S ribosomal gene fingerprinting using SSCP technique. It was observed a high variability in total bacteria and phytoplankton counts along time for the studied Sites. Pre-treatments efficacies were evaluated in terms of microorganisms removal. For this study our data revealed that microfiltration (MF) followed by slow sand filtration (SSF) are more efficient than the conventional dual media filtration.

Genetic fingerprinting (CE-SSCP) was successfully applied to monitor bacterial diversity changes in different compartments and also showed on a temporal scale that bacterial community structure changed according to sampling date in one full-scale desalination plant (Site D). Furthermore for a certain time point (Nov/2007) we show that community composition from water after retreatment was quite similar to the corresponding open intake water. The identity of the major groups/species present in raw and RO feed waters was accessed by clone libraries, which revealed some ubiquitous groups in both water samples (e.g., SAR11 clade) but also some sample specific ones. In agreement with the SSCP data, cloning and sequencing revealed that bacterial communities were similar in OI and after DMF (Site D) suggesting that pre-treatment did not remove the major bacterial groups but just reduced the total bacterial load. The experimental strategy used in this study to characterize the microbial content of seawater at different compartments of a FSDP proved to be appropriate.

1.3 DEVELOPMENT OF PARTICULATE FOULING INDICES

A reliable index to predict the fouling potential of RO feedwater is important in preventing and diagnosing fouling at the design stage of RO plants and for monitoring the performance of pre-treatment during plant operation. The silt density index (SDI) is widely used to measure the fouling potential of RO feed water; however, fouling problems have been reported even with very low SDI values, i.e., SDI < 1. The modified fouling index (MFI), developed by Schippers and Verdouw in 1980, has many advantages over the SDI including: i) a linear relation between the concentration of colloidal particles and MFI ii) cake filtration is assumed to be the dominant filtration mechanism when employing the MFI while the SDI is not based on any filtration mechanism, and iii) since the MFI is based on the occurrence of cake filtration, flux decline in an RO plant can be predicted using cake formation or pressure increase. Both the SDI and MFI operate using constant pressure and a nominal pore size of 0.45 μm (Schippers and Verdouw, 1980). However, particles smaller than 0.45 μm are not captured by the 0.45 μm membrane and thus are not measured in the MFI test. It is widely believed that particles smaller than 0.45 μm are responsible for fouling of RO membranes. Consequently, the MFI using membranes with a pore size of 0.05 μm was introduced in 1981 (Schippers *et al.*, 1981).

In 2003, Boerlage *et al.* developed the MFI-UF test to capture smaller particles using a polyacrylonitrile UF membrane with a MWCO of 13 kDa. Furthermore, Boerlage *et al.* also introduced the MFI at constant flux (Boerlage *et al.*, 2003a,b). In recent times the concept of MFI-NF was proposed to capture even smaller particles, but the obtained MFI-NF values for seawater with NF membranes (0.5 kDa) were similar to the MFI-UF values obtained with UF membranes (30 kDa) (Khirani *et al.*, 2006). Thus, MFI-NF is not applicable to measure the fouling potential of seawater. In 2008, the cross flow sampler (CFS) was introduced to simulate the hydrodynamic conditions occurring in a (cross flow) RO process (Adham and Fane, 2008).

1.3.1 SDI and MFI

In order to predict the fouling of the reverse osmosis membranes, pretreated water have to be characterized using fouling indexes. Up to now the most applied index has been the Silt Density Index (SDI) whose ASTM test is widely used today. However, the SDI already showed limits and inconsistent results. That is why there is a need to develop and study more relevant indexes, in order to improve the characterization of pretreated water.

Two kinds of pre-treated water are assessed in this study: Dual Media Granular Filtration (DMF) and Membrane Filtration (UF). SDI is compared to another fouling index, the Modified Fouling Index (MFI), in terms of average value at the outlet of each pre-treatment and of variation profile during a filtration cycle.

1.3.1.1 Tools and methods

All the results for these two indexes are gathered with the online SDI & MFI-2200 analyzer provided by the MABAT® Company.

SDI: The SDI & MFI analyzer automatically calculates SDI according to the ASTM D4189-07 method (American Society for Testing Materials, 2007). The time required to filtrate 500 ml of water is measured twice: at the beginning of the test (t_0) and after a given filtration period (t_f), fixed at 15 min in this case as only pre-treated water are characterized.

SDI is obtained by the following relation: $SDI = (100/15)*(1 - t_o/t_f)$.

MFI: MFI is calculated nearly the same way as SDI. The difference lies in the fact that the water flow rate through the 0.45 μm filter is measured during the entire test. MFI value is then calculated by plotting the t/V ratio (t: filtration time, V: cumulated filtrate volume) against V and calculating the slope of the linear part which corresponds to the cake filtration. MFI value is then corrected with water temperature.

MFI is determined on site directly with the online SDI & MFI-2200 analyzer: water flow rate through the 0.45 μm filter is determined each 500 ml until the linear part of the t/V vs. V curve is reached.

1.3.1.2 Evolution of the fouling indexes during a DMF cycle

The objective was to find the most relevant index between SDI and MFI for the characterization of the pre-treated seawater quality. To do so, variations of these two indexes were measured during some DMF filtration cycles (with pre-coagulation). Figure 1.12 depicts the profile of SDI and MFI values along one the filtration run.

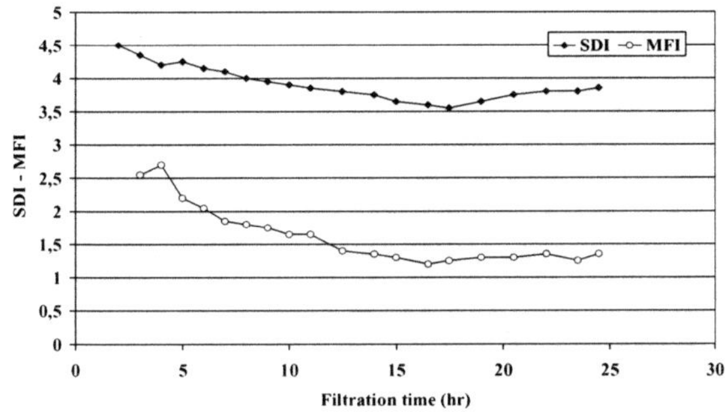

Figure 1.12 Typical evolution of SDI and MFI vs filtration time during a DMF cycle

The values of the two indexes decrease in the course of time in parallel to granular filter maturation but SDI obviously presents a lower range of variation than MFI, as confirmed in Table 1.8, which shows that MFI is a much better indicator of water quality variations than SDI.

Table 1.8 Range of variation of SDI and MFI during a DMF cycle

	Max	Min	Variation (%)
SDI	4.5	3.8	**17**
MFI (s/L^2)	2.7	1.3	**52**

Furthermore, SDI cannot be measured during the first two hours of the dual media filtration after backwash since 0.45 μm membrane clogs before reaching 15 minutes of filtration (SDI > 6). Nevertheless, this drawback is not observed for MFI since it can be measured from the beginning of the filtration cycle, as shown in Figure 1.13. Whereas SDI remains quite constant during this filtration cycle, MFI values present a 90% decrease.

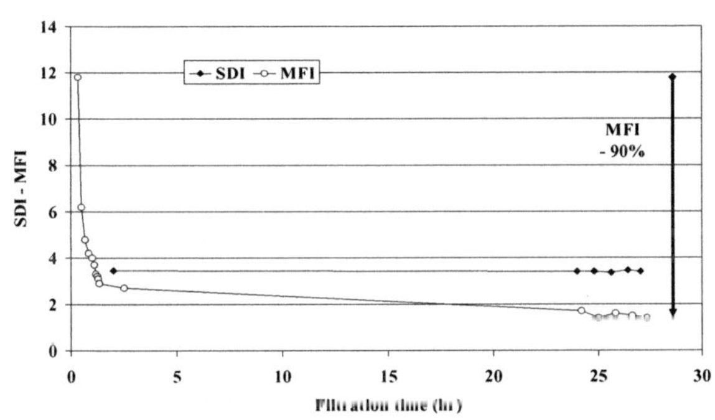

Figure 1.13 Evolution of SDI and MFI during a DMF cycle

1.3.1.3 Comparison granular filtration/membrane filtration

A study was performed to compare the performance of two pre-treatments (pre-coagulation + granular vs. low-pressure membrane without pre-coagulation) according to the two fouling indexes. Membrane pre-treatment was performed with an UF membrane (average pore size = 0.02 μm). The values for the UF pre-treatment were compared with fouling index values at the outlet of the DMF pre-treatment after more than 10 hours of filtration (between 10 hours of filtration and the end of the filtration cycle), when the filter is mature and pre-treated water quality is fairly stable.

The results are presented in the following table (Table 1.9) in terms of minimum, average and maximum values for each index (5 SDI measurements and 7 MFI measurements for DMF, 28 SDI/MFI measurements for UF).

It appears that average SDI and MFI are significantly lower at the outlet of the UF pre-treatment compared to the DMF pre-treatment which shows that, according to these two fouling indexes, membrane pre-treatment provides a better pre-treated seawater quality than granular pre-treatment even without pre-coagulation.

The range of variation from the maximum value for the UF pre-treatment is respectively 21% for SDI and 43% for MFI. As noted in the previous paragraph about DMF, the ranges of variation for MFI is higher than for SDI, even at the outlet of the UF pre-treatment which should provide quite a constant water quality. This confirms that MFI is a more relevant index than SDI to capture the variations of seawater quality.

Table 1.9 Average values of SDI and MFI after the two pre-treatments

	DMF pre-treatment	UF pre-treatment
SDI	Min = 3.50	Min = 1.82
	Avg = 3.70	Avg = 2.01
	Max = 3.95	Max = 2.32
MFI (s/L^2)	Min = 1.08	Min = 0.31
	Avg = 1.59	Avg = 0.43
	Max = 1.82	Max = 0.54

In parallel to the operation of these two pre-treatments, two RO units were also continuously fed by the DMF pre-treated seawater and the UF pre-treated seawater. After four months of continuous operation, the fouling rate of the RO membrane fed the UF unit was found to be significantly higher than the fouling rate of the RO membrane fed by the DMF unit. It is hence worthwhile to assess whether the MFI index could show this worse water quality at the outlet of UF membrane or at least could show a lower difference in terms of water quality between these two pre-treatment processes than expressed by SDI.

For this purpose, the following figure (Figure 1.14) presents the parallel evolution of SDI and MFI for a DMF cycle and the UF pre-treatment operated at the same time.

It clearly appears that each index is lower at the outlet of the UF pre-treatment than at the outlet of the DMF pre-treatment. The difference percentage between the DMF pre-treatment and the UF pre-treatment is respectively 41% for

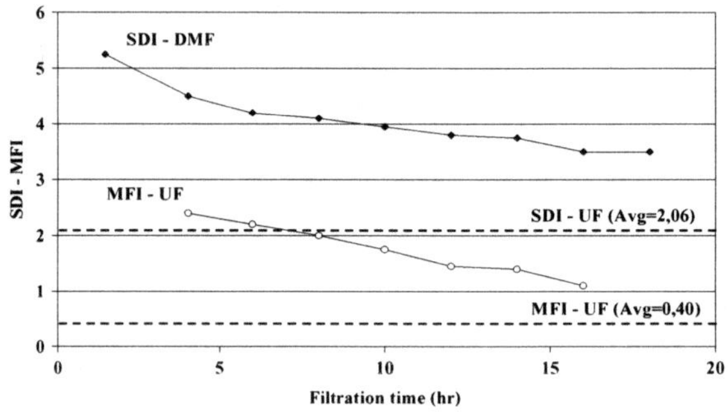

Figure 1.14 SDI and MFI profiles during a DMF cycle with average SDI and MFI values after UF pre-treatment

SDI and 63% for MFI which shows that the difference of pre-treatment efficiency in terms of water quality is somewhat higher for MFI than for SDI. This means that UF pre-treated seawater has a better quality than DMF pre-treated seawater according to the two indexes, with MFI showing the highest degree of difference between the two pre-treatment processes, which contradicts the results gathered from the RO membranes evaluation.

Even if more data are required for MFI, these results tend to show that none of these two particulate fouling indexes can explain the difference of fouling for the two RO membranes.

1.3.1.4 Conclusions

MFI is an interesting index for evaluating the fouling potential of seawater as it shows a wider range of variation than SDI during a DMF cycle or at the outlet of a UF pre-treatment, which enables it to be much more sensitive than SDI to potential changes of pre-treated seawater quality changes.

Nevertheless, as for SDI, MFI remains an index for the evaluation of the particulate content of seawater. Consequently, it do not provide information about the organic content of pre-treated seawater or its biofouling potential, which is crucial for the prediction of the fouling extent on the downstream RO system.

1.3.2 Modified fouling index ultrafiltration at constant flux

Since the introduction of the MFI-UF by Boerlage *et al.* (2001, 2003a), applications have mainly been limited to fresh water sources and the MFI-UF has not yet been tested and evaluated for seawater.

1.3.2.1 How to measure the MFI-UF at constant flux?

A filtration set-up was developed to work at constant flow as illustrated in Figure 1.15. The key components of the set-up are: pump, membrane holder, pressure sensor, thermometer, computer and a three-way valve.

Two piston pumps were tested (shown in Figure 1.15). The maximum pressure they can operate is the difference. The small pump has a maximum capacity of 1.2 bar while the bigger pump has a capacity of 3.5 bar. This makes the small suitable for working at low flux rates while the big pump can work at any flux rate and with any membrane pore size.

Membranes are commercially available. They should be flat filters. The diameter of the filters depends on the size of the pressure holder (e.g., 25, 45 mm). The membrane holder (porous support) was modified so the flow distribution towards the membrane is as uniform as possible.

The pressure transmitter is commercially available and suitable to work with seawater. The operational pressure range is 0–4 bar with a maximum deviation of 0.04%. The three-way valve is used to connect the syringe with the membrane holder and at the same time with the pressure transmitter.

Before any filtration test, the set-up is checked to avoid the presence of air trapped in the tubes and inside the membrane holder.

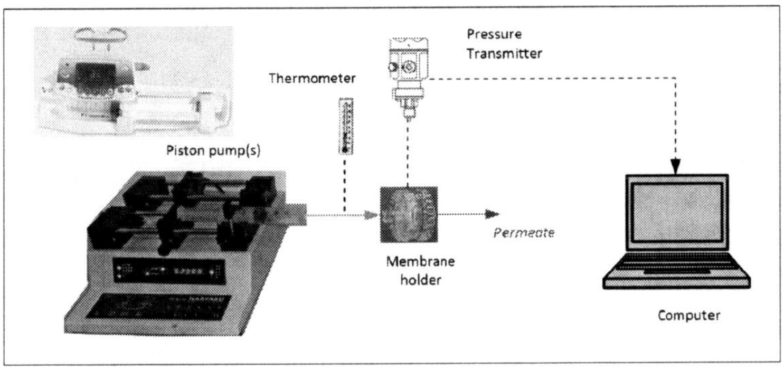

Figure 1.15 Constant flux filtration set-up

The procedure for the test is as follows:

1. The membrane resistance is measured with ultra pure water.
2. The membrane is placed into the membrane holder.
3. The required flow (flux) is controlled manually in the pump by defining the flow rate in ml/hr. Membrane area is known.
4. All the tests should be stopped after:
   a. cake filtration is reached,
   b. a minimum I value is observed or the *fouling index vs. time* slope shows no change in time; and
   c. let run at least for 35 minutes.
5. The fouling index I is calculated by dividing the slope of the *pressure vs. time* line over square flux and water viscosity ($I = Slope/J^2\eta$). MFI-UF is calculated considering the minimum I values.

$$MFI = \frac{\eta_0 \cdot I}{2 \cdot \Delta P_0 \cdot A_0^2} \tag{1.1}$$

6. In order to keep MFI-UF values comparable with $MFI_{0.45}$, the MFI-UF values are standardized to reference conditions namely: viscosity at temperature of 20°C ($\eta_{20°C}$), pressure of 2 bar (ΔP_0) and surface of area of a MFI 0.45 µm micro filter (A_0) as shown in Eq. 1.

1.3.2.2 Which pore size of membrane to use?

The range of ultrafiltration membranes is large. It can vary from a few micrometers to nanometres. As the MFI-UF test works in a dead-end configuration, all the particles bigger than the pore size of the membranes are retained. This means that the smaller the membrane pore size the more particles will be captured creating a thicker and less porous cake. At the same time, the fouling potential of the water is proportional with the concentration of particles in the water; this means that the MFI value with a smaller pore size membrane will be higher that with a looser membrane. This is illustrated in Figure 1.16 where North Sea water was tested with various pore sizes.

It can be observed that the measured MFI value depends strongly on the pore size of the membrane used in the test. It also shows that the MFI value is a function of the particles concentration present in water.

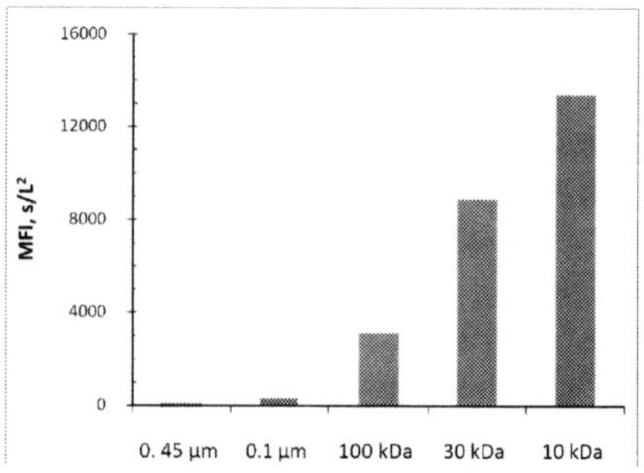

Figure 1.16 MFI values for North Seawater as function of pore size

Nevertheless, the important question remains: what does this value mean for the reverse osmosis operation? It is clear that the small particles will reach the RO units, but can they be directly translated into fouling of the RO?

For solving this question, a prediction model was developed to predict the rate of particulate fouling considering the measured MFI values and considering some other important factors: flux effect on the test, amount of particles depositing in the RO unit. This is illustrated later on 3.2.6.3 section.

1.3.2.3 Which membrane material to use?

There are several materials used in ultrafiltration such as: PES, RC, PAN, PVDF. PVDF is produced mainly for loose UF membranes. PAN and PES are more likely to find in hollow fibre membranes and in various pore sizes. For this research, PES and RC membranes were tested as the range of pore size available was wider.

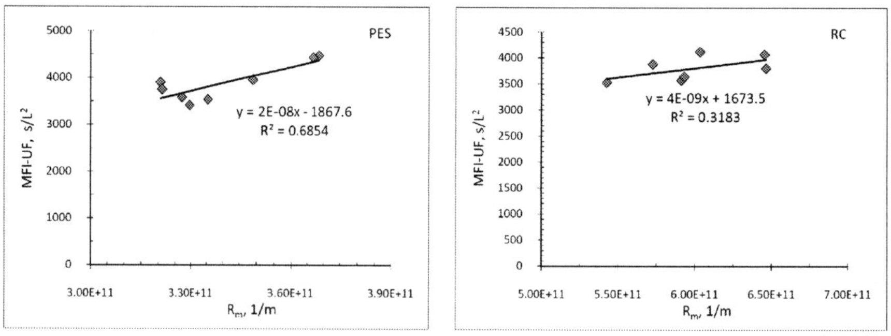

Figure 1.17 MFI values for Delft canal water measured with 100 kDa PES (left) and 100 kDa RC (right) at 100 L/m^2-hr

Figure 1.17 shows the measured MFI values for the same solution (canal water) using a whole set of new membranes. For the PES membranes the average was 3880 s/L^2 ± 395 (10.3%), and for the RC membranes the average was 3800 s/L^2 ± 235 (6.3%).

Both membrane materials have the average value close to each other. RC membranes are slightly more uniform than the PES membranes.

1.3.2.4 At which flux should the test be performed?

In principle, the MFI-UF test can be performed at any filtration flux. However, at high flux rates two effects may play an important role in the results: flux effect on particles arrangement in the cake and cake compression effect.

Figure 1.18 shows the MFI-UF values for raw seawater (left) and for UF permeate and dual media filtration effluent (right) measured at various flux rates from ~50 L/m^2-hr up to 350 L/m^2-hr. It was observed a direct relation between applied flux and measured MFI-UF value.

The applied flux may depend on the use of the measurements. In case of pre-treatment performance comparison, the MFI values can be measured at a high flux as the test duration would be short (e.g., 30 minutes) while for predicting RO operation the test can last longer as the filtration rate should be as close as possible to the RO one (e.g., 15 L/m^2-hr).

1.3.2.5 Has salinity an effect in the MFI-UF measurements?

Salinity may affect the MFI-UF test in two ways: to the particles in the solution and to the membrane itself.

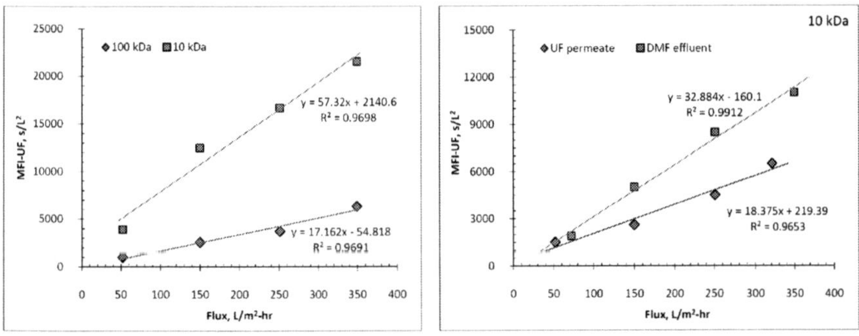

Figure 1.18 MFI-UF values as function of filtration flux

In Figure 1.19 the MFI-UF values for Delft canal water (diluted 4 times) and for RO feed water (diluted from 35 to 23 g/L) are presented. The salinity of the initial solution was altered by adding concentrated solution of NaCl (99.9999%). For both samples an increase of MFI-UF with salinity was observed.

For a salinity level and an increase similar to a 40% SWRO recovery, in case of canal water the increase was about 10% and in case of RO feed water the increase was about 20%.

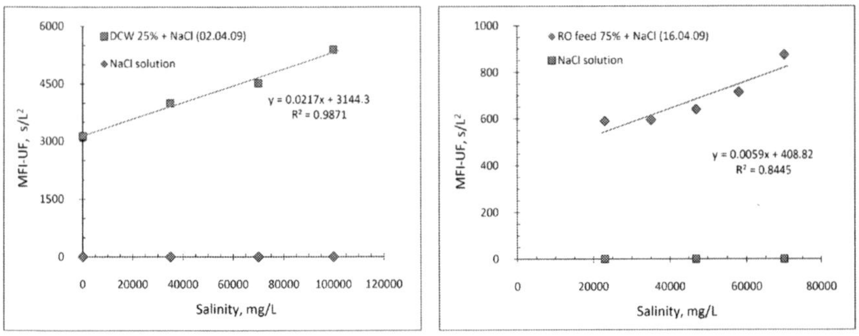

Figure 1.19 Salinity effect on particles – MFI-UF of Delft canal water diluted to 25% (left) and SWRO feed (right)

The observed effects suggest that salinity may play a role when comparing RO feed and Ro concentrate waters.

The effect of salinity on the membrane was studied by measuring the MFI-UF value of NaCl solution. The measured values were low enough to be considered the effect minimal.

1.3.2.6 Applications

1.3.2.6.1 Raw water comparison

When performed the test with the same conditions (membrane material, pore size and flux rate) water from different locations and after different pre-treatments can be compared.

Figure 1.20 shows the MFI-UF values of raw seawater from different locations. In many cases, the MFI values are 4–10 times different. These values give an indication of the particle content that the pre-treatment need to remove before reaching a RO unit.

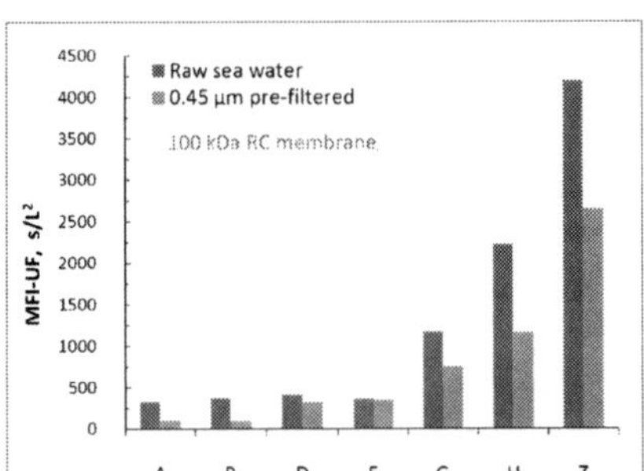

Figure 1.20 MFI values for raw seawater from different locations. Measured with a 100 kDa RC membrane at 250 L/m²-hr

1.3.2.6.2 Plant profiling

Figure 1.21 shows the MFI-UF values measured with 100, 50 and 10 kDa membranes at 250 L/m^2-hr along a SWRO plant treating brackish water.

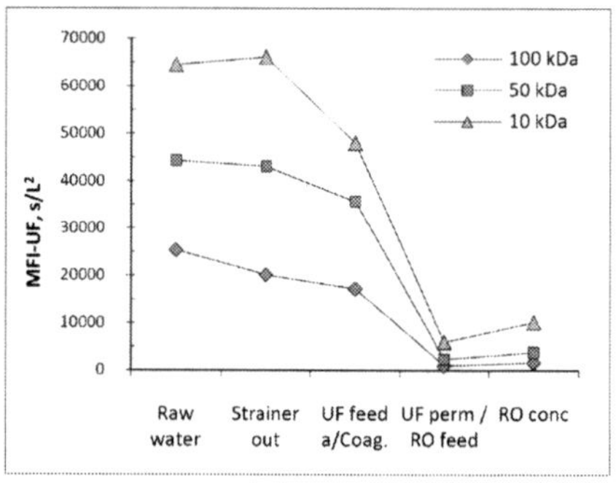

Figure 1.21 Plant profiling with MFI-UF values measured with 100, 50 and 10 kDa at 250 L/m^2-hr

The percentages in reduction of MFI values after water passing through the ultrafiltration units were 94.3%, 93.4% and 87.6% for 100, 50 and 10 kDa respectively.

Table 1.10 MFI-UF (100 kDa) values in s/L^2 and percentage removal

Date	Raw water	UF feed	UF perm	Removal
23.04.09	4310	2935	190	94%
28.04.09	4840	4295	125	97%
16.06.09	3800	3650	395	89%
02.07.09	2950	2285	203	91%
06.07.09	2840	2450	200	92%
10.05.10	25340	17190	980	94%

In Table 1.10 are presented the MFI measurements at various dates. Although the raw water varied, the percentage decrease of MFI-UF value was in all cases more than 90%.

1.3.2.6.3 RO particulate fouling prediction

By using the cake filtration model to predict particulate fouling on the RO elements (see Eq. 1.2) it is possible to estimate the time for a defined net driving pressure increase (15% was considered).

$$t_r = \frac{(\Delta NDP)}{I \cdot \psi \cdot \Omega \cdot J_0^2 \cdot \eta_r} \qquad (1.2)$$

Where: ΔNDP is the net driving pressure increase (bar) which typically is the criteria for cleaning in RO systems; I is the fouling index (m^{-2}) and is calculated from MFI filtration test; Ω is the deposition factor (–) and it is measured on-site; ψ is the cake ratio factor (–).

For the projection the following conditions were considered: T = 10.5 °C, P_{feed} = 58.5 bar, NDP = 22.2 bar, % increase = 15% (3.32 bar), Recovery = 40%, deposition factor (Ω) = 1.

Table 1.11 Predicted cleaning frequency in RO unit for particulate fouling

Sample	MWCO, kDa	$MFI_{@ 250 lmh}$, s/L^2	$MFI_{@ 15 lmh}$, s/L^2	Ω	$t_{r@ 250 lmh}$, months	$t_{r@ 15 lmh}$, months
RO feed	100	980	80	1	7.7	94.6
	50	2350	340	1	3.2	22.3
	10	5975	850	1	1.3	8.9
Raw water	100	25340	2816	1	0.30	2.7
	50	44285	4921	1	0.17	1.5
	10	64500	7167	1	0.12	1.1

The projected values for cleaning frequency for RO feed water and for Raw water are presented in Table 1.11. For the RO feed water, in the worse case it would take ~9 months to reach a 15% increase in NDP. Considering that Raw water would be fed directly into the RO units, the predicted times for 15% increase in NDP are 1.1. These values are considering that the deposition factor is 1 (all particles would accumulate on the surface of the membranes).

1.3.3 Crossflow sampler modified fouling index ultrafiltration

1.3.3.1 The rationale of CFS-MFI$_{UF}$

RO membrane systems are mostly operated in crossflow mode, whereas the common fouling existing indices are all measured under dead-end mode. None of them has tried to simulate the actual hydrodynamic flow conditions in fouling index measurement. During crossflow filtration, large particles with larger back transport velocities usually move away from the membrane surface due to the convective flow and particles back-transport mechanisms. These hydrodynamic conditions in crossflow lead to different cake composition and structure when compared to dead-end filtration. A novel fouling index called the Crossflow Sampler – Modified Fouling Index Ultrafiltration (CFS-MFI$_{UF}$) was developed in order to simulate the hydrodynamic conditions occurring in a (crossflow) RO process (Sim *et al.*, 2010, 2011a). This fouling index was performed in a typical crossflow filtration module followed by a dead-end MFI measure device (Figure 1.22). The aim of the crossflow sampler (CFS) is to simulate the hydrodynamic conditions of crossflow cell, so the membrane in crossflow sampler should be "non-retentive" that is permeable to any foulants that flow near to it. The membrane used in the dead-end unit has MWCO of 10kDa to capture as much fine components in the feed as possible. The measurement is operated under the constant flux mode to replicate the balance of permeate and tangential velocities experience in reverse osmosis units.

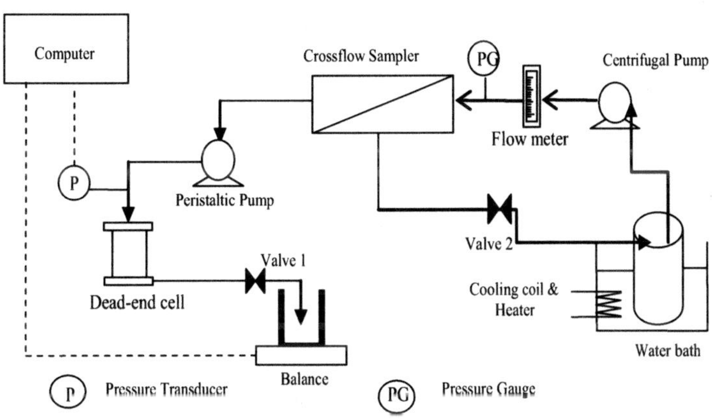

Figure 1.22 Schematic diagram of CFS-MFI$_{UF}$ device

1.3.3.2 Fundamental study of CFS-MFI$_{UF}$

Colloidal silica and humic acid were selected as representatives of a colloidal foulant and macromolecular foulant respectively for the fundamental investigation of CFS-MFI$_{UF}$. Similar to MFI-UF, CFS-MFI$_{UF}$ was found to have a linear relationship with the feed concentration, suggesting that CFS-MFI$_{UF}$ is not only sensitive toward colloids, but also to organic matter.

CFS-MFI$_{UF}$ was also found to be sensitive to the crossflow velocity. The crossflow velocity in the sampler unit influences the particle concentration and the particle size distribution in its permeate. CFS-MFI$_{UF}$ was found decreased gradually as crossflow velocity increased, however when reached a certain crossflow velocity, CFS-MFI$_{UF}$ started to increase and plateau off. Clear minimum values of CFS-MFI$_{UF}$ were observed for both model feed and raw seawater. It was also found that CFS-MFI$_{UF}$ has a lower value when a spacer was placed in the channel. This indicates that CFS-MFI$_{UF}$ can successfully capture the hydrodynamic behaviour during the crossflow filtration.

Values of CFS-MFI$_{UF}$ and MFI-UF$_{const.flux}$ for the same feed were also compared. MFI-UF$_{const.flux}$ tends to predict a higher fouling propensity of the feed, approximately 10–40% higher than CFS-MFI$_{UF}$. This discrepancy suggests that the original feed has been hydrodynamically fractionationed by the crossflow sampler. This can be further supported by the particle size distribution analysis, at which the particle size distribution in permeate of CFS has shifted significantly to smaller size range when compared to the original feed. The results highlighted the existence and the importance of the hydrodynamics on particle capture. Hence, it is crucial to incorporate in the fouling test.

1.3.3.3 Effect of CFS and dead-end permeate flux ratio

CFS-MFI_{UF} was found to be sensitive to the permeate flux of both CFS and dead-end unit. The increase of permeate flux in either CFS or dead-end unit will increase CFS-MFI_{UF} values. Thus, to better simulate the hydrodynamic conditions of a RO unit, the fluxes used for both units in CFS-MFI_{UF} measurement should be the same and comparable to the RO flux (~20 L/m^2-h). However, this may prolong the duration of the test substantially. A flux correction factor was then proposed to streamline the test. In order to accelerate the fouling rate in CFS-MFI_{UF}, the area of dead-end cell was reduced to 1/6 of the area of CFS. As such, the dead-end flux J_2 was increased to 120 L/m^2.h while maintaining the CFS flux J_1 at 20 L/m^2.h. The accelerated CFS-MFI_{UF} cannot be used for quantitative fouling rate prediction in RO due to the effect of flux difference on particle deposition. As a result, a flux correction factor that correlates the CFS-MFI_{UF} measured at dead-end flux of 120 L/m^2.h to CFS-MFI_{UF} measured at dead-end flux of 20 L/m^2.h was determined. Different types of feed solution were used in a series of experiments to investigate if the flux correction factor exists for specific operation conditions regardless of the types of feed. For each type of feed, two experiments with different flux ratios (J_1 = 20 L/m^2.h and J_2 = 20 L/m^2.h vs J_1 = 20 L/m^2.h and J_2 = 120 L/m^2.h) were conducted. Figure 1.23 shows a good correlation (R^2=0.92) for different feeds with a wide range of MFIs. Thus, a flux correction factor of about 0.67 can be applied when converting CFS-MFI_{UF} 6:1 to CFS-MFI_{UF} 1:1. This flux correction factor is specific for the hydrodynamic conditions. The changes in either crossflow velocity or the flux may lead to a different value of this factor. As a result, a new calibration on flux correction factor must always be conducted in accordance to the real RO plant operating flux and crossflow velocity.

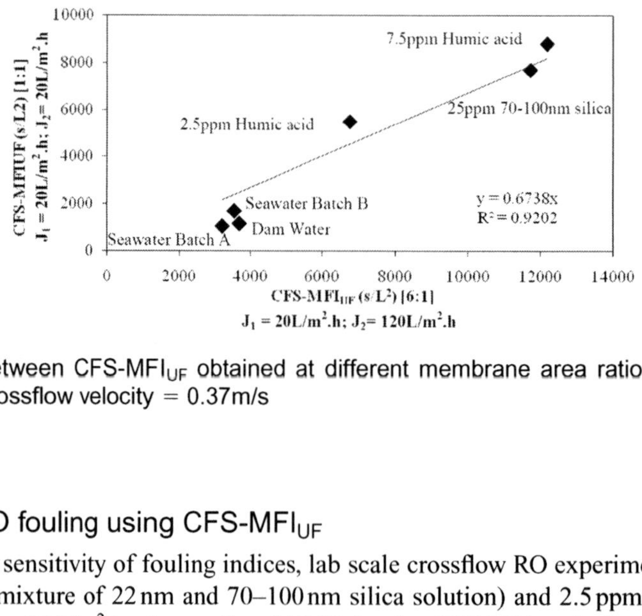

Figure 1.23 Relationship between CFS-MFI_{UF} obtained at different membrane area ratio [1:1 and 6:1] using different feed solutions. Conditions: crossflow velocity = 0.37m/s

1.3.3.4 Prediction of RO fouling using CFS-MFI_{UF}

To validate the accuracy and sensitivity of fouling indices, lab scale crossflow RO experiments were carried out. For RO filtration of 100 ppm silica (mixture of 22 nm and 70–100 nm silica solution) and 2.5 ppm humic acid mixture solution in the absence of salt at flux 20 L/m^2.h, the predicted TMP by the CFS-MFI_{UF} deviated only 11% from the actual RO TMP profile (Figure 1.24). Although MFI-$UF_{const.flux}$ showed relatively good prediction at the early stage of the filtration, MFI-$UF_{const.flux}$ predicted trend deviated significantly from the actual fouling trend as the filtration proceeded, with the deviations of more than 50% occurring over time. This might be due to the difference in cake compositions and cake structure between crossflow RO filtration and dead-end MFI-$UF_{const.flux}$ test. Better fouling prediction was observed when deposition factor was taken into account. However, there is still at least a 30% deviation from the actual fouling rate. This implies that even with the additional measurement, the accuracy of the fouling prediction cannot be guaranteed. The changes of cake formed on RO over time may be difficult to be represented by this single value.

To demonstrate the impact of salt concentration, an experiment using higher salt concentration feed solution was conducted, shown in Figure 1.25. CFS-MFI_{UF} predicted trend was found unable to match the actual RO fouling behaviour, suggesting that another fouling mechanism, known as cake-enhanced osmotic pressure (CEOP) effect, may come into play due to the stagnant foulant layer on the membrane surface. When incorporating the mass information from the CFS-MFI_{UF} test to account for the CEOP effect, the prediction showed good agreement with the TMP profile of the RO system. The results indicated that the CFS-MFI$_{UF}$ test which includes the CEOP effect is a very promising technique to provide an estimation of RO colloidal fouling profile. However, the predicted trend did not indicate two stages of the extended fouling profile that is present in the experimental data (Sim *et al.*, 2011b). By further adding to the model the potential changes in the cake porosity, the predicted TMP profile clearly generated two stages of fouling profile which agreed well with the experimental data. The results highlighted the observed impact of the CEOP may not only dependent on the load of foulant but also the structure of the deposit such as cake thickness and porosity.

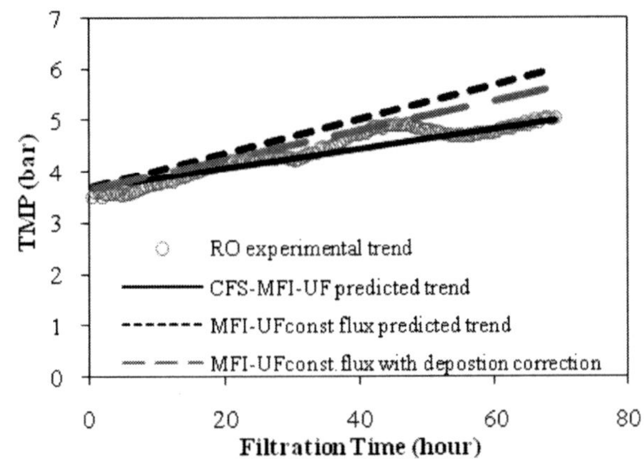

Figure 1.24 Predicted RO fouling trend and actual RO fouling behaviour. Conditions: RO flux = 20 L/m^2.h; Crossflow velocity = 0.37 m/s; Feed = 50 ppm of 22 nm and 50 ppm of 70–100 nm silica colloidal mixed with 2.5 ppm of Humic acid.

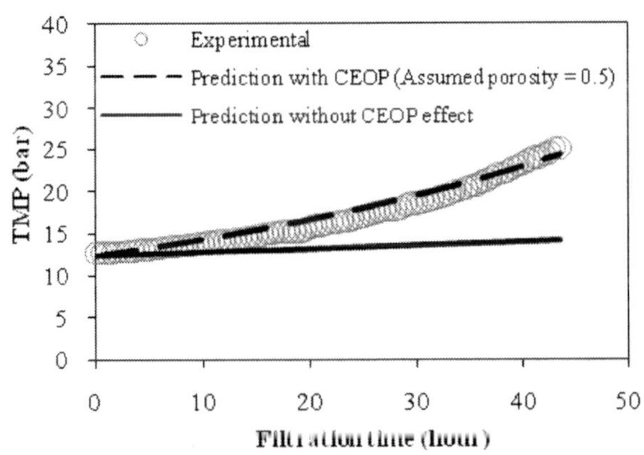

Figure 1.25 CFS-MFIUF predicted TMP profile (ε = 0.5) with and without CEOP effect and the actual RO fouling behaviour. Conditions: RO flux = 20 L/m^2.h; crossflow velocity = 0.22 m/s; feed = 50 ppm of 22 nm and 50 ppm of 70–100 nm silica colloidal and 6 g/L NaCl.

1.4 ASSESSMENT OF BIOFOULING POTENTIAL OF SALT AND BRACKISH WATER FOR RO DESALINATION PLANTS – BDOC/LC-OCD

1.4.1 Method

To characterise the biofouling potential of marine and brackish waters an advanced analytical method based on the combination of the parameters Biodegradable Dissolved Organic Carbon (BDOC) coupled with liquid chromatography was developed and evaluated. To characterise and quantify organic carbon components in salt and brackish water the approach known as Liquid Chromatography – Organic Carbon Detection (LC-OCD) (Huber, 2007; Huber and Frimmel, 1991 and Huber and Frimmel, 1994) was modified to allow the quantification of organic carbon down to the low-ppb range in salt water matrices. The results of a reproducibility test (Table 1.12) showed that parallel samples had variations in the range of ±20–30 μg/L for summaric DOC and few μg/L for individual fractions. Adaptation of the LC-OCD technique to marine waters included: *a*) the prolongation of the chromatographic run by coupling two standard columns (TSK HW 50(S), Toso, Japan) in series to improve the resolution and *b*) the increase of the flow rate for acidification from 0.3 ml/min to 0.5 ml/min. The latter was required to convert potentially present nanocrystalline carbonate and magnesium/calcium bicarbonate ("Dolomite") to carbonic acid in the inorganic carbon (IC) removal step.

The quantification of the amount of biodegradable dissolved organic carbon (BDOC) in salt water was performed by applying the BDOC assay based on bacteria attached to sand. Biodegradation was higher using attached bacteria rather than suspended bacteria, as published by Volk *et al.* (1994). Authors had investigated the role of several operational conditions and their effect on biodegradation effectiveness. They found that bacterial inocula, incubation time and the ratio of sand volume to water (sample) volume were of great importance for BDOC results. As stated by the authors, the ratio of sand to water was optimal with 300 g of sand and 600 ml of water sample resulting in an optimal biodegradation. The BDOC method in general is operationally defined. It is common to all modifications that the bacteria used are sessile on sand or gravel. No higher organisms should be present. BDOC can be carried out with incubation times from 14 to 28 days at a defined temperature with sand or gravel as substrate. How suspended bacteria become sessile bacteria, which populations are used, and how they are able to degrade components is specific for each experimental approach (Volk *et al.*, 1994).

Table 1.12 Reproducibility test

	DOC	BIO-polymers	DON	Humic Subst. (HS)	DON	Aromaticity (SUVA-HS)	Mol-Weight (Mn)	Building Blocks	Neutrals	Acids	Inorg. Colloid.
Approx. Molecular Weights in g/mol:		>>20.000		~1000 (see separate HS-Diagram)				300-500	<350	<350	
			(Norg)		(Norg)						SAC
	ppb-C	ppb-C	ppb-N	ppb-C	ppb-N	L/(mg*m)	g/mol	ppb-C	ppb-C	ppb-C	(m^{-1})
	% TOC	% TOC	--	% TOC	--	--	--	% TOC	% TOC	% TOC	--
Sample 1	**1474**	**259**	9	**435**	11	**0,68**	**494**	**162**	**500**	**117**	0,00
	100,0	17,6	--	29,5	--	--	--	11,0	33,9	8,0	--
Sample 2	**1418**	**249**	8	**411**	11	**0,95**	**493**	**138**	**507**	**113**	0,17
	100,0	17,6	--	29,0	--	--	--	9,7	35,7	8,0	--
Sample 3	**1394**	**255**	9	**393**	9	**1,07**	**495**	**136**	**497**	**112**	0,13
	100,0	18,3	--	28,2	--	--	--	9,7	35,7	8,0	--
Sample 4	**1470**	**274**	10	**406**	12	**0,95**	**485**	**144**	**533**	**113**	0,10
	100,0	18,6	--	27,6	--	--	--	9,8	36,2	7,7	--
Sample 5	**1454**	**272**	11	**400**	15	**0,94**	**472**	**146**	**524**	**113**	0,17
	100,0	18,7	--	27,5	--	--	--	10,0	36,1	7,7	--
Sample 6	**1390**	**259**	12	**392**	13	**0,93**	**474**	**136**	**486**	**117**	0,19
	100,0	18,6	--	28,2	--	--	--	9,8	35,0	8,4	--
average	**1434**	**261**	10	**406**	12	**0,92**	**486**	**144**	**508**	**114**	0,13
standard deviation	38	10	1	16	2	0,13	10	10	18	2	0
half confidence interval											
in ppb	**40**	**10**	1	**17**	2	**0,14**	**11**	**11**	**18**	**3**	0,07
in %	2,8	3,9	14,3	4,2	18,7	14,9	2,2	7,4	3,6	2,2	57,6

In our experiments, the newly developed BDOC assay for salt and brackish water was performed in a batch assay using "DOC-free" two-litre glass flasks. Each flask contained 300 ml of sand grains with a diameter of 0.8 to 1.5 mm, resulting in a total number of approximately 70,000 grains with an average surface area of 3,000 cm^2 ± 500 cm^2. An important requirement for BDOC experiments is "DOC free" sand which was obtained by annealing the sand (950° C) at the beginning of each BDOC experiment resulting also in sterility of the sand and the flask. BDOC assay preparations started by adding DOC-free and particle-free (filtered over 0.2 μm pore size), sterile water to the sand, shaking the sand/water mixture and decantation to remove remaining particles.

1.4.2 Sampling

To perform BDOC/LC-OCD experiments, several flasks with water sample have to be collected at site:

1. 100 ml for LC-OCD analysis which has to be sent immediately for analysis to the laboratory ("BDOC start")
2. 2 × 1,000 ml of raw water sample which should be characterised for BDOC experiments.
3. 2 × 1,000 ml of RO feed or biofilm from RO membrane for preparation of the bacterial inoculum.

1.4.3 Test procedure

BDOC experiments started by preparing the bacterial inoculum. For this, sample No 3 in case of RO feed was concentrated e.g. by centrifugation. The bacterial pellet was then re-suspended and washed with sterile sodium chloride solution (3%), then diluted to a concentration of at least 5×10E+05 bacteria/ml. However, higher bacterial concentrations up to 10E+07 bacteria/ml are desirable, but often difficult to obtain, because the bacterial concentration in RO Feed is low. 100 ml of this bacteria suspension was added to the sand with the aim to allow bacteria to adhere to the sand surface. It is known that higher cell densities will lead to relatively more bacteria attached and will probably result in better biodegradation rate. As the decrease in DOC may be relatively small, it is recommended to run two assays in parallel. After an incubation time of at least 48 hours the surplus bacterial suspension was decanted from the sand. Further preparation has to make sure that very low concentration of DOC is present in the assay before water sample is added.

Thereafter the BDOC assay started by adding 600 ml of the original water sample. The remaining air volume in the flask was 1,200 ml; flasks were closed tightly. During the incubation time the flasks were positioned horizontally and were agitated slightly on a daily basis. Incubation temperature was 20°C. Incubation time was 14 days or 21 days. The amount of biodegraded matter was determined by analysing DOC before and after incubation with the LC-OCD method. As illustrated by Huber (2007), LC-OCD splits DOC compounds in five fractions as described in section 1.2.1.

1.4.4 Results

Water samples have been collected within the MEDINA project in the years 2007 to 2009 from four different desalination plants around the world. They were characterised by LC-OCD and their BDOC was determined. BDOC assays were performed and evaluated at IWW Water Centre, Germany and LC-OCD analyses were carried out by DOC Labor Dr. Huber (Germany).

BDOC of water samples mentioned above was determined as difference (start minus end) by calculating the decrease of DOC for each of the fractions detected by LC-OCD. Thus, in addition to summaric DOC values, more information is obtained for individual fractions. Specific LC-DOC fractions could be identified which most contributed to biofouling potential on the basis of their biodegradability (Figure 1.26).

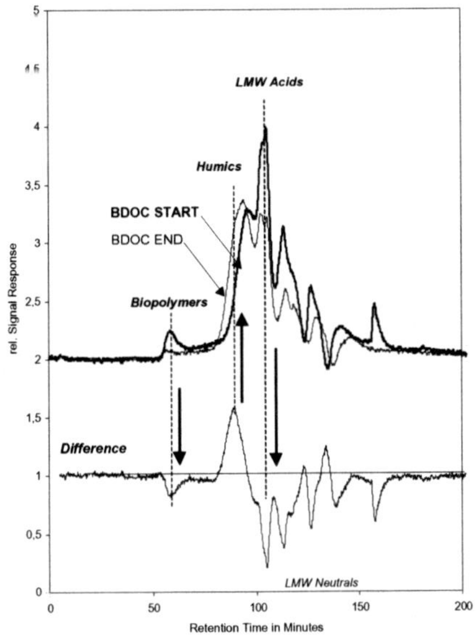

Figure 1.26 LC-OCD chromatograms of a BDOC assay (Site A, DMF effluent, before and after incubation for 21 days)

For some samples studied it was found that the fraction called Biopolymers (>>20,000 g/mol) contributed to BDOC whereas the other fractions contributed less. For Humic Substances with molecular weights of about 1,000 g/mol/L the results were not conclusive, in some cases Humic matter (and "LMW Acids") decreased, in other cases humic matter (and "Building Blocks" increased. It may be speculated that humic matter is produced by bacteria as final, refractory end product of microbial activity. In some tests this surplus humic matter was probably adsorbed on the sand substrate while in other tests it did not or did less. For other sample sites studied, no or only little degradation of biopolymers was found, which is most likely related to the composition of in these waters: The amount in biopolymers was very low and no or only little degradation was observed. Clearly, the tests performed within MEDINA project are regarded as snapshots as the method had to be developed and evaluated but many more tests were performed since.

1.4.4.1 Behaviour of antiscalants

In one sample characterised by BDOC/LC-OCD, the fraction "Humics" was very high which could be explained by presence of an antiscalant which overlapped with the humic matter fraction. The fraction was partly biodegradable and enhanced BDOC drastically. Further antiscalants and other water additives were investigated with LC-OCD (Figure 1.27). It could be demonstrated that all of them have specific retention times and in one case a strong response for UV at 254 nm. Typically, antiscalants with molecular weights of 5,000 g/mol elute between the fraction of "Biopolymers" and "Humics". This was confirmed for 4 products, while 2 products co-eluted with the humic fraction.

Therefore in conventional BDOC tests based on summaric DOC values it is not possible to distinguish between biodegradation of natural compounds and synthetic additives. They may contribute to DOC as well as to BDOC. This was investigated within MEDINA in chapter 3.

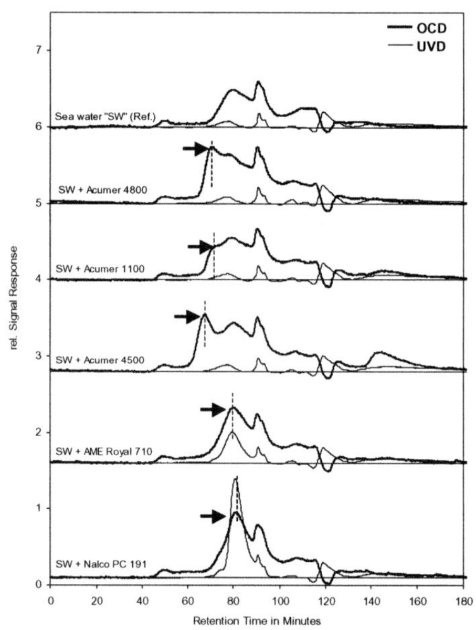

Figure 1.27 LC-OCD chromatograms of a raw water spiked with about 1 mg/L (as mass) of polymer additives of 6 different manufacturers

1.5 DETERMINATION OF THE CONCENTRATIONS OF AOC AND ATP IN SEAWATER

1.5.1 Introduction

1.5.1.1 Biofouling and substrate utilization by bacteria

Biofouling of spiral-wound RO membranes used for desalination of seawater and brackish water and in water treatment is a main cause of operational problems, e.g., increase of pressure drop in the feed channel and a decrease of the membrane flux (Khedr, 2000 and Schippers *et al.*, 2004). Biofouling processes include microbial attachment and growth which lead to biomass accumulation in the feed channel (Ridgway and Flemming, 1996). Initial attachment of microorganisms to the water-exposed surface usually starts within a few minutes and may be reversible only for a short period of time Irreversible attachment is the next phase, caused by the production of extracellular polymeric substances (EPS) which adsorb strongly onto the surface (Marshall *et al.*, 1971 and Zobell, 1943). Simultaneously, multiplication proceeds by the uptake of growth substrates from the water.

Many natural organic compounds (NOM), e.g., amino acids, proteins, carbohydrates, and carboxylic acids are favorite sources of carbon and energy for bacterial biomass production (growth), but also certain inorganic compounds, e.g., ammonia, can promote growth. The growth rate of bacteria is characterized by the time period needed for one division of the bacterial cell, the generation time (G, h) or doubling time, which is constant during exponential growth. The growth rate (V, h^{-1}), which is 1/G, of a specific micro-organism depends on the concentration and nature of the growth-promoting compound(s) and temperature, provided that sufficient other nutrients (e.g., phosphorous, nitrogen, oxygen) are present and inhibiting compounds (e.g., disinfectants) are absent. The relationship between V and the concentration (S) of the growth-limiting substrate is given by the Monod equation, viz. $V = V_{max} \times S/(S + K_s)$, where V_{max} is the maximum growth rate (h^{-1}), and K_s is the value for S at which $V = 0.5V_{max}$. Values for K_s determined with pure cultures isolated from treated water can be as low as a few micrograms per litre (Van der Kooij and Hijnen, 1984 and Van der Kooij and Hijnen, 1988). Experiments with mixtures of substrates have further shown that individual compounds present in a mixture can be utilized at a level below 0.1 μg C/L (Van der Kooij and Hijnen, 1988). The production of biomass is directly related to the amount of utilized substrate. Complete utilization of 1 μg of C yields 4×10^6 to 2×10^7 bacterial cells. Differences in these yield values are explained by (i) difference in cell size and/or (ii) difference in biomass production on different organic compounds, e.g. oxalate and acetate (Van der Kooij and Hijnen, 1984).

1.5.1.2 Microbial growth potential of feed water

The nature and concentration of growth-promoting (biodegradable) compounds in feed water depends on (i) quality of the raw water, (ii) effects of pre-treatment processes and (iii) effects of chemicals added to the feed water. Information about effects of raw water composition, water treatment and addition of chemicals on the concentration of growth-promoting compounds is needed for taking decisions regarding prevention and control of biofouling. A variety of methods is available for determining the microbial growth potential (MGP) of treated fresh water. These techniques include (i) the AOC test (Van der Kooij, 1992 and Van der Kooij *et al.*, 1982), (ii) the BDOC test, which determines the concentration of biodegradable dissolved organic carbon in water exposed to sand with attached bacteria (Joret and Lévi, 1986), (iii) determination of the biofilm formation rate (BFR) of the water using a biofilm monitor (van der Kooij *et al.*, 1995). The AOC test is based on determining the maximum level of growth of two pure bacterial cultures, *Pseudomonas fluorescens* strain P17 and *Spirillum* sp. strain NOX, in pasteurized samples of water to be tested. The maximum level of growth (N_{max}, CFU/ml) of the pure cultures is used to calculate the concentration of utilized compounds by using the yield of these strains for acetate. The result of the AOC test is expressed as acetate-carbon equivalents/L. The AOC test was developed for fresh water and cannot be applied for seawater, because one of the used bacteria does not grow at high salt concentrations.

1.5.1.3 Microbial biomass

Microbial biomass present in the feed water can contribute to biofouling of the feed channel of spiral-wound membranes. Bacterial cells constitute an important fraction of the biomass in seawater. Heterotrophic plate counts represent only a small fraction of the total number of bacterial cells, that can be determined using microscopy (Hobbie *et al.*, 1977). In addition to this method, the analysis of the concentration of ATP in water also gives information about the total amount of active biomass (Holm-Hansen and Booth, 1966 and Karl, 1980). High salt concentrations hamper the ATP analysis.

1.5.1.4 Study objectives

The following objective were defined: (i) selection of one or more marine bacteria suited for use in an AOC test for seawater; (ii) application of this test to seawater and feed water prepared from seawater and (iii) adaptation of the ATP analysis for use in seawater.

1.5.2 Materials and methods

1.5.2.1 Bacterial strains and precultivation

The following strains were selected based on a study of the literature: *Halomonas cupida* (DSM No. 4740), *Alteromonas macleodii* (DSM No 6062), *Marinomonas communis* (DSM No. 5604). *Pseudoalteromonas atlantica* and *Vibrio* species had been isolated from tap water supplemented with NaCl (30 g/L), enriched with acetate (100 μg C/L) and incubated at 25°C after inoculation with seawater. The identity of the strains was either confirmed or established by sequence analysis of the gene encoding 16S rRNA.

The bacterial strains were grown in 100 ml of tap water, supplemented with 30 g/L of NaCl and enriched with 1 mg/L of either acetate or glucose, incubated at 25°C. Growth tests in either seawater or tap water with added NaCl (30 g/L) were conducted in thoroughly cleaned glass-stoppered Erlenmeyer flasks with a volume of 1 L. Tap water (600 ml) with added NaCl contained in the Erlenmeyer flasks was heated to 60°C in a water bath and maintained at this temperature for 30 min. The procedures for cleaning and sample treatment have been described in a previous publication (Van der Kooij, 1992). Samples of seawater were membrane filtered (UF membrane, polysulfone, MWCO 30 kD; Fresenius Medial Care, Bad Homburg, Germany). Selected organic compounds from heat treated freshly prepared stock solutions were added to the water in the Erlenmeyer flasks prior to the heat treatment. These compounds included 20 different amino acids (AA), 19 different carboxylic acids (CA), 19 carbohydrates and polyalcohols (CH) and 11 aromatic acids (AR), which were present as mixtures. Duplicate flasks were incubated at 25°C and bacterial growth was assessed with the plate count method using Lab Lemco Agar (Oxoid Ltd) supplemented with 30 g/L of NaCl. Plates were incubated at 25°C for 48 to 72 hours, depending on the used strain. Growth tests with sea water were done after UF treatment to remove suspended biomass, followed by pasteurization to inactivate indigenous bacteria. All samples were incubated at 25°C.

1.5.2.2 ATP analysis

The measurement of ATP is based on the production of light in the luciferine-luciferase assay. ATP was released from suspended cells with nucleotide-releasing buffer (NRB, Celsis). The intensity of the emitted light was measured in a luminometer (Celsis AdvanceTM) calibrated with solutions of free ATP (Celsis) in autoclaved tap water following the procedure given by the manufacturer. The detection limit was 1 ng/L of ATP.

1.5.2.3 Results

1.5.2.3.1 Selection of test strains

Growth of 5 different strains was measured in seawater sampled from the North Sea, close to the shore. The maximum level of growth, (N_{max}, CFU/ml) which was reached within one week of incubation, ranged from 2.6×10^3 CFU/ml (PA) to 1.6×10^4 CFU/ml (HC) for the tested strains. After confirmation that N_{max} had been reached, a mixture (total mixture, TM) of 73 easily biodegradable low molecular-weight compounds was added to the flasks and growth measurements were continued. Also under these conditions, clear differences were observed between the N_{max} values, with strain HC reaching the highest value (Figure 1.28A). No decline in plate counts was observed with strain HC, of which colonies on the solid medium were clearly visible after 24 hours. Furthermore, this organism had been characterized as highly versatile (Baumann *et al.,* 1972). Therefore, strain HC was selected for further testing.

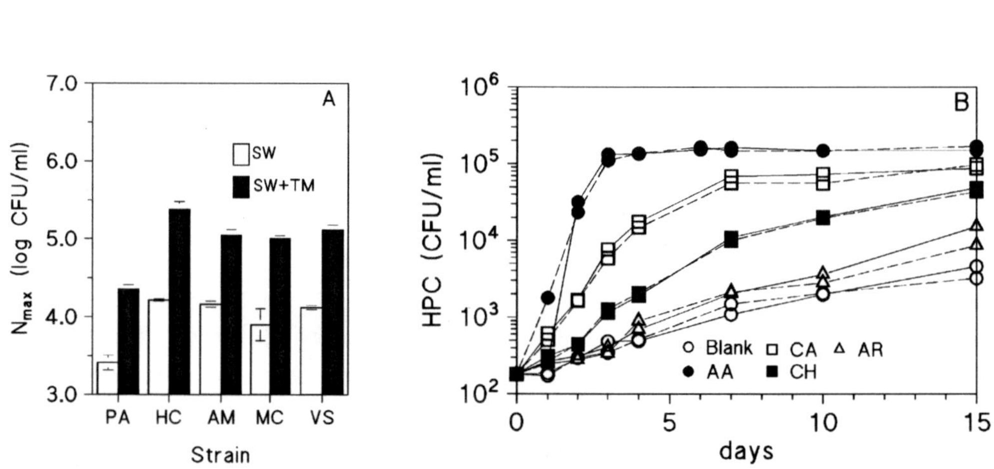

Figure 1.28 A: Maximum colony counts (N_{max}, CFU/ml) of seleted strains grown in pasteurized seawater (SW) without and with the total mixture (TM). PA, *Pseudoalteromonas atlantica*; AM, *Alteromonas macleodii*; HC, *Halomonas cupida*; MC, *Marinomonas communis*, and VS, *Vibrio species*. B: Growth of strain HC with mixtures of substrates at individual concentrations of 1 μg C/L in tap water supplemented with NaCL (30 g/L). AA, amino acids; AR, aromatic acids; CA, carboxylic acids and CH, carbohydrates.

1.5.2.3.2 Nutritional versatility of strain HC and its growth yield for acetate

Growth tests were conducted in tap water with NaCl (30 g/L) supplemented with mixtures of amino acids (AA), aromatic acids (AR), carboxylic acids (CA) and carbohydrates (CH) for determining the nutritional versatility of strain HC at low concentrations (micrograms/L) Growth was most rapidly with AA and CA where N_{max} was reached within one week of incubation (Figure 1.28B).

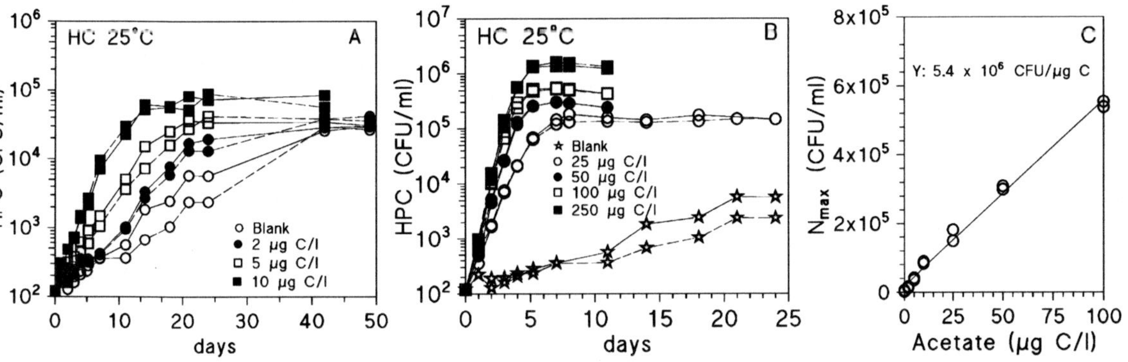

Figure 1.29 A, B: Growth of strain HC in tap water supplemented with NaCl (30 g/L) and different concentrations of acetate. C: relationship between N_{max} and acetate concentration for calculation of yield

Growth tests with individual carboxylic acids at a concentration of 10 μg of C/L revealed that most of these compounds promoted growth at this concentration (results not shown). Growth was most pronounced ($N_{max} > 10^5$ CFU/ml) with DL-lactate, malate, succinate, propionate, acetate, malonate and pyruvate. Growth tests at a range of acetate concentrations were conducted to determine the relationship between the acetate concentration and the N_{max} value (yield). Figure 1.29 shows that acetate concentrations as low as of a few μg of C/L promoted growth. The maximum colony counts of strain HC were linearly related with the acetate concentration and a yield of 5.4 (±0.1) × 10^6 CFU/μg of C was calculated. In a similar way the yield for the mixture of 20 amino acids was obtained, viz. 8.4 (±0.3) × 10^6 CFU/μg of C.

1.5.2.3.3 AOC in North Sea water and feed water

Strain HC multiplied in samples of seawater collected from the North Sea (DOC 0.8 to 1.1 mg/L) at different distances to the coast. These samples were UF filtered prior to pasteurization and inoculation. In a few of these samples maximum growth was reached within one week of incubation, corresponding with an AOC concentration of about 10 μg ac-C equivalents/L (Figure 1.30A). Growth of strain HC in most samples was relatively slow, and the AOC concentrations calculated from the colony counts remained below 1 μg ac-C equivalents/L within one week of incubation. This observation suggests that the concentration of compounds readily available to strain HC, e.g., amino acids, was less than 1 μg of C/L. AOC tests using strain HC in seawater used for desalination (site A) showed an elevated level, but growth was relatively slow. Feed water, after pre-treatment with either microfiltration or sand filtration showed very low AOC concentrations (Figure 1.30B). These observations show that growth measurements with *Halomonas cupida* strain HC can be used to obtain information about the microbial growth potential of sea water, as affected by low molecular weight compounds.

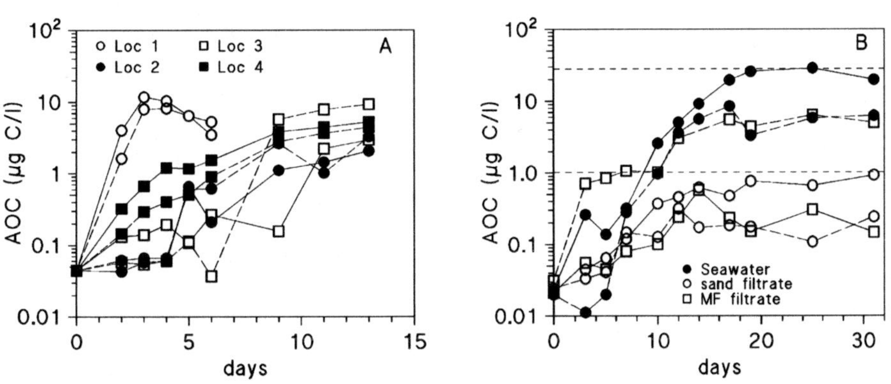

Figure 1.30 Growth of *Halomonas cupida* strain HC in water from the North Sea, treated with UF

1.5.2.3.4 Determination of the ATP concentration of seawater

The impact of the concentration of salts on the determination of ATP was tested in tap water supplemented with ATP (c. 800 ng/L) and with different concentrations of either NaCl or Red Sea Salts (RSS). Salt concentrations clearly below concentration in seawater strongly reduced the production of light (RLU, relative light units) in the ATP assay (Figure 1.31). No difference was observed between the effects of NaCl and RSS. At concentrations <1 g/L, the reduction was less than 20%. Figure 1.31B shows that the conductivity of the water is linearly related to the salts concentration. Therefore, the following method for measuring the ATP concentration of seawater is proposed: (i) dilution of seawater with ATP free (autoclaved) demineralized water to a salts concentration of 1 to 2 g/L; (ii) analysis of the ATP concentration in the diluted seawater, (iii) determination of the conductivity of the diluted seawater and (iv) calculation of the ATP concentration of undiluted seawater by taking into account the reduced yield of the ATP analysis and the dilution factor.

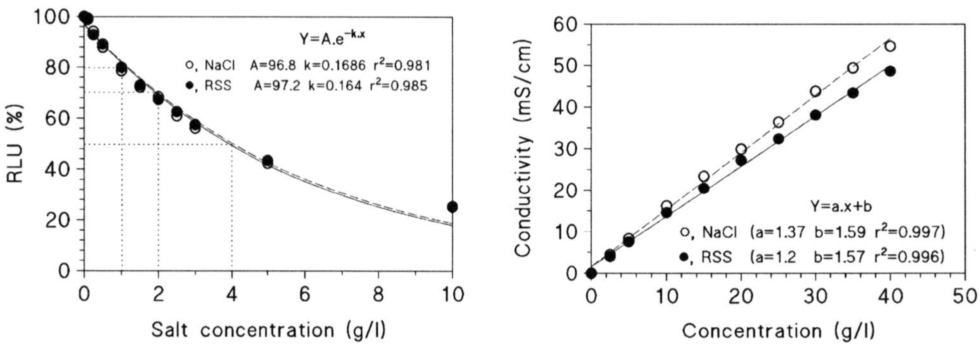

Figure 1.31 A. Impact of salts on the ATP analysis; B, Relationship between salts concentration and conductivity. RSS = Red Sea salts (Red Sea Salts Fish Farm Ltd, Eilat, Israel)

1.5.2.4 Discussion and conclusions

1.5.2.4.1 AOC

The presented data confirms that bacteria present in marine environments are able to utilize very low concentrations of easily biodegradable compounds, which is no surprise.A single organism cannot represent the metabolic potential of natural bacterial assemblages, but a single strain gives information about the presence of specific types of compounds. For strain *Halomonas cupida* (HC), these compounds include the amino acids, the carboxylic acids and also a number of carbohydrates (Baumann *et al.*, 1972). However, *Halomonas cupida* does not hydrolyze biopolymers such as polysaccharides and proteins. Inclusion of strains with such capacities and one or more other bacteria may lead to a more complete assessment of the microbial growth potential of sea water. Still, the use of strain HC, which has a number of favorable properties, including a very large nutritional versatility and rapid growth on solid media, gives information about the presence of compounds which are commonly related to biological activity, viz. amino acids and carboxylic acids. Low N_{max} values of growth of strain HC therefore indicate low activity and high values indicate a high growth potential. The observed values are much lower than the AOC value (about 280 μg of C/L) reported for filtered seawater collected at the coast of Singapore (Ng *et al.*, 1997). The difference may be explained by differences in water quality and in sample treatment.

A recent study showed that the threshold value for the concentration of an easily biodegradable compound (acetate) to induce biofouling problems in spiral-wound membranes is only a few micrograms/L (Hijnen *et al.*, 2009). At a concentration of 15 μg of acetate-C/L, the rate of biofouling was half of the maximum value. AOC concentrations observed in UF-treated North Sea water were less than 10 μg C/L (Figure 1.30). The slow growth in most samples from the North Sea suggests that the concentration of low molecular-weight compounds which are readily available to strain HC is very low (<1–2 μg of C/L), i.e., close to the threshold value for biofouling. Establishment of a relationship between AOC concentrations obtained with growth tests using strain HC and the rate and extent of biofouling of RO membranes requires further study. However, other compounds, which remain undetected with the AOC assay with strain HC, e.g., biopolymers derived from biomass, may be present and induce biofilm formation. These compounds are part of the BDOC concentration and can be detected with LC-OCD (Schaule *et al.*, 2010).

Other challenges in conducting tests for the microbial growth potential of feed water include sampling and sample treatment. Sampling should be conducted at a representative location, preferentially at a continuously flowing tap. The water should be collected in thoroughly cleaned glass containers to prevent introduction of easily biodegradable compounds from the container. Subsequently, during storage and transportation the sample should be kept at a low temperature (<5°C) to limit microbial activity in the sample, which may lead to uptake or production of AOC. Finally, the use of UF for the removal of suspended biomass from the samples needs further attention.

1.5.2.4.2 ATP

The observations presented in Figure 1.31 confirm that high salts concentrations seriously hamper the ATP analysis of seawater and brackish water. Dilution is one way to overcome this problem. Another method is based on sample filtration to collect the biomass. Subsequently, the collected biomass is inserted in the test system and ATP is measured. Advantages of this filtration method are (i) salts do not hamper the analysis and (ii) the detection limit is clearly lower than in the dilution method. A disadvantage is ATP leakage from cells during filtration. A comparison of methods is needed to determine which approach gives the best results. In addition, determination of the total number of cells in the water, e.g., using microscopy or flow cytometry (Manes *et al.*, 2010) is an alternative approach for quantifying the concentration of bacteria in (sea)water establishing the other.

REFERENCES

Adham S. and Fane A. (2008). *Cross Flow Sampler Fouling Index*. National Water Research Institute, California, USA.

American Society for Testing Materials. (2007). *D4189 – 07 Standard Test Method for Silt Density Index (SDI) of Water*. ASTM.

Amy G. and Her N. (2004). Size exclusion chromatography (SEC) with multiple detectors: a powerful tool in treatment process selection and performance monitoring. *Water Science and Technology: Water Supply*, **4**, 19–24.

Baumann L., Baumann P., Mandel M. and Allen R. D. (1972). Taxonomy of aerobic marine bacteria. *J. Appl. Bacteriol.*, **110**, 402–429.

Boerlage S. F. E. (2001). *Scaling and Particulate Fouling in Membrane Filtration Systems*. Swets & Zeitlinger Publishers, Lisse.

Boerlage S. F. E., Kennedy M. D., Aniye M. P. and Schippers J. C. (2003a). Applications of the MFI-UF to measure and predict particulate fouling in RO systems. *Journal of Membrane Science*, **220**, 97–116.

Boerlage S. F. E., Kennedy M. D., Tarawneh Z., Abogrean E. and Schippers J. C. (2003b). The MFI-UF as a water quality test and monitor. *Journal of Membrane Science*, **211**, 271–289.

Her N., Amy G., Foss D. and Cho J. (2002). Variations of molecular weight estimation by HP – size exclusion chromatography with UVA versus on-line DOC detection. *Environmental Science and Technology*, **36**, 3393–3399.

Her N., Amy G., Park H. and Von-Gunten V. (2004). UV absorbance ratio index with size exclusion chromatography (URI-SEC) as a NOM property indicator.

Hijnen W. A. M., Biraud D., Cornelissen E. R. and Kooij D. V. D. (2009). Threshold concentration of easily assimilable organic carbon in feed water for biofouling of spiral-wound membranes. *Environmental Science & Technology*, **43**, 4890–4895.

Hobbie J. E., Daley R. J. and Jasper S. (1977). Use of nucleopore filter for counting bacteria by fluorescence microscopy. *Appl. Environ. Microbiol.*, **33**, 1225–1228.

Holm-Hansen O. and Booth C. R. (1966). The measurement of adenosine triphosphate in the ocean and its ecological significance. *Limnol. Oceangr.*, **11**, 510–519.

Huber S. (2007). *LC-OCD applications* [Online]. DOC-Labor Dr. Huber (online). Available: http://www.doc-labor.de/ (accessed 01 August 2007).

Huber S. A. and Frimmel F. H. (1991). Flow injection analysis of organic and inorganic carbon in the low-ppb range. *Anal. Chem.*, **63**, 2122–2130.

Huber S. A. and Frimmel F. H. (1994). Direct gel chromatographic characterization and quantification of marine dissolved organic carbon using high-sensitivity DOC detection. *Environmental Science and Technology*, **28**, 1194–1197.

Joret J. C. and Lévi Y. (1986). Méthode rapide d'évaluation au carbone éliminable des eaux par voie biologique. *Trib. Cebedeau*, **510**, 3–9.

Karl D. M. (1980). Cellular nucleotide measurements and implications in microbial ecology. *Microbiol. Rev.*, **44**, 739–796.

Khedr M. (2000). Membrane fouling problems in reverse osmosis desalination applications. *Desalination & Water Reuse*, **10**, 8–17.

Khirani S., Ben Aim R. and Manero M.-H. (2006). Improving the measurement of the modified fouling index using nanofiltration membranes (NF-MFI). *Desalination*, **191**, 1–7.

Manes C., West N. J. and Lebaron P. (2010). Enumeration, identification and diversity of marine microorganisms including bacteria and phytoplankton. In: *Medina Project*, IWA (ed.), IWA.

Marshall K. C., Stout R. and Mitchell B. (1971). Mechanism of the initial events in the sorption of marine bacteria to surfaces. *J. Gen. Microbiol.*, **68**, 337–348.

Mcknight D. M., Boyer E. W., Westerhoff P. K., Doran P. T., Kulbe T. and Andersen D. T. (2001). Spectrofluorometric characterization of dissolved organic matter for indication of precursor organic material and aromaticity. *Limnol. Oceanog.*, **46**, 38–48.

Ng W. J., Ong S. L., Hu J. Y. and Wang L. (1997). Study on biological stability of water. *J. Chinese Inst. Environ. Eng.*, **9**, 37–42.

Ridgway H. F. and Flemming H. C. (1996). Membrane biofouling. In: *Water Treatment Membrane Processes*, J. Mallevialle, P. E. Odendaal and M. R. Wiesner (eds.), McGraw-Hill, New York, pp. 6.1–6.60.

Schaule G., Petrovski K. and Huber S. (2010). Assessment of the biofouling potential of salt and brackish water for RO desalination plants –BDOC/LC-OCD. In: *Medina Project*, IWA (ed.), IWA.

Schippers J. C., Hanemaayer J. H., Smolders C. A. and Kostense A. (1981). Predicting flux decline or reverse osmosis membranes. *Desalination*, **38**, 339–348.

Schippers J. C., Taylor J. S., Kruithof J., Hofman J. and Nederlof M. (2004). *Integrated Membrane Systems*. American Water Works Research Association, Elbert, Colorado.

Schippers J. C. and Verdouw J. (1980). The modified fouling index, a method of determining the fouling characteristics of water. *Desalination*, **32**, 137–148.

Sim L. N., Ye Y., Chen V. and Fane A. G. (2010). Crossflow sampler modified fouling index ultrafiltration (CFS-MFIUF) – an alternative fouling index. *Journal of Membrane Science*, **360**, 174–184.

Sim L. N., Ye Y., Chen V. and Fane A. G. (2011a), Comparison of MFI-UF constant pressure, MFI-UF constant flux and Crossflow Sampler – Modified Fouling Index Ultrafiltration (CFS-MFIUF), *Water research*, **45**(4), 1639–1650.

Sim L. N., Ye Y., Chen V. and Fane A. G. (2011b) Investigations of the coupld effect of cake-enhanced osmotic pressure and colloidal fouling in RO using crossflow sampler-modified fouling index ultrafiltration, Desalination, doi: 10.1016/j. desal.2011.01.059

Van Der Kooij D. (1992). Assimilable organic carbon as an indicator of bacterial regrowth. *J. American Water Works Association*, **84**(2), 57–65.

Van Der Kooij D. and Hijnen W. A. M. (1984). Substrate utilization by an oxalate-consuming Spirillum species in relation to its growth in ozonated water. *Appl. Environ. Microbiol.*, **47**, 551–559.

Van Der Kooij D. and Hijnen W. A. M. (1988). Nutritional versatility and growth kinetics of an Aeromonas hydrophila strain isolated from drinking water. *Appl. Environ. Microbiol.*, **54**, 2842–2851.

Van Der Kooij D., Veenendaal H. R., Baars-Lorist C., Van Der Klift D. and Drost Y. C. (1995). Biofilm formation on surfaces of glass and teflon exposed to treated water. *Water Research*, **29**, 1655–1662.

Van Der Kooij D., Visser A. and Hijnen W. A. M. (1982). Determining the concentration of easily assimilable organic carbon in drinking water. *J. American Water Works Association*, **74**, 540–545.

Volk C., Renner C., Robert C. and Joret J. C. (1994). Comparison of two techniques for measuring biodegradable dissolved organic carbon in water. *Environmental Technology*, **15**, 545–556.

Zobell C. E. (1943). The effect of solid surfaces upon bacterial activity. *J. Bacteriol.*, **46**, 39–56.

Chapter 2

Evaluation and comparison of seawater and brackish water pre-treatment

Y. Orem[1], R. Messalem[1], E. Ben-David[1], M. Herzberg[1], A. Kushmaro[1], X. Ji[2], G. Di Profio[2], E. Curcio[2], E. Drioli[2,3], J. F. Laroche[4], P.-J. Remize[4], J. Leparc[4], S. Vigneswaran[5], K. J. Chinu[5], A. H. Johir[5], J. J. Lee[5], H. K. Shon[5], J. Kandasamy[5], Y. Ye[6], L. N. Sim[6], B. Herulah[6], V. Chen[6], A. G. Fane[6], C. Tansakul[7], S. Laborie[7] and C. Cabassud[7]

[1]Ben-Gurion University (Israel)
[2]University of Calabria (Italy)
[3]ITM-CNR (Italy)
[4]Anjou-Recherche, Veolia Environment (France)
[5]University of Technology, Sydney (Australia)
[6]University of New South Wales (Australia)
[7]INSA Toulouse (France)

The main objective of WP2 was to conduct a comprehensive evaluation on available pretreatment approaches for seawater, brackish water, and estuarine/bay water for membrane protection and recovery maximization.

During the project promising results were obtained by UNICAL and BGU on MBR as a means for reducing organic content from seawater as a step towards reducing fouling potential. It seems that the pretreatment in a MBR system is effective at improving the water's quality (removal of TOC and EPS) for further process, but further economical and implemental examinations are necessary in order to adopt this treatment as an alternative pretreatment.

The removal of particulate matter and organics from seawater by the use of biofiltration was investigated by UTS. Granular activated carbon (GAC) and anthracite were tested as a filter medium at two different velocities. The filtrate quality was evaluated by measuring the silt density index (SDI), modified fouling index (MFI) and turbidity removal. Both biofilters demonstrated similar behavior in terms of SDI and MFI. In addition, fiber filter as pre-treatment of SWRO in terms of MFI, SDI_{10}, pressure drop, turbidity and molecular weight distribution (MWD) was investigated with and without precoagulation. Submerged microfiltration coupled with physcio-chemical processes such as adsorption on powder activated carbon was investigated as well.

The objectives of UNSW were to investigate the seawater pretreatment using the submerged hollow fibre without the aid of coagulant. Three different MF/UF membranes are investigated based on the fouling performance. The effect of operation parameters (filtration duration, backwash duration, backwash flux etc.) on the TMP and fouling mechanisms are investigated. Based on the finding, the system operation optimization would be available. Also the composition of the feed, permeate and the waste water collected after backwash are analysed for the better understanding of the fouling mechanisms. For the seawater in Sydney area, a PVDF 0.04 um (ID/OD: 1300/650) hollow fibre membrane has the best performance during the filtration with periodical backwash when compared with the other two hollow fibre membranes. Backwash modes were investigated. It was found that the existence of the bubble during backwash did not remove more percentage of foulant from the cake. However, the bubble does influence the structure of the remaining cake and then resulting in the different fouling rate during the filtration cycle.

The objective of INSA Toulouse was to study different methods at laboratory scale including hybrid membrane systems for the pre-treatment of seawater before RO. The methods studied include UF alone or combined with a physico-chemical pretreatment such as coagulation and adsorption.

Several pre-treatment procedures have been studied in pilot scale modes by AR. These include various configurations of granular media filtration, microfiltration and a combination of various pre-treatment steps. According to the results collected over two distinct testing periods, whereas membrane pretreatment (MF/UF) provided a better SDI abatement than granular media filtration, the comparatively lower removal of organic matter through MF/UF appears to have induced a higher extent of organic fouling on the RO membrane. Also, this study demonstrated the advantages and interest for the use of advanced seawater characterization (i.e. NOM characterization in pretreated water and MFI measurement) as well as for the use of membrane autopsies during pilot-scale studies aiming at comparing the performance of pretreatment processes. It was shown that identifying a single water quality parameter to fully characterize the performance of pretreatment processes and predict the fouling rate on RO units is not a realistic objective. While SDI, MFI, particle counts provide good information related to the risk of particulate fouling, these parameters provide very limited information on organic fouling, and even less information on the risk of biofouling. This is why a multi-parametric approach is yet recommended for the monitoring of pretreatment performance.

2.1 REMOVAL OF ORGANIC MATTER AND MICROBIAL CHARACTERIZATION IN PRETREATMENT MBR SYSTEM FOR RO SEAWATER DESALINATION

2.1.1 Introduction

One of the acute problems that the world in general, and the state of Israel in principle is dealing with, is the lack of potable and agriculture use water. Seawater desalination is one of the attractive ways to deal with this problem. In the desalination process reversed osmosis (RO) membranes are used to separate the salts from the water.

There are four main types of fouling that appear on the membrane surface and damage its performance: scaling, organic fouling, colloidal fouling and biofouling. While the first three types had been intensively investigated and the methods to remove them by chemicals are relatively simple, biofouling is defined as the most problematic one, with the least research knowledge. Biofouling defined as accumulation of biomass on a surface by growth and/or deposition to such a level that it is causing operational problems (Ridgway and Flemming, 1996 and Vrouwenvelder *et al.*, 2000). Investigations into the beginning of the fouling process reveal that the microorganisms settle on the membrane during the first hours of the operation but only when a certain threshold of interference is exceeded does biofouling take place (Flemming, 2002).

Biofilm caused by interaction between microorganisms and the membrane by several steps: the first step in biofilm formation is a rapid adsorption of organic molecules to the membrane surface. This layer enhances subsequent microbial adhesion. The second step is the adhesion of microorganisms to the conditioned surface, this step followed by continued microbial adhesion, growth of adhered cells and production of extracellular polymer. The third step is attachment of microbial cells to the surface and the accumulated biomass is recognized as a biofilm (Ahmad and Ubaisy, 2000). As a result there is an increase of total organic carbon (TOC) concentration. This attachment of microorganisms to the membrane enables them to passively increase the nutrient consumption from the surrounding water (Rittschof, 1999).

An effective way to decrease TOC level in the feed water and thus to prevent fouling is by removing organic substances that cause the fouling before the contact with the RO membrane, as a result there is a need to use a pretreatment system prior to the RO. In this research a membrane bioreactor (MBR) system will be examined for the potential removal of organic matter and various nutrients. The MBR system is composed two processes: biological process and ultrafiltration. In the biological process microorganisms are nourished by the organic matter present in the seawater, by that reducing TOC level and other nutrients such as nitrogen sources, phosphorous, iron etc.

The high potential of MBR in improving water quality and TOC level reduction in the feed water, prior to the membrane, can enable the removal of organic matter and minimize the fouling problem. Prior experiments that were held, using a MBR system as a pretreatment to RO discovered it has short term impacts, which enabled work with higher flows relative to non treated water, and long term effects including removal of biological matter from the feed water and biofouling control (Visvanathan *et al.*, 2002).

Seawater is relatively poor in organic matter and microorganisms. TOC concentration in the water is 1.5–19 mg/L (Frebrero, 1984). During this research we examined the influence of the MBR and its efficiency in the removal of organic matter and nutrients from the seawater. In addition, we identified the microorganisms in the feed water, in the MBR system (as suspended in water, and as attached to the UF membrane surface), which take part in the removal of organic matter and nutrients.

2.1.1.1 The laboratory-scale MBR system

The installation includes a seawater feed tank (1,000 L) that feeds continuously the MBR tank, (30 L). Inside the MBR, in which the micro organisms' activity is carried out, a submerged UF membrane module (ZW-1, Zenon-GE, Canada) and an air diffuser are installed. The filtered water is collected in a filtration tank and the surplus flows to the sewage (Figure 2.1).

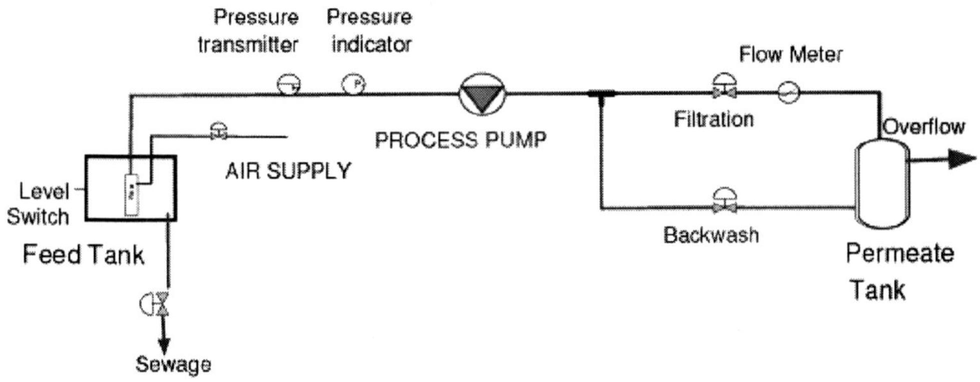

Figure 2.1 Layout of the MBR lab-scale system

Feed-water filtered through a strainer, enters the feed tank through a float valve (FV1). In production mode, feed water passes through the feed valve (SV-1) into the membrane tank, under gravity. The water level in the membrane tank is set by the level in the feed tank to maintain water above the membrane fibres. The process pump speed is set by the PLC based on the permeation flow set-point. Permeate flow is directed to the Back-pulse tank (TK2) where it overflows to the outlet. There is an option to have an aeration step at the end of each production cycle also (this is normally set to 0 seconds).

After initial operation, continuous flow was reached, seawater was fed into the system. The main goal was to accumulate biomass in the MBR tank and to enable bacteria acclimation.

2.1.1.2 Aeration and Sludge removal

The permeate flux varied between 50–60 l/m^{2}h and the TMP is about 0.3–0.5 bar. In general, biomass accumulation was very limited since total organic mater did not exceed more than 1.25 ppm, thus no sludge production was observed in the MBR system. The unit was under constant aeration, both during filtration and during backwash, as opposed to the normal conditions of operation of conventional MBR with municipal wastewater.

2.1.2 Results

2.1.2.1 TOC removal

An effective way to decrease TOC level in the feed water and thus to prevent fouling is by removing organic substances that cause the fouling. In this research we examined different retention times (Tables 2.1 to 2.3) to find out the optimal retention time with maximal decrease of TOC concentration in permeate, relative to the feed water. We characterized the microorganisms and examined species diversity in the seawater relative to those developed in the MBR system and on the UF membrane. From the analysis of the TOC results, it appears that the most effective retention time from those examined in the research is 0.5 day (Table 2.1). In this retention time, 55% average decrease in TOC level was achieved in the MBR systems permeate in comparison to feed water. This retention time is the longest examined and apparently enabled long contact time between microorganisms and the organic matter, which serves as nutrient in the MBR system.

Table 2.1 Influence of MBR and its efficiency in the removal of organic matter and nutrients from the seawater at retention time: 0.5 day

Retention time: 0.5 day			
% reduction of TOC	**TOC OUT (ppm) (permeate water)**	**TOC IN (ppm) (feed water)**	**Date of sampling**
47	0.36	0.68	06.12.08
69	0.33	1.07	30.12.08
41	0.39	0.66	07.01.09
64	0.27	0.75	12.01.09

Table 2.2 Influence of MBR and it's efficiency in the removal of organic matter and nutrients from the seawater at retention time: 0.3 day

Retention time: 0.3 day			
% reduction of TOC	**TOC OUT (ppm) (permeate water)**	**TOC IN (ppm) (feed water)**	**Date of sampling**
10	0.3	0.37	4.2.09
6	0.3	0.32	11.2.09
30	0.21	0.3	16.2.09
50	0.62	1.24	23.2.09
−13	0.37	0.33	26.2.09

Table 2.3 Influence of MBR and it's efficiency in the removal of organic matter and nutrients from the seawater at retention time: 0.1 day

Retention time: 0.1 day			
% reduction of TOC	**TOC OUT (ppm) (permeate water)**	**TOC IN (ppm) (feed water)**	**Date of sampling**
15	0.29	0.34	16.3.09
25	0.42	0.56	30.3.09
−35	0.27	0.2	20.4.09
−46	0.35	0.24	5.5.09
−23	0.37	0.3	20.5.09

2.1.2.2 Microbial diversity

From the microorganisms tests conducted in the research in various locations, we obtained the following results: in most of the samples the dominant group in the phylum level belonged to *Proteobacteria* (Figure 2.2). In all genomic libraries *Alphaproteobacteria* and *Gammaproteobacteria* were dominant, these are heterotrophic bacteria nourished from organic substances present in the MBR system that originate from the feed (sea water). Photosynthetic microorganisms from the phylum *Cyanobacteria* appeared in high percentages in feed water, but were almost non-present in the MBR system (for example compare Figure 2.2B with Figure 2.2A). The percent of sequences identified as Unclassified Bacteria on the UF membrane filament (Figure 2.3) was much higher (23%) than the percent accepted in all other samples from feed water and the MBR system water (up to 15%, see Figure 2.2). It is possible that these unclassified microorganisms acclimatized and developed in the system even though they are less common in seawater.

We compared changes (by amplification and comparison of 16S rRNA genes sequences) of microbial diversity in MBR system (0.5 day retention time) with feed water. Water in MBR system was sampled after 90 days filtration activity. Microbial population which developed in the MBR displayed shifting in microbial diversity in comparison to the seawater (feed water). Sequences of 16S rRNA genes of photosynthetic cyanobacteria (17%) retrieved in feed water completely, disappeared from the MBR system. Most abundant phylum Proteobacteria was obtained in seawater as well as in MBR system, with increasing from 50% to 67% in later (Figure 2.2). The abundance of Proteobacteria in seawater was supported by other study (Acinas *et al.*, 1999).

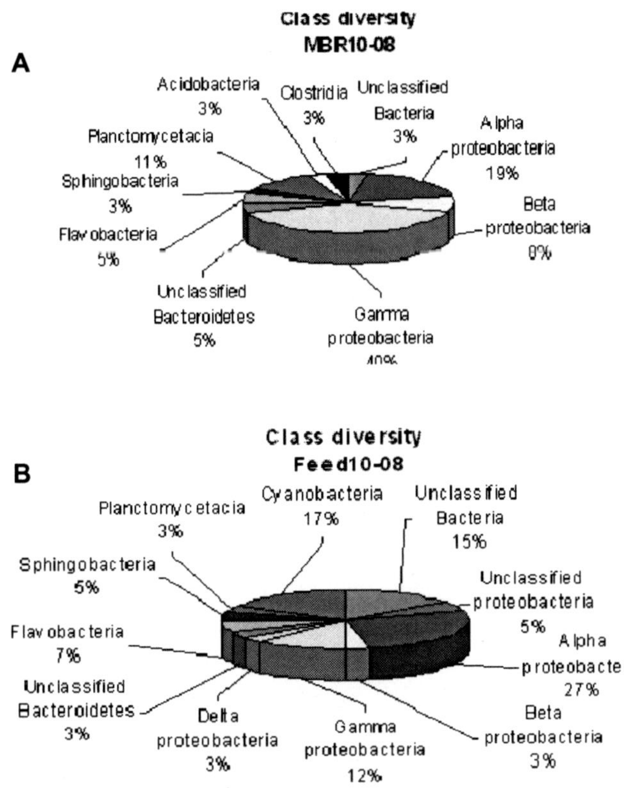

Figure 2.2 Bacterial diversity in MBR system (A) and in feed-water (B)

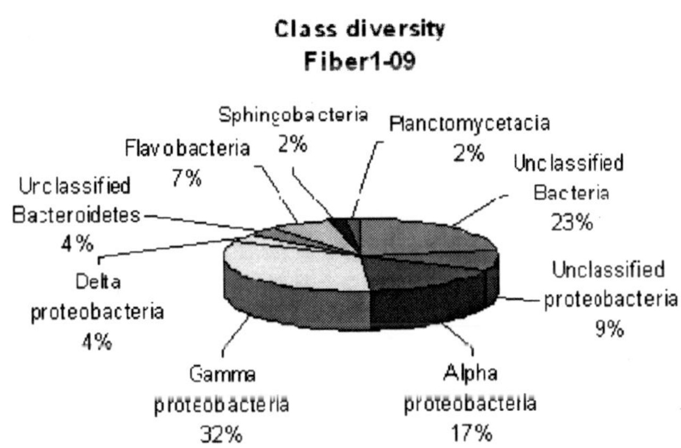

Figure 2.3 Microbial diversity on UF fiber in MBR system

A hump-shaped species diversity is formed and depends on the magnitude of the changes of the environmental gradients along which the community parameters are measured. In microhabitats with a modest supply of resources, groups of species can coexist, and each group will be at a relatively lower dominance level, but species richness will be higher. After biomass production reaches a critical range, competition becomes sufficient to eliminate less competitive species from the community. Our results displayed a decrease in bacterial diversity (at both 90% and 97% sequence similarity) in MBR system where TOC concentration was higher then in feed water. This result implies a 'negative' response to increase in TOC concentration.

The overall diversity of cloned sequences was analyzed on two levels (90% and 97% similarity) using cluster analysis by the DOTUR program (Schloss and Handelsman, 2005). A total of 27 OTUs were observed from feed water samples at the 97% similarity level (roughly representing species level), compared to about 22 OTUs obtained in MBR system. This result was repeated for the Chao1 richness estimator and Shannon diversity index. A similar image was obtained for the 90% similarity level (roughly representing phylum level diversity). A total of 20 OTUs was observed in feed water, compared to average 16 OTUs in MBR samples. A similar trend was obtained for the Cahol and Shannon indices, indicating a decrease in diversity at the phylum-order level.

2.1.2.3 Biofouling potential

The reduction in the amount of protein and polysaccharides in permeate of the MBR that feeds the RO system is highly important. We carried out an experiment to check the fouling potential of the MBR system in comparison to UF system only, in different time scales, (7 days and 28 days). From the fouling potential examination, it appears that the total amount of live and dead microorganisms (Figure 2.4) and the amount of protein (data not shown) and polysaccharides (Figure 2.5) on the RO membrane (on which permeate from these systems came in contact with) from the MBR was much lower than from the UF system's permeate. There is a great importance regarding the system's ability to serve as a pretreatment for desalination process and to reduce the rate of biofilm formation on the RO membranes.

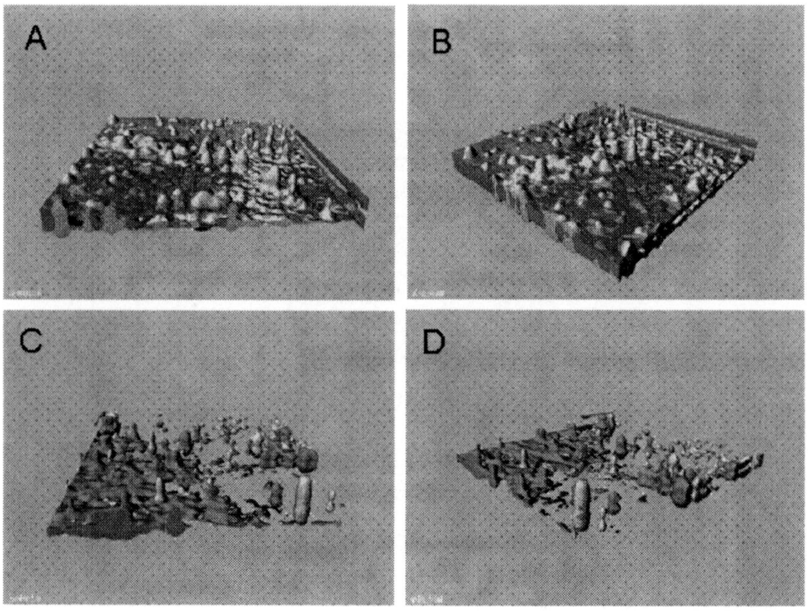

Figure 2.4 Biofouling of live (green), dead (red) cells and EPS (blue) on RO membrane of the MBR and the UF system's permeate after 7 days. A and B UF system permeate and C and D- MBR + UF system's permeate. This figure is available in colour at http://www.iwapublishinq.com/template.cfm?name=isbn9781843393214

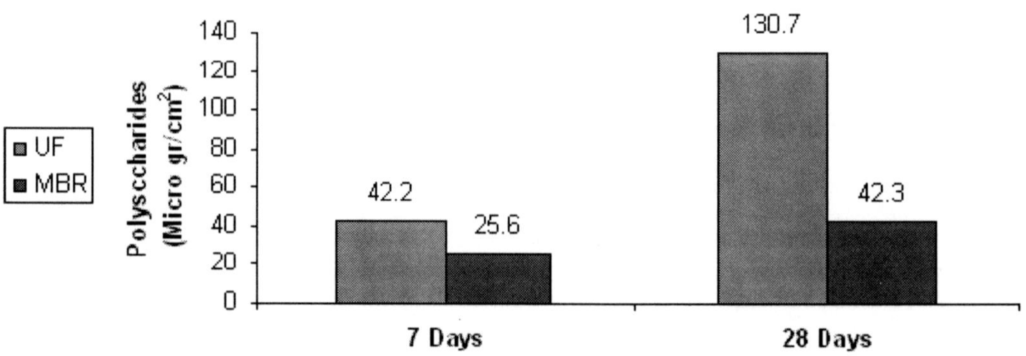

Figure 2.5 Changes in quantities of polysccharides on RO membrane's surface of the MBR and UF system's permeate

In conclusion, it seems that the pretreatment in a MBR system is effective at improving the water's quality (removal of TOC and EPS) for further process, but further economical and implemental examinations are necessary in order to adopt this treatment as an alternative pretreatment.

2.2 SUBMERGED ULTRAFILTRATION FOR REDUCING NOM IN SWRO DESALINATION

2.2.1 Introduction

The aim of the present work was to carry out experimental research addressed to the optimization of operational conditions addressed to use MBR technology to reduce/remove organic matter from marine feed water. The final goal was to obtain (1) a further improvement of the desalination process by lowering membrane fouling and (2) to reduce/ eliminate organics matter which might affect the crystallization stage operated on the brines of the NF and/or of the RO units.

Experimental test carried out by UNICAL relied mainly on the study of the optimal combination of operative parameters, like trans-membrane pressure and flux, aeration rate and mode and back flushing operations, for submerged ultrafiltration processes. This activity would provide the hardware that, in conjunction with experiments carried out in the frame of the project on biomass growth in marine environment, would complete the concept of a membrane bioreactor application as seawater pretreatment in desalination practice.

The experimental activity has been anticipated by a critical literature survey on current state-of-the-art of membrane bioreactor technology in traditional municipal and industrial wastewater treatment processes, its benefits and main limitations wit respect to conventional purification processes, and the valuation of the possibility to use this novel technology as pre-treatment in seawater desalination. This survey has been concretized with the preparation of the deliverable D 2.1.2 "A report on the current advantages and drawbacks of MBR as a pre-treatment step in the SW and BW desalination". Figure 2.6 shows the integrated system studied.

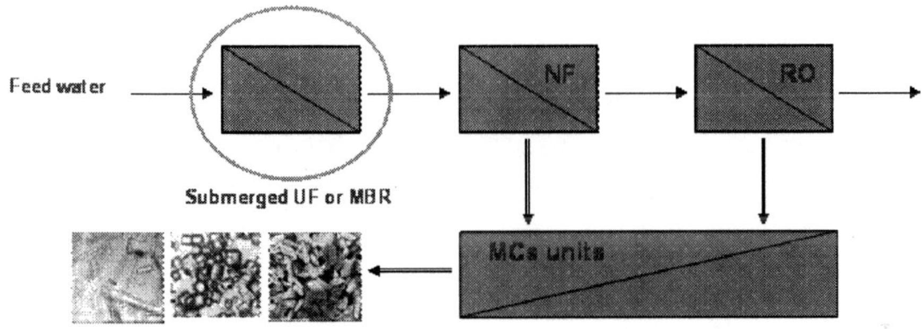

Figure 2.6 The flow sheet of an integrated membrane system for seawater desalination studied in the frame of the MEDINA project with MBR as pre-treatment

2.2.2 Experimental activity

2.2.2.1 Experimental plants

The first objective of the present work was the design and the realization of experimental submerged membrane bio-reactor (sMBR) laboratory plants. The design of the experimental plants has been carried out on the bases of the literature survey on current state of the art membrane bioreactors configurations, including membrane type, performances and operation mode. The general scheme of the plant is composed by (Figure 2.7): a feed water tank, containing the raw seawater; a feed pump (P1); a membrane tank, with a specific capacity; a membrane module (1); an aerator or bubbler (2); an air compressor; a suction pump (P2); an permeate water tank; a back-flushing pump (P3). Controllers for mass flow (FM), temperature (T), turbidity (Tu), dissolved oxygen (DO), pH (pH), conducibility (I) were inserted in the plant. An air line was inserted in the plant to allow air back-pulse to check, when necessary, module integrity. According with the described configuration, two experimental plants with different capacity have been assembled.

2.2.2.1.1 Experimental plant N. 1

Figures 2.8 and 2.9 show the first membrane bioreactor plant assembled. It consist of a membrane tank with an overall capacity of 230 L and a permeate flow rate range of 20–80 L/h. This plant might contain commercial modules of different size and shape, in both hollow fibers and flat sheet membrane configuration. It includes several control apparatus to monitor temperature, pH, turbidity, conductivity, DO. This configuration has been designed by taking into account the utilization of the plant as pre-treatment unit to be used for the integrated membrane system, which is the objective of the work package 8 (WP8) of the MEDINA project.

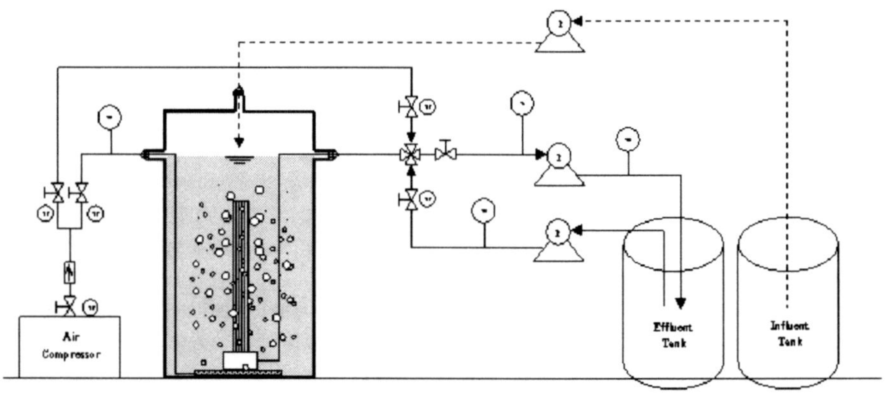

Figure 2.7 Scheme of the laboratory-scale MBR hardware used as pre-treatment step for sweater desalination in the integrated membrane system proposed in the MEDINA project

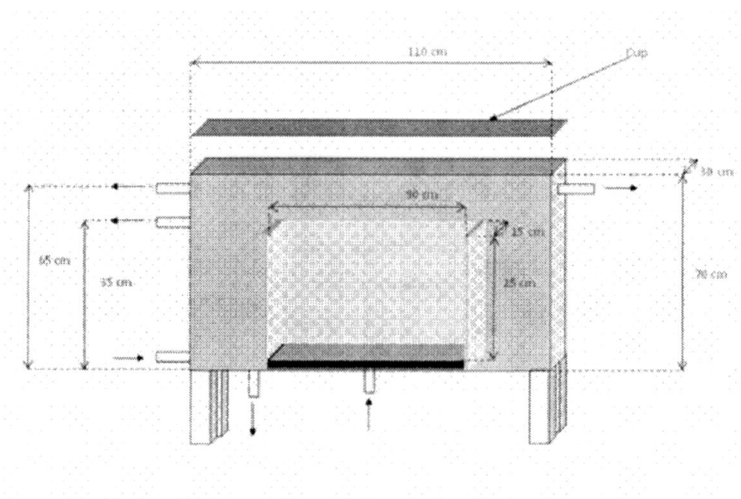

Figure 2.8 Design of the experimental plant N. 1 used in the integrated membrane system

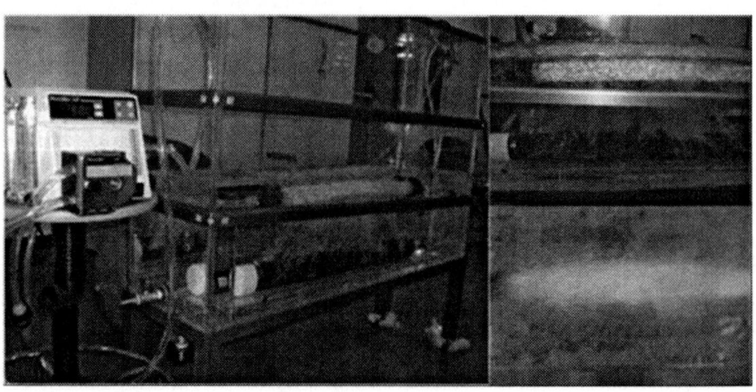

Figure 2.9 Image of the plant N.1 during an experiment

Tests have been carried out to study the effect of DOC removal from synthetic seawater in the submerged configuration.

2.2.2.1.2 *Experimental plant N. 2*

Figures 2.10 and 2.11 display the second membrane bioreactor plant assembled. It consists of a membrane tank with the full capacity of 20 L and permeates flow rate range of 20–80 L/h. This plant contains membrane modules, with hollow fibers membranes in vertical arrangement, assembled directly in laboratory and includes control apparatus to monitor temperature, pH, turbidity, conductivity, DO. This smaller lab-scale plant has been used to test real seawater instead of the bigger one because of the limitate availability of seawater in the location of the University, which is 35 km far from the seacoast.

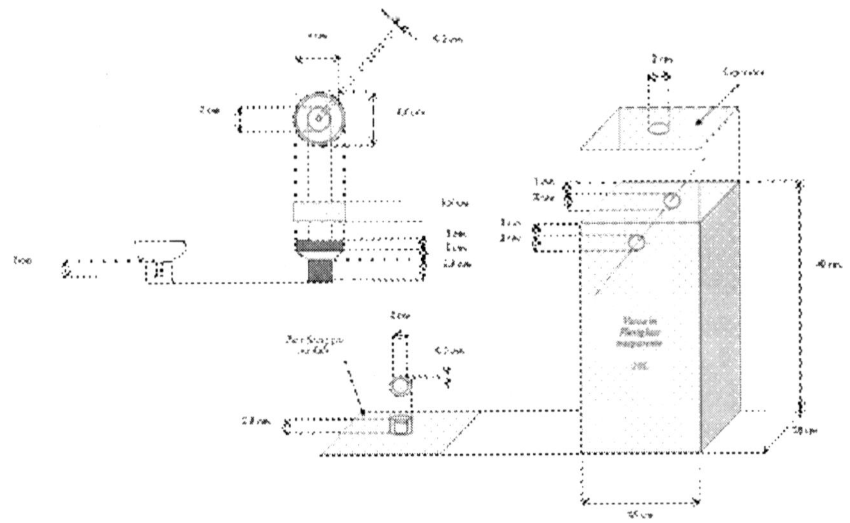

Figure 2.10 Design of the experimental plant N. 2 to be used in the integrated membrane system

Figure 2.11 Image of the plant N.2 during an experiment

2.2.2.2 Membranes

For the plant N. 1, ultrafiltration right fibers polyvinilidenefluoride (PVDF) membrane module, in horizontal position, has been used. For experiments in plant N. 2, polyethersulfone (PES) hollow fiber membranes (UltraPes, from Membrana GmbH, Germany; 70 *kDa* MWCO; nominal permeance as reported by the manufacturer: $L_p > 0.65\,mL/cm^2sbar - 3.9\,L/m^2hkPa$), microfiltration polyvinilidenefluoride (PVDF) hollow fibers (pore size 0.09 μm) and polypropylene (PP) hollow fibers (Accurel, from Membrana GmbH, Germany; pore size 0.2 μm) were used to assemble the one-side potted modules with the total membrane area A of 0.068 m^2. Modules were prepared in non-cauterized configuration to allow more freedom for fibers motion to reduce fouling under aeration conditions. Filtration was carried out in outside-to-inside filtration mode. Air sparing, through a diffuser, was operated from the bottom of the modules, by using an air compressor, at 8 L*PM*.

2.2.2.3 Water samples

2.2.2.3.1 *Synthetic seawater*

The synthetic seawater has been prepared according with the following composition, by dissolving the opportune amount of the following salts in de-ionized water: Na_2SO_4, NaCl, KBr, $MgCl_2$, $CaCl_2$, $CaSO_4$, $MgSO_4$. The required amount of dissolved organic carbon (DOC) was provided by adding humic acid. pH has been corrected by using NaOH and HCl solutions.

2.2.2.3.2 Real seawater

Seawater from the Tyrrhenian coast (location: Belvedere Marittimo, Southern Italy), without any initial pre-treatment, was used to feed the plant. Real seawater has been obtained from three location indicated in Figure 2.12. Water was uptake 2–3 meters far, on the sea side, from the estuarine of little rivers or streams in which the effluents from traditional municipal wastewater depuration plants are discharged directly in the sea after treatment. Therefore, the amount of DOC in these coastal zones is over the range of typical DOC concentration for the other areas of the cost (see Table 2.4).

Figure 2.12 Location where the real seawater has been checked for DOC content

Table 2.4 Characteristics of real seawater from the three locations, used in this work

	Location 1	Location 2	Location 3
DOC (mg/L)	2.9	3.1	2.7
pH	8.0	7.9	8.0

Water characterization was carried out by ion chromatography (Metrohm IC861 compact system) for ion composition and by a TOC analyzer (Shimadzu V_{CSN}) for TOC measurements according to standard procedure for seawater.

2.2.3 Results and discussions

The systems have been tested initially as submerged Microfiltration/Ultrafiltration unit in order to check performance and optimal operative conditions in the logic to integrate it in the overall integrated membrane system, which is the objective of WP8. Tests have been carried out to study the effect of DOC removal from synthetic seawater in the submerged configuration. Rejection experiments to DOC in plant N. 2 have been carried out by using either microfiltration (pore size 0.2 µm) polypropylene, microfiltration (pore size 0.09 µm) polyvinylidene fluoride and ultrafiltration (molecular cutoff 70 kDa) polyethersulphone hollow fibers modules arranged in vertical configuration. Results demonstrated that Microfiltration hollow fibers (both polypropylene and polyvinylidenefluoride) were not effective in removing DOC, as the rejection was always lower than 10% in every condition. In preliminary testing of PES (plant N. 2) and ultrafiltration PVDF (plant N. 1) membranes a significative DOC removal was observed. Accordingly PES membranes were used fior the subsequent experimental activity.

2.2.3.1 Systems' efficiency

Experiments by using the plant N. 2 have been carried out by using real seawater.

Figure 2.13 shows the $J - TMP$ relation for constant pressure filtration experiments of pure and seawater. It is evident that the two curves diverge since the lower values of pressure, due to the adsorption of foulants at the beginning of the test. This is consistent with a weak-form of critical flux which can be individuated already for low values of J. From the figure, the $J - TMP$ curve starts to be not linear already for $J > 26\,L/m^2\,h$, this value representing J_{cw}. Accordingly, filtration experiments with at the flux of $26\,L/m^2\,h$ were carried out in the subsequent tests at constant J.

In order to study the effect of aeration/suction operation mode on process performances, different experimental sections were performed: (i) continuous aeration and suction, (ii) continuous suction and intermitted aeration (10 min ON/10 min OFF), (iii) intermitted suction (8 min ON/2 min OFF) and aeration. Figure 2.14 compares fluxes against *TMP* for SW filtration in the different modes.

The curves in the figure demonstrate that in continuous aeration conditions slightly lower fluxes were observed with respect to intermitted conditions. Accordingly, system performance in intermitted aeration conditions does not substantially decrease with respect to continuous aeration mode, with consequent gain, however, in terms of energy consumption. However, for both intermitted suction and aeration conditions the highest limiting flux J_∞ was obtained, resulting in the more efficient operation mode in terms of energy consumption provided effluent production requirements.

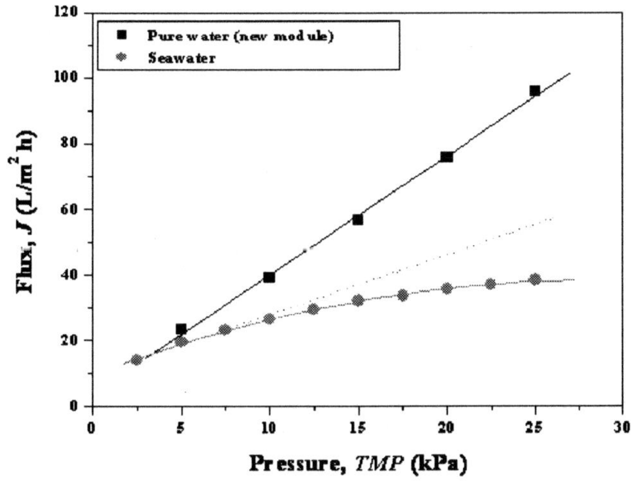

Figure 2.13 Flux (*J*) as function of the trans-membrane pressure (*TMP*) for pure water (squares) and seawater (circles) filtration test at increasing pressure steps

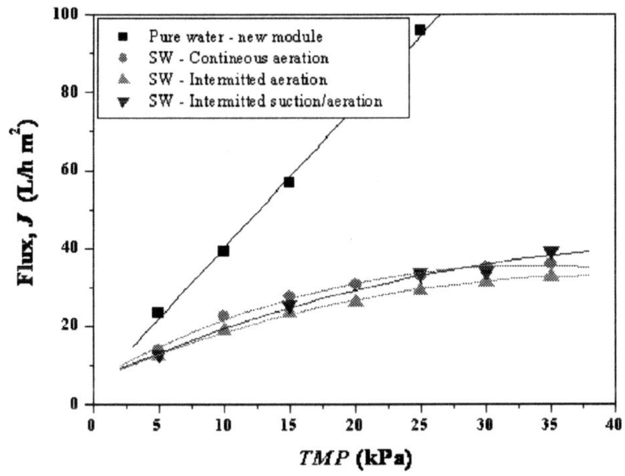

Figure 2.14 Trans-membrane flux *J* as function of *TMP* during filtration of SW in continuous and intermitted aeration/ suction regimes

On what concerns TOC removal, in Figure 2.15 the rejection of TOC after membrane filtration with respect to *J* is reported. As shown in the figure, for the lowest flux the maximum value of 63% in TOC rejection was achieved. Increasing *J* TOC rejection decreased although some scattered points are present in the figure. This is due to the fact that as *J* increases polarization phenomena, which encourage small organic particles to permeate the membrane, raise as well with consequent reduction in TOC removal.

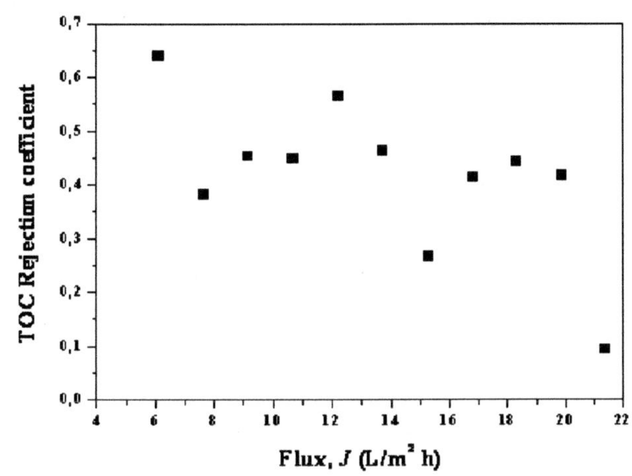

Figure 2.15 Trans-membrane flux *J* and TOC removal in seawater filtration test as function of *TMP*

2.2.4 Conclusions

This work explores the use of submerged membrane operations as pre-treatment step in the logic of integrated membrane desalination systems to reduce natural/polluting organic matter existing in feed seawater. In this logic, by achieving this purpose would allow to reduce significantly RO membranes bio-fouling and to eliminate exogenous molecules that will affect/hinder crystallization kinetics in membrane crystallizers.

Results demonstrated that irreversible particles adsorption on/in the membrane surface occurs during the early stages of filtration while reversible polarization phenomena/fouling become dominant for higher fluxes. The plant can be operated for more than 60 hours without any chemical cleaning, before of significant loss of performance. At this stage, simple cleaning with NaOCl allows the almost complete recovery of module permeability.

As small organic molecules permeates the membranes and deposit also on the lumen side, back-flushing with pure water (or even more with the effluent with contains particles not retained!) induces compaction of these deposits enhancing the overall resistance to mass transport. Therefore back-flushing is not recommended when treating seawater, which contains a significant amount of small molecular size organic particles permeating the 70 kDa MWCO membrane.

Intermitted aeration/suction operation represent the best compromise between process efficiency and energy consumption, provided the requirements of effluent production.

TOC removal up to 63% for the sole filtration stage can be achieved for low values of flux by using 70 kDa MWCO membranes, although polarization phenomena supported the decrease in TOC rejection for higher trans-membrane fluxes. In this respect, the increase of TOC removal by stimulating biodegradation before membrane filtration in a full MBR system would increase the overall efficiency of the process.

2.3 COMPARISON OF GRANULAR MEDIA FILTRATION AND LOW-PRESSURE MEMBRANE FILTRATION FOR SEAWATER PRETREATMENT

2.3.1 Introduction

Desalination by reverse osmosis (RO) has become the most cost-effective process to convert seawater into fresh water for potable use at large scale. However, a pretreatment is necessary to ensure that feed water will not cause fouling of the RO membranes. The main goal of this study was to evaluate whether microfiltration/ultrafiltration (without pre-coagulation) could provide a higher level of protection against fouling for the RO membrane as compared to a conventional granular filtration with pre-coagulation. In addition to the conventional performance parameters (i.e. turbidity, particle counts, SDI), "Modified Fouling Index" (MFI) measurements were performed according to the method developed by (Schippers *et al.*, 1980). Indeed, despite the extended use of the Silt Density Index as fouling indicator by the desalination industry, studies have shown that this index is not the most appropriate parameter for measuring the fouling potential of seawater (Kaakinen and Moody, 1985 and Potts *et al.*, 1981). Two RO units each fed by the two pretreatment trains (coagulation + dual-media filtration, and microfiltration/ultrafiltration) were operated over two periods of four months each in order to monitor the impact of the two pretreatment trains on the hydraulic performance of the RO units. During this pilot testing, seawater samples were also collected along the two pretreatment trains for quality characterisation so as to identify the best water quality parameters that could be used to quantify the performance of the pretreatment processes and their capacity to minimize fouling risks on RO membranes.

Two pilot testing phases were conducted as follows:

- Phase 1: Comparison of "coagulation + dual-media filtration" and "0.04 μm MF/UF"
- Phase 2: Comparison of "coagulation + dual-media filtration" and "0.02 μm MF/UF"

2.3.2 Pilot-scale experimental set up

Tests were performed on a site located in the South of France (Mediterranean Sea) and pilot plants were fed through an open intake. Figures 2.16 and 2.17 depict the process trains used for Phase 1 and Phase 2, respectively.

2.3.2.1 Granular media filtration with pre-coagulation/flocculation

Phase 1: After pH correction from 8.2 to 6.8, the conventional pretreatment process was composed of a coagulant injection (ferric chloride), a 15 min flocculation (injection of polymer) and a granular dual-media filtration through sand and anthracite.

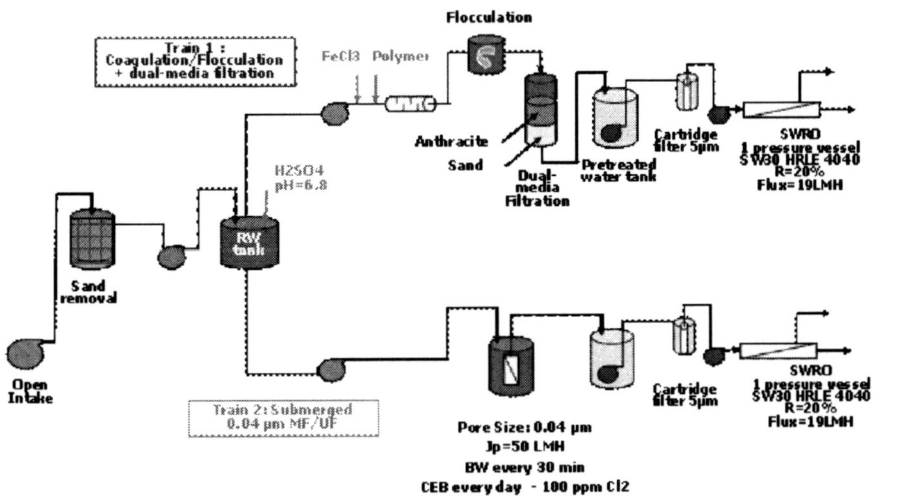

Figure 2.16 Pilot-scale units used for pretreatment comparison – Phase 1 – May to October 2007

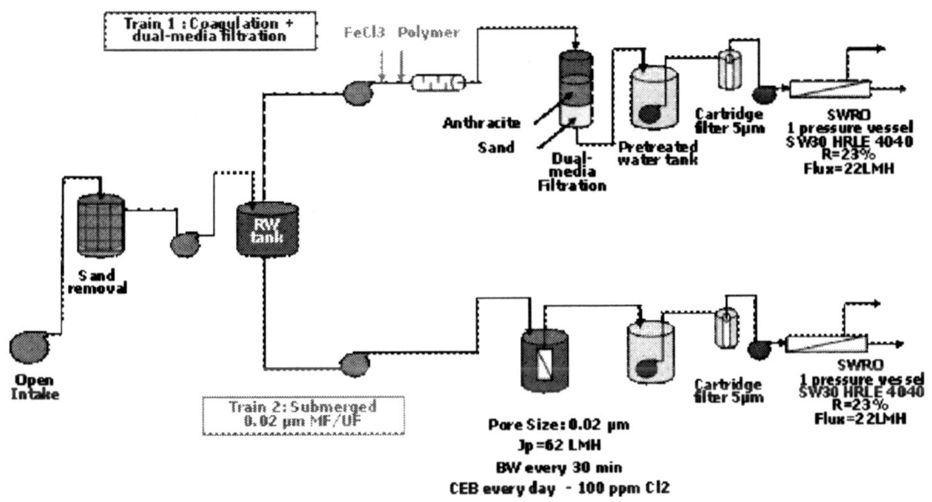

Figure 2.17 Pilot-scale units used for pretreatment comparison – Phase 2 – June to November 2009

Phase 2: As the flocculation stage was deemed not necessary for the raw seawater quality, it was decided to use only in-line coagulation (static mixer) upstream of the dual media filter for Phase 2. The type and dosing of coagulant and polymer was the same as for Phase 1, but pH correction to 6.8 was stopped (pH depression for the pretreatment is now often stopped at full-scale SWRO plants).

2.3.2.2 Membrane pretreatment process

Phase 1: After pH correction to 6.8, seawater was pretreated by a MF/UF pilot plant which consisted in an immersed dead-end membrane filtration with an out/in membrane module made up of hollow-fibres with a 0.04 µm nominal pore size. The operating conditions were as follows:

- Filtrate flux: 50 $L.h^{-1}.m^{-2}$.
- Backwashes every 30 minutes, operated with an aeration at the bottom of the membrane tank (3.5 $m^3.h^{-1}$), and a water backwash at 2 $m^3.h^{-1}$ (flux: 72 $L.h^{-1}.m^{-2}$)
- Chemically enhanced backwashes (CEB) performed once a day at 100 ppm chlorine.

Phase 2: The same MF/UF pilot unit was used for Phase 2. The 0.04 µm membrane was replaced by a 0.02 µm membrane. The operating conditions were as follows:

- Filtrate flux: 62.5 $L.h^{-1}.m^{-2}$.
- Backwashes every 30 minutes, operated with an aeration at the bottom of the membrane tank (3.5 $m^3.h^{-1}$), and a water backwash at 2 $m^3.h^{-1}$ (flux: 77 $L.h^{-1}.m^{-2}$)

- Chemically enhanced backwashes (CEB) performed once a day at 100 ppm chlorine. Some CEB with citric acid (pH = 2.5) were also performed when deemed necessary (about 1 acid CEB per month).

2.3.2.3 Reverse osmosis pilot units

Each pretreatment process fed a RO pilot unit. Each RO pilot unit consisted of a 5 μm cartridge filter followed by a RO pressure vessel with a single 4 inch Dow Filmtec SW30HR LE-4040 RO membrane module. The two RO pilot plants were operated under the same conditions:

Phase 1: conversion rate of 20% and permeate flux of 19 $L.h^{-1}.m^{-2}$
Phase 2: conversion rate of 23% and permeate flux of 22 $L.h^{-1}.m^{-2}$

2.3.3 Water quality parameters and fouling indices

During both phases, regular samples were collected on the raw and pretreated seawaters. The following analyses were performed:

- Characterization/quantification of organic matter
  - TOC, Total Nitrogen – for more details, see reference (Rapenne *et al.*, 2005)
  - UV254
  - Algal pigments such as chlorophyll-a and pheophytin-a – for more details, see reference (Neveux and Lantoine, 1993)
  - Liquid Chromatography with organic carbon detection (LC-OCD), which allows quantification of the main fractions of the organic matter (biopolymers, humic substances, low-molecular weight organic acids and neutrals – for more details, see references (Hubert and Frimmel, 1994 and Vincelet and Jacquemet, 2008)
- Characterization/quantification of the bacteria and algal content
  - Total bacteria and picophytoplankton counts through flow cytometry, which is an individual, qualitative and quantitative characterization technique for particles (cells, bacteria. . .) in a liquid field, which allows discrimination between picophytoplankton and heterotrophic bacteria populations according to their signals of diffusion and fluorescence. More details are given in reference (Lebaron *et al.*, 2001 and Courties *et al.*, 1994)
- Characterization/quantification of particulate matter
  - Turbidity
  - Particle counts
  - Fouling indexes
    - SDI_{15}-0.45 μm
    - MFI-0.45 μm

2.3.4 Results and interpretation

2.3.4.1 Raw seawater quality

Tables 2.5 and 2.6 give the raw seawater quality during Phase 1 and Phase 2.

Table 2.5 Raw seawater quality from May 2007 to October 2007 – Phase 1

	Min	Average	Max
T (°C)	14.3	19.5	23.6
Turbidity (NTU)	0.1	0.3	1.3
$UV_{254\,nm}$ (m^{-1})	0.5	1.0	1.6
TOC ($mg.L^{-1}$)	1.0	1.2	1.5
DOC ($mg.L^{-1}$)	0.9	1.0	1.1
Biopolymers ($mg.L^{-1}$)	0.06	0.07	0.08
Total Bacteria (mL^{-1})	$1.4.10^5$	$1.6.10^5$	$1.9.10^5$
Total Picophytoplankton (mL^{-1})	$0.8.10^4$	$4.2.10^4$	$7.8.10^4$
Pigments > 0.7 μm (μg.L^{-1})	0.9	1.1	1.3
$SDI_{3\,min}$	9.3	19.8	31.5

During both periods, the raw seawater presented similar quality. Overall, this seawater is characterised by a low particle content and a low natural organic matter content as average turbidity was around 0.3–0.4 NTU and average DOC around 1.0 $mg.L^{-1}$.

Table 2.6 Raw seawater quality from July 2009 to November 2009 – Phase 2

	Min	Average	Max
T (°C)	15.1	21.6	26.3
Turbidity (NTU)	0.1	0.4	3.0
UV_{254nm} (m^{-1})	0.2	0.8	1.5
TOC (mg.L^{-1})	0.8	1.0	1.2
DOC (mg.L^{-1})	0.9	1.0	1.2
Polysaccharides (mg.L^{-1})	0.08	0.11	0.13
Total Bacteria (mL^{-1})	$1.5.10^5$	$2.0.10^5$	$2.4.10^5$
Total Picophytoplankton (mL^{-1})	$2.0.10^4$	$2.4.10^4$	$2.9.10^4$
Pigments > 0.7 μm (μg.L^{-1})	0.30	0.40	0.60
SDI_{3min}	11.9	16.9	30.8

2.3.4.2 Seawater quality at the outlet of each pretreatment

2.3.4.2.1 Conventional analytical parameters (turbidity, particle counts, and SDI)

Table 2.7 shows the average seawater quality at the outlet of both pretreatment trains in terms of SDI, turbidity and particle counts. The SDI_{15} values for the Granular Media Filtration correspond to SDI_{15} data recorded after at least 10 hours of filtration following the end of the backwash.

Table 2.7 Seawater quality at the outlet of Granular Media Filtration and MF/UF

	$SDI_{15} < 3$ (%)	$SDI_{15} < 3.5$ (%)	Average SDI_{15}	Turbidity (NTU)	Particle > 1 μm count (mL^{-1})
Granular Filtration (Phase 1)	25	55	3.5	0.03	160
0.04 μm MF/UF (Phase 1)	94	99	2.5	0.03	70
Granular Filtration (Phase 2)	4	28	3.8	0.05	50
0.02 μm MF/UF (Phase 2)	86	96	2.4	0.05	40

During both phases, the MF/UF process provided a significantly lower SDI level as compared to the Granular Media Filtration (GMF), with an average SDI_{15} around 2.5 after MF/UF and an average SDI_{15} around 3.5–3.8 after GMF. This observation is consistent with the previous studies on seawater membrane pretreatment reporting low SDI at the outlet of MF/UF membranes. During Phase 1, the 0.04 μm MF/UF-treated seawater also presented a lower particle count than the water at the outlet of GMF. This difference was not seen during Phase 2, and it is important to note that even a particle count (>1 μm) of 160/mL remains a very acceptable level for RO feed waters.

2.3.4.2.2 MFI measurements

Figure 2.18 compares typical SDI_{15} and MFI profiles recorded at the outlet of both pretreatment processes (GMF and 0.04 μm MF/UF). MFI at the outlet of the MF/UF membranes showed a steady level over time, while the MFI at the outlet of the GMF continuously decreased over the filtration cycle. Table 2.8 shows the average SDI_{15} and MFI levels recorded during both testing phases. As it was observed for the SDI_{15}, the MFI level after MF/UF is significantly lower as compared to the MFI after GMF.

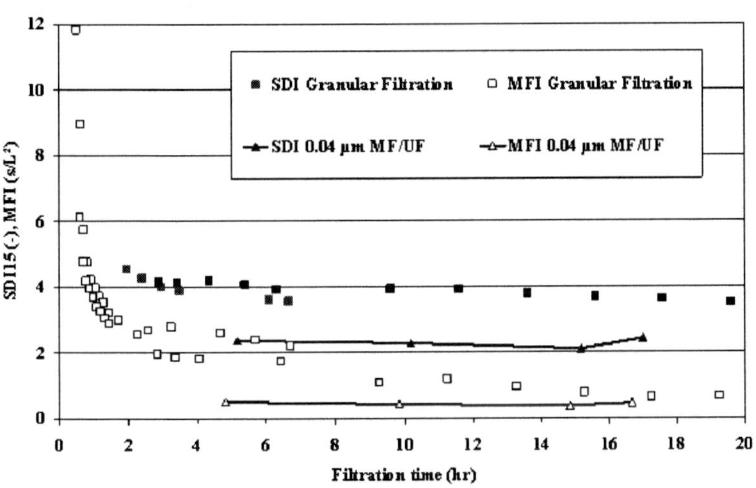

Figure 2.18 SDI and MFI profiles after MF and granular media filtration profiles

Table 2.8 Average values of SDI and MFI at the outlet of Granular Media Filtration and MF/UF

	GMF – Phase 1	0.04 μm MF/UF – Phase 1
SDI	3.5	2.5
MFI (s/L²)	1.5	0.6
	GMF – Phase 2	**0.02 μm MF/UF – Phase 1**
SDI	3.8	2.4

2.3.4.2.3 Advanced analytical parameters – organic and bacteria content

Table 2.9 shows the average seawater quality at the outlet of both pretreatment trains in terms of removal for DOC, biopolymers, total bacteria, total picophytoplankton, and algal pigments.

Table 2.9 Average removal after Granular Media Filtration and MF/UF

	DOC (%)	Biopoly. (%)	Bacteria (Log)	Picoplankton (Log)	Pigments > 0.7 μm (%)
Granular Filtration (Phase 1)	19 ± 10	58 ± 12	0.5	1.2	>94
	n = 3	n = 3	n = 3	n = 3	n = 4
0.04 μm MF/UF (Phase 1)	<5	26 ± 10	2.1	2.8	>97
	n = 3	n = 3	n = 3	n = 3	n = 4
Granular Filtration (Phase 2)	13 ± 10	32 ± 6	0.5	0.8	>90
	n = 4	n = 4	n = 4	n = 4	n = 3
0.02 μm MF/UF (Phase 2)	<5	20 ± 20	1.5	>4	>90
	n = 4	n = 4	n = 4	n = 3	n = 3

This table shows that the membrane pretreatment provided a better microorganism removal than granular media filtration in terms of bacteria and picophytoplankton content. On the other hand, the dissolved organic matter removal was lower through MF/UF. The level of biopolymer removal through MF/UF shows a similar trend (lower through MF/UF than through GMF). Both pretreatment processes efficiently removed algal pigments.

2.3.4.2.4 Impact of granular media filtration and MF/UF pretreatment on the RO process

Phase 1 – comparison between granular media filtration and 0.04 μm MF/UF

Figure 2.19 depicts the profile of the temperature-corrected feed-brine pressure drop recorded during the testing period on both RO units. No significant increase in the feed-brine pressure drop was noticed. This graph shows that no significant fouling (and notably biofouling) occurred on either RO units during the testing period.

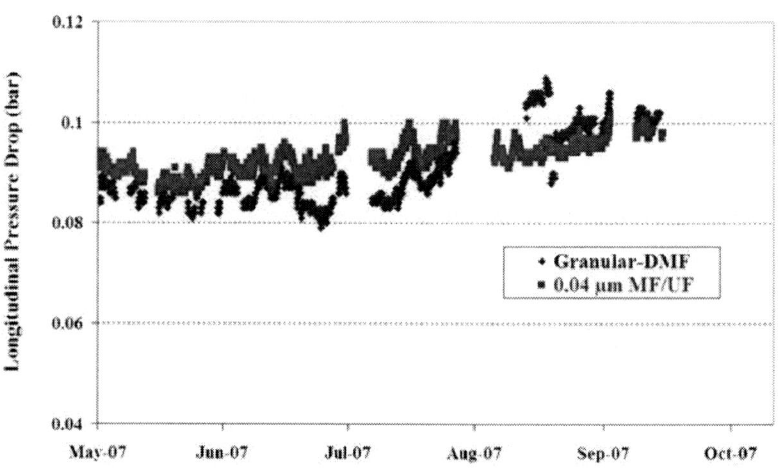

Figure 2.19 Temperature-corrected feed-brine pressure drop on RO units fed by GMF and 0.04 μm MF/UF – Phase 1

Figure 2.20 presents the profile of the normalized permeate (NPF) flow for each RO unit. The NPF of the RO membrane fed by MF/UF decreased by 30%, whereas NPF only decreased by 15% for the RO membrane fed by GMF. Therefore, despite a better seawater quality at the outlet of MF/UF as quantified with SDI, MFI and microorganism removal, the RO unit fed by MF/UF showed a more pronounced decline in performance as compared to the RO unit fed by GMF.

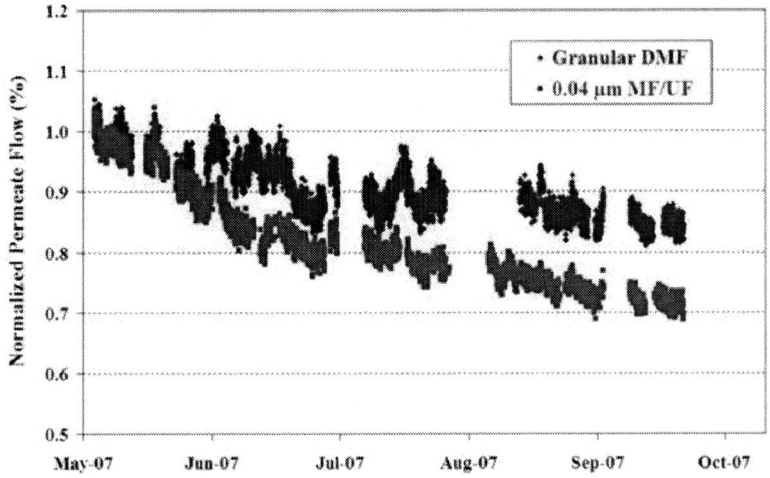

Figure 2.20 Profile of the normalized permeate flow of the RO membranes – Phase 1

To better understand this difference in NPF profile, the two membranes were autopsied at the end of the four-month study.

Figure 2.21 depicts the LC-OCD analysis performed on the deposit which was extracted at the surface of each RO membrane. The analysis of the deposit on each RO membrane surface by Liquid Chromatography revealed that the concentration of Dissolved Organic Carbon was 5 times higher on the deposit on the MF/UF fed- RO membrane ($1.0\,\mu g.cm^{-2}$) than on the RO membrane fed by GMF ($0.2\,\mu g.cm^{-2}$). The deposit on the granular filtration fed membrane was mainly made of organic molecules of high molecular weight (>50,000 Da) like polysaccharides. The deposit on the MF/UF fed membrane was composed of high molecular weight organic compounds but also by organic compounds with a lower molecular weight (<350 Da) which were not found on the RO membrane fed by granular filtration.

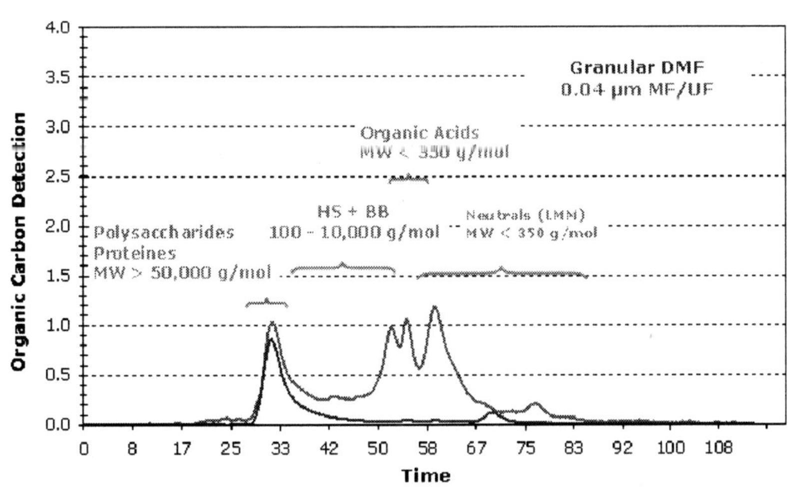

Figure 2.21 LC-OCD analysis of the RO membrane deposits – Phase 1

Figure 2.22 shows the bacteria content (total and active bacteria) within both membrane deposits. No significant difference is noticeable, and these levels of bacteria density prove the absence of major biofouling on both RO membranes.

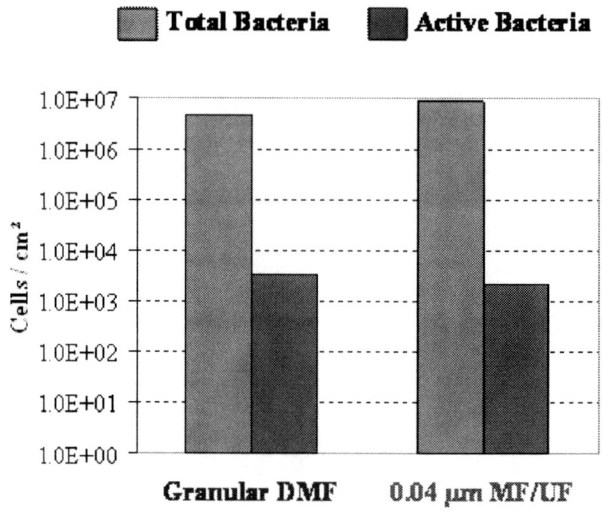

Figure 2.22 Bacteria content of the RO membrane deposits – Phase 1

Phase 2 – comparison between granular media filtration and 0.02 µm MF/UF

Figure 2.23 depicts the profile of the normalized permeate (NPF) flow for each RO unit. During this testing period, the NPF of the RO membrane fed by MF/UF decreased by 30%, whereas NPF only decreased by 20–25% for the RO membrane fed by GMF. This result tends to be somewhat consistent with what was observed during Phase 1, even though the difference between the granular filtration and the membrane filtration appears slightly less pronounced during Phase 2 as compared to Phase 1. The feed-brine pressure drop could not be compared because of problems encountered on the pressure-drop transducer installed on the RO unit fed by the MF/UF unit.

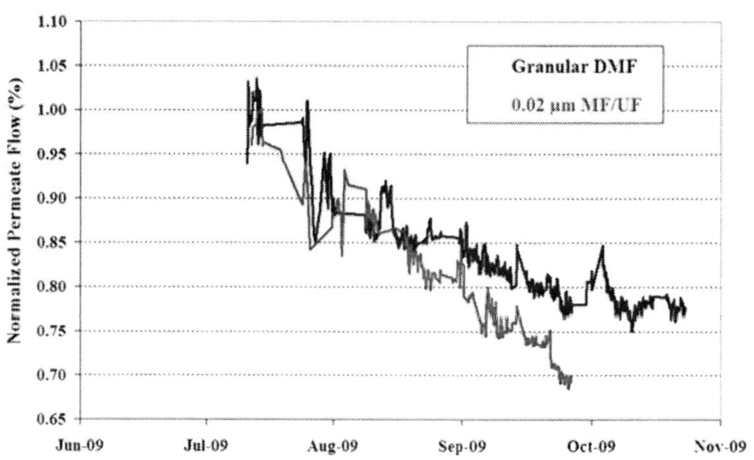

Figure 2.23 Profile of the normalized permeate flow of the RO membranes – Phase 2

2.3.5 CONCLUSIONS

According to the results collected over two distinct testing periods, whereas membrane pretreatment (MF/UF) provided a better SDI abatement than granular media filtration, the comparatively lower removal of organic matter through MF/UF appears to have induced a higher extent of organic fouling on the RO membrane. Also, this study demonstrated the advantages and interest for the use of advanced seawater characterization (i.e. NOM characterization in pretreated water and MFI measurement) as well as for the use of membrane autopsies during pilot-scale studies aiming at comparing the performance of pretreatment processes.

This study also demonstrates that identifying a single water quality parameter to fully characterize the performance of pretreatment processes and predict the fouling rate on RO units is not a realistic objective. Indeed, while SDI, MFI, particle counts provide good information related to the risk of particulate fouling, these parameters provide very limited information on organic fouling, and even less information on the risk of biofouling. This is why a multi-parametric approach is yet recommended for the monitoring of pretreatment performance.

2.4 COMPARISON OF DIFFERENT PRETREATMENT METHODS FOR RO DESALINATION

The suitability of the combined processes of coagulation/flocculation/adsorption and MF/UF as pretreatment to RO was evaluated experimentally at Chowder Bay, Sydney using semi-pilot scale experiments. The brief findings are as follows.

2.4.1 Biofilter as pretreatment to membrane based desalination: evaluation in terms of fouling index

2.4.1.1 Objectives

The objective of this study was to compare the performance of two biofilters as a pretreatment method to SWRO in terms of the reduction in membrane fouling. In this study, the reduction of the fouling potential of raw seawater arising from GAC and Anthracite biofilters was studied and related to modified fouling index (MFI), headloss, silt density index (SDI), turbidity removal.

2.4.1.2 Experimental setup

In this study, biological filter experiments were conducted using transparent acrylic filter columns. These columns have an internal diameter of 9.5 cm and length of 150 cm with outlet pipes along and at the bottom of the column. Prepared filter media comprising GAC and anthracite were packed in 4 columns up to 80 cm. The column operated in downflow mode at two different velocities of 5 and 10 m/h. Feeding water was pumped from a feeding water tank to the top of the columns and passed through the filter bed. These four columns ran for nearly 8 weeks. Effluent samples were collected at the bottom of the column for analysis (see Figure 2.24).

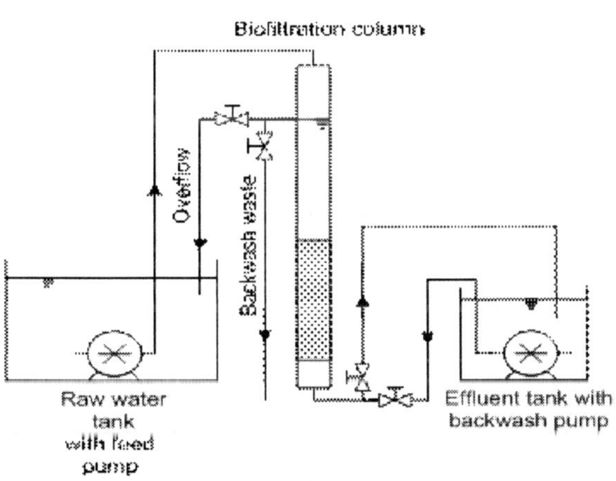

Figure 2.24 Schematic diagram of biofiltration column

2.4.1.3 Variation of seawater characteristics during experiments

Biofilter experiments took place in Chowder Bay, Sydney from April, 2008 to July 2008. The variation of seawater turbidity with temperature was monitored during this period. The average temperature found during the experimental time was between 15 to 19°C. Apart from a period of rain, seawater temperature was fairly uniform and close to 15°C. The turbidity was consistent at between 0.5–0.7 NTU. Turbidity increased during the rainy period up to 1.10 NTU. Apart from the rainy period seawater the average turbidity was close to 0.5 NTU during most of the test period.

The SDI_{10} of raw seawater was also monitored during the test period. SDI_{10} fluctuated during the test period and the value was found to be between 7.5–9.2. It was observed an increase of SDI values with the turbidity. The SDI_{10} value was consistent with the value of 7.5–8 during the experimental period when the turbidity of seawater was between 0.5–0.7 NTU. SDI_{10} increases up to 9.2 when the turbidity was high at a value of 1.10 NTU.

2.4.1.4 Effect of filtration velocity to turbidity removal

Consistency in performance of the two long term biofilters (GAC and Anthracite) was closely monitored during the experimental period for two different velocities of 5 and 10 m/h. Parameters that were investigated regarding their effect of filtrate quality were filtration velocity and filter medium. Initially GAC and Anthracite filter column operated at a filtration velocity of 10 m/h gave a filtrate turbidity which fluctuated between 0.28–0.35 NTU.

At a filtration velocity of 5 m/hr, at the initial stage, the filtrate from the anthracite biofilter gave a lower turbidity than the GAC biofilter. The filtrate turbidity value was 0.2–0.3 NTU and 0.28–0.31 NTU for the anthracite and GAC biofiltration respectively. From Figure 2.25, it is observed that a higher filtration velocity gave a worse the filtrate quality. After 55 days of operation the turbidity reduced to a value of 0.2–0.22 NTU for both GAC and Anthracite biofilter columns.

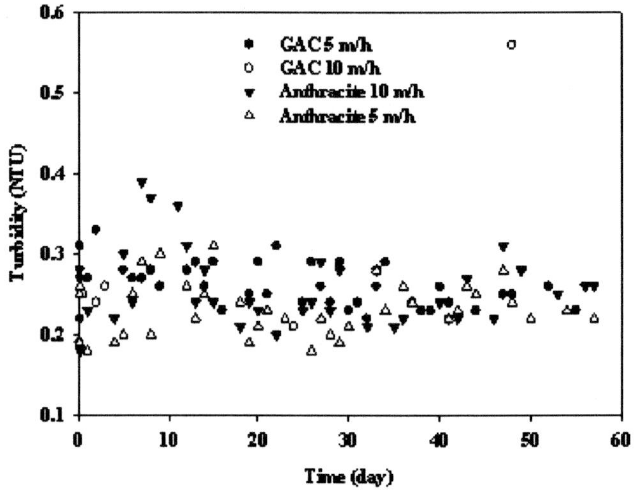

Figure 2.25 Effect of filtration velocity on filtrate turbidity

2.4.1.5 SDI_{10} and MFI

Figure 2.26 (a, b, c and d) shows the values of MFI and SDI_{10} with GAC and Anthracite biofilter at different filtration velocities (5 and 10 m/h). The average SDI_{10} and MFI for seawater were 7.5–9.2 and 250 s/l^2 respectively. A total of four biofilter columns comprising GAC and Anthracite media run at two different velocities were continuously monitored in terms of SDI_{10} and MFI. SDI_{10} fluctuated during initial period for both GAC and anthracite biofilter columns. The initial SDI_{10} value was 7.5–5.5 for Anthracite for both velocities (Figure 1.3 a. and b). The fluctuation of SDI_{10} was more pronounced when it was operated at 10 m/h velocities than 5 m/h.

For GAC biofilter column SDI10 fluctuated in the range of 8.8–5.5. After the maturation stage both biofilter column become steady and SDI10 values reduced to 3.5–4 and 3–4 for the GAC and anthracite column respectively. A similar result was found for both Anthracite and GAC biofilter columns. It is notable that the lower filtration velocity (5 m/h) gave a better quality filtrate as a result lower value of SDI.

In case of MFI, it is observed from Figure 2.26 c) and d) that MFI value decreased and became stable after 10 days of operation for both GAC and anthracite biofilters. The MFI during the initial period was found to be very high at a value of 312 and 176 s/L^2 for anthracite biofilters run at velocities of 10 and 5 m/h respectively. With the continuous operation of 1 day the MFI values stabilize and reduced to 5 and 10 s/L^2 for 5 m/h and 10 m/h velocities. A similar trend was noted for the GAC biofilters but both columns (5 and 10 m/h) stabilized after 10 days of operation. It was also observed that the MFI values decreased and was stable up to 55 days of operation for both GAC and anthracite biofilters. The decrease in the MFI values was notable. This suggests that it is possible to decrease fouling potential of the RO membrane with a biofilter columns operated at slow filtration velocities. After reaching ripening time/maturity time biofilters gave steady MFI and SDI results.

2.4.1.6 Head build up

The head loss development was monitored as a function of filtration time for GAC and anthracite biofilter columns. A higher filtration velocity and finer grain size gave rise to a faster headloss development.

In particular, it was observed that for GAC and anthracite head development was higher when the filtration velocity was 10 m/h. This is due to the increase of particle deposition at a higher filtration velocity. When GAC and anthracite biofilters were operated at 5 m/h the head development was not significant. The head loss for a anthracite biofilter was smaller than the GAC biofilter as the grain size of the GAC media is smaller anthracite (Boller and Kavanaugh, 1995).

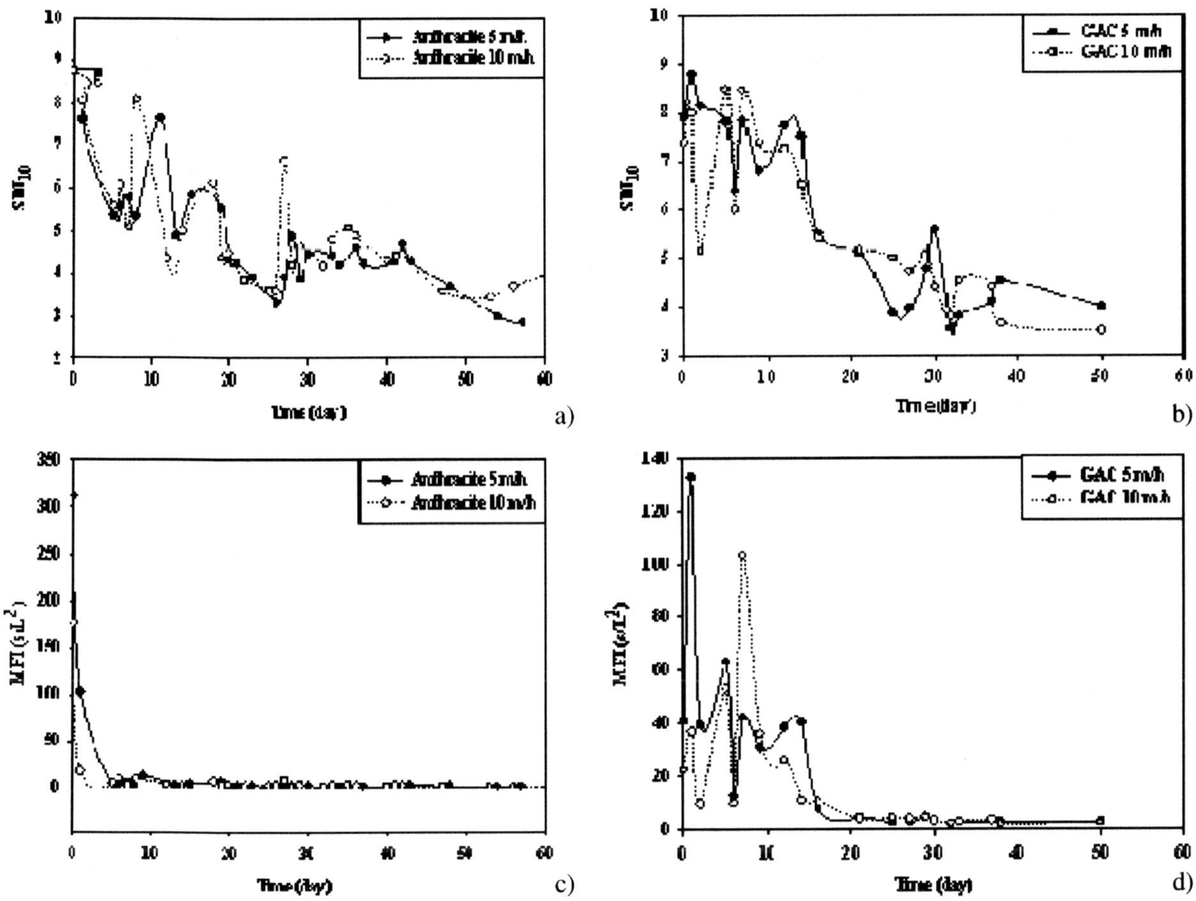

Figure 2.26 Variation of SDI and MFI for Anthracite and GAC biofilters run at 10 m/h and 5 m/h

2.4.1.7 Summary

The removal of particulate matter and organics from seawater by the use of biofiltration was investigated. Granular activated carbon (GAC) and anthracite were tested as a filter medium at two different velocities. The filtrate quality was evaluated by measuring the silt density index (SDI), modified fouling index (MFI) and turbidity removal. Seawater reverse osmosis (SWRO) also conducted as a post treatment. Both biofilters demonstrated similar behavior in terms of SDI and MFI. MFI values decreased to 10 s/L^2 and was stable up to 55 days of operation for both GAC and anthracite. The filtrate turbidity was steady after 10 days and the value was 0.2–0.3 NTU and 0.28–0.31 NTU for anthracite and GAC biofiltration respectively. Furthermore, when two biofilter columns operated at different velocity, it was observed that slower velocity column had a lower head development than the higher velocity.

2.4.2 Fibre media filtration as a pretretment for seawater

2.4.2.1 Objectives

The objectives of this study were to investigate the performance of the fibre filter as pre-treatment to SWRO desalination. In-line coagulation and different packing densities of fibre filter and operating velocities of the filter were evaluated in terms of MFI, and SDI_{10}.

2.4.2.2 Materials and methods

2.4.2.2.1 Seawater

Seawater was drawn from Chowder Bay where the Sydney Institute of Marine Science (SIMS), Australia is located. The average turbidity of the seawater ranged from 0.5 to 3 NTU. Dissolved Organic Carbon (DOC) during the operation of fibre filter was below 1 mg/L. The pH was about 8.

2.4.2.2.2 Coagulation

Ferric chloride ($FeCl_3 \cdot 6H_2O$) was used as the in-line coagulant of influent seawater to fibre filter. Coagulation was carried out using the optimum dose (1 mg/L) of ferric chloride ($FeCl_3 \cdot 6H_2O$) predetermined by standard jar tests. Ferric chloride was chosen in these experiments as it is capable of removing colloidal organic matter in seawater.

2.4.2.2.3 Fibre filter

A lab-scale fibre filter was installed at SIMS. The schematic and photographs of fibre filter are shown in Figure 2.27. The experiment was performed using a vertical column in up-flow mode. The height and diameter of the fibre filter were 1,000 mm and 30 mm, respectively. A bundle of micro-fibre was used as a the filter media.

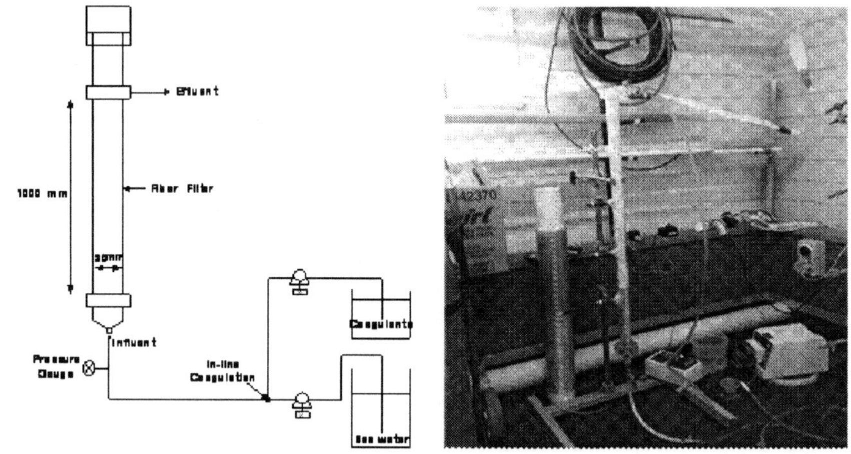

Figure 2.27 Schematic and on-site picture of fibre filter

2.4.2.3 Operational conditions

Experiment were conducted with and without in-line coagulation to a fibre filter with a packing density 115 kg/m^3 operated at a filtration velocity of 20 m/h to examine the effect of in-line coagulation.

Another set of experiments was carried out to investigate the effect of different packing densities (105 and 115 kg/m^3) and filtration velocities (40 and 60 m/h) of the fibre filter.

2.4.2.4 Results and discussion

2.4.2.4.1 Effect of in-line coagulation

Table 2.10 shows the values of MFI and SDI_{10} with and without in-line coagulation of seawater influent to the fibre filter. The SDI_{10} and MFI values of initial seawater were 8.75 and 214 s/L^2, respectively. Here, it should be noted that seawater characteristics vary from season to season and place to place. Following fibre filtration at a velocity of 20 m/h and packing density of 115 kg/m^3 without in-line coagulation, the SDI_{10} and MFI values were 7 and 30 s/L^2 respectively. When 1 mg/L of $FeCl_3$ was used as the in-line coagulant with the same fibre filter operating condition, the SDI_{10} and MFI values were decreased down to 6 and 4 s/L^2, respectively. The SDI_{10} values decreased slightly in contrast to the MFI values which decreased remarkably. This suggests that in-line coagulation decreases the fouling potential of the membrane.

Table 2.10 MFI and SDI_{10} values with and without in-line coagulation (MFI of seawater influent = 214 s/L^2, SDI_{10} of seawater influent = 8.75)

Filtration velocity (m/h)	Packing density (kg/m^3)	Coagulants (mg/L)	MFI (s/L^2)	SDI_{10}
20	115	0	30	7
		1	4	6

2.4.2.4.2 Effect of different packing densities and filtration velocities

Table 2.11 shows the values of MFI and SDI_{10} with fibre filter different packing densities and filtration velocities. A dosage of $FeCl_3$ (1 mg/L) was used as the in-line coagulant for all conditions. As stated above, the SDI_{10} and MFI values of untreated seawater were 8.75 and 214 s/L^2, respectively. When fibre filter was operated at different packing densities and filtration velocities, the SDI_{10} and MFI values decreased down to 4 and 2 s/L^2 respectively. The decrease in the MFI values was notable. This suggests that it is possible to decrease fouling potential of the membrane with a fibre filter of low packing density operated at a high filtration velocity.

Table 2.11 MFI and SDI_{10} values with different packing densities and filtration velocities (MFI of seawater influent = 214 s/L^2, SDI_{10} of seawater influent = 8.75)

Filtration velocity (m/h)	Packing density (kg/m^3)	MFI (s/L^2)	SDI_{10}
40	105	2	4
	115	2	4
60	105	2	4
	115	2	4

2.4.2.5 Pressure drop and turbidity

2.4.2.5.1 Effect of in-line coagulation

The pressure drop (ΔP) with and without in-line coagulation of fibre filter operated at a velocity of 20 m/h and packing density of 115 kg/m^3 was monitored. The ΔP without in-line coagulation was 4 mbar, suggesting that the retentions of suspended particles were marginal. The ΔP with in-line coagulation of filtration increased to 33 mbar. Table 2.12 shows the increase of ΔP from the initial filtration to the end of the filtration period of 90 minutes and turbidity with and without in-line coagulation. The average of turbidity was about 1.75 NTU during experiments which coincided a rainy period. Turbidity with and without in-line coagulation of fibre filtration decreased from 1.75 NTU (raw seawater) to 0.16 NTU and 0.49 NTU, respectively. It can be concluded that the fibre filtration with in-line coagulation removes 91% of turbidity.

Table 2.12 ΔP and turbidity with and without in-line coagulation (turbidity of initial seawater: 1.75 NTU)

Filtration velocity (m/h)	Packing density (kg/m^3)	Coagulants (mg/L)	ΔP (mbar)	Turbidity (NTU)
20	115	0	4	0.49
		1	33	0.16

2.4.2.5.2 Effect of different packing densities and filtration velocities

ΔP with different fibre filter packing densities and filtration velocities was also measured. The ΔP with a packing density of 105 kg/m^3 and 115 kg/m^3 operated at filtration velocities of 40 m/h was 38 mbar and 42 mbar, respectively. The ΔP with a packing density of 105 kg/m^3 and 115 kg/m^3 operated at filtration velocity 60 m/h was 41 mbar and 49 mbar, respectively. Table 2.13 shows the increase of ΔP from its initial filtration to the end of the filtration period of 90 minutes and turbidity with different fibre filter packing densities and filtration velocities. The increase of ΔP is proportional to that of the packing density and filtration velocity. The lowest turbidity was 0.14 NTU for a packing density of 115 kg/m^3 and filtration velocity of 40 m/h. The values of turbidity at other operating conditions were between 0.22 and 0.23 NTU. Regardless of packing density and filtration velocity, turbidity remained low until the end of the filtration period.

Table 2.13 ΔP and turbidity with different packing densities and filtration velocities (Average turbidity of seawater: 1.75 NTU)

Filtration velocity (m/h)	Packing density (kg/m^3)	ΔP (mbar)	Turbidity at 30 min of operational time (NTU)	Turbidity at 90 min of operational time (NTU)
40	105	38	0.35	0.22
	115	42	0.28	0.14
60	105	41	0.27	0.23
	115	49	0.32	0.23

2.4.2.6 Summary

Fiber filter as pre-treatment of SWRO in terms of MFI, SDI_{10}, pressure drop (ΔP), turbidity and MWD was investigated. The following conclusions were obtained:

1. The use of in-line coagulation gave the lowest values of SDI_{10} and MFI.
2. When fiber filter was operated with different packing densities (105 and 115 kg/m^3) and different filtration velocities (40 and 60 m/h), the SDI_{10} and MFI values were 4 and 2 s/L^2, respectively.
3. The pressure drop (ΔP) with and without in-line coagulation was 33 and 4 mbar respectively. Turbidity with and without in-line coagulation decreased from 1.75 NTU to 0.16 NTU and 0.49 NTU, respectively. The increase of ΔP was proportional to the fiber filter packing density and filtration velocity. Regardless of fiber filter packing density and filtration velocity, turbidity remained low until the end of the filtration period.

2.4.3 Submerged microfiltration coupled with physico-chemical processes as pre-treatment to sea water desalination

2.4.3.1 Background

In this study we used a submerged membrane system as a pre-treatment process to seawater desalination with and without pre-flocculation and pre-adsorption. Hollow fibre membrane module was used and its characteristics are summarized in Table 2.14. Seawater was pumped into the submerged membrane reactor using a feeding pump and the effluent (permeate) was collected at a predetermined flux using a suction pump. A pressure gauge was used to measure the transmembrane pressure (TMP) and a soaker hose air diffuser was used to provide air for membrane scouring.

Table 2.14 Characteristics of the hollow fiber membrane module used

Item	Characteristics
Material	Hydrophilic polyethylene
Nominal pore size	0.1 μm
Outer diameter	0.41 mm
Inner diameter	0.27 mm
No. of fibre	320 (16 × 20)
Length of fibre	12 cm
Surface area	0.05 m^2
Membrane packing density	9858 m^2/m^3
Membrane manufacturer	Mitsubishi-Rayon, Tokyo, Japan

Adsorption and flocculation pre-treatment were performed using a lab-scale batch reactor equipped with mechanical stirrers. Powdered activated carbon (PAC, wood based) (Table 2.15) was used as adsorbent and ferric chloride ($FeCl_3$) was used for flocculant. After the adsorption and flocculation pretreatment, the treated seawater was settled down before passing it through submerged membrane system.

Table 2.15 Characteristics of powdered activated carbon (PAC) used

Specification	PAC-WB
Iodine number (mg/g min)	900
Ash content (%)	6 max.
Moisture content (%)	5 max.
Bulk density (kg/m^3)	290–390
Surface area (m^2/g)	882
Nominal size	80% min finer than 75 micron
Type	Wood based
Mean pore diameter (Å)	30.61
Micropore volumn (cc/g)	0.34
Mean diameter (μm)	19.71
Product code	MD3545WB powder

2.4.3.1.1 Critical flux experiments with seawater

Critical flux is defined as the maximum flux at which the membrane system can operate without any accumulation of foulants on the membrane. The flux step method is a common experimental technique for determining critical flux. Flux is held constant for the specified filtration time (1 hour in this study) during which the evolution of trans-membrane pressure (TMP) is monitored with time. The critical flux is the lowest set-point flux at which the suction pressure is noted to increase with time due to fouling rather than remaining constant.

Figure 2.28 shows the variation of TMP value for constant filtration flux using the seawater without and with pre-flocculation. Flocculation was provided in batch mode in the following conditions: addition of $FeCl_3$ (1 mg/L), agitation during 20 minutes and settling for 20 mins prior to application of the supernatant to submerged membrane system. The critical flux was around 5 L/m^2.h for the seawater without pre-flocculation (Figure 2.28a). It increased to 6.7 L/m^2.h when a pre-treatment of flocculation was provided (Figure 2.28b). The pre-flocculation helped in capturing and agglomerating colloids, thereby reducing the membrane fouling by colloids. According to the results, flocculation had more considerable effect on the critical flux than adsorption.

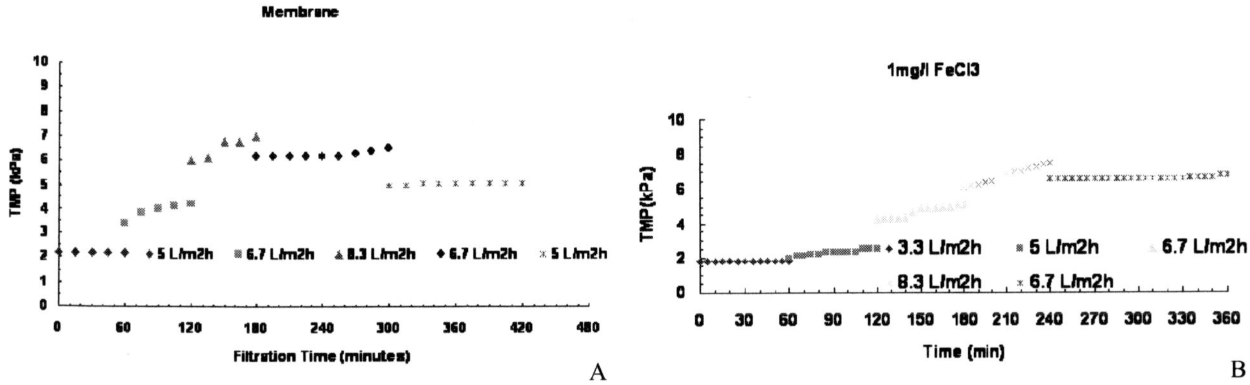

Figure 2.28 Constant filtration flux experiments with seawater (with-B and without-A pre-flocculation) (membrane pore size 0.45 μm; $FeCl_3$ dose = 1 mg/L; air flow rate 5 m^3/m^2 membrane area/hr)

The effect of adsorption as pre-treatment on critical flux is shown in Figure 2.29. Here two different doses of PAC (0.05 and 1 g/L) were added in the seawater and mixed for 1 hour before the PAC was settled down for 1 hour. With the pre-treatment of adsorption with 0.05 g/L PAC, the critical flux increased from 5 L/m^2.h to 6.7 L/m^2.h. When a dose of 1 g/L PAC was used the critical flux increased from 5 L/m^2.h to 13.3 L/m^2.h. The results indicated that the pre-treatment of flocculation and adsorption was able to remove large and small molecular weight organic matter respectively while enhancing the filtration flux of submerged membrane system and dissolved organic removal.

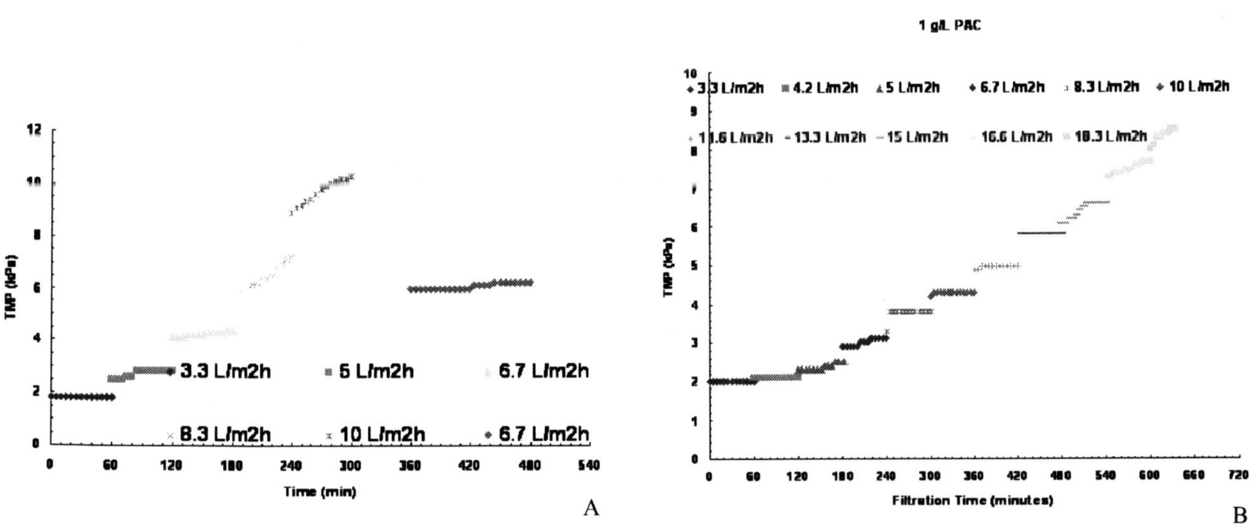

Figure 2.29 Constant filtration flux with seawater after adsorption. A- 50 mg/L PAC, B- 1 g/L PAC, (membrane pore size = 0.45 μm; PAC dose = 2 g/L)

2.4.3.1.2 Effect of pre-treatment on fouling reduction

The relative merits of different pre-treatments were studied in terms of fouling potential. Modified fouling index using ultrafilter membrane of 30 kDa (UF-MFI) and SDI_{15} (using 0.45 microfilter membrane) measurements were made to compare the different pre-treatments. UF-MFI and SDI_{15} was measured using dead end cell. The UF-MFI and SDI_{15} of raw seawater were 4531 s/L^2 and 5.8, respectively. After the pre-treatment of flocculation with $FeCl_3$ (of 1 mg/L), the UF-MFI was decreased to 2305 s/L^2 and 4.5. This may be due to the formation of flocs of different sizes during flocculation which lead to reduce the fouling. On the other hand, a pre-treatment with PAC adsorption with 0.05 g/L gave rise to higher organic removal and also resulted in a very low UF-MFI and SDI_{15} value of 1329 s/L^2 and 3 respectively.

2.4.3.2 Summary

1. The pre-treatment of PAC adsorption with a dose of 1 g/L showed an improvement in the critical flux up to 13.3 L/m^2.h. The pre-flocculation could marginally increase the critical flux up to 6.7 L/m^2.h from 5 L/m^2.h.
2. The performance of these pre-treatments was determined in terms of modified fouling index using ultrafilter membrane (UF-MFI) and silt density index (SDI). UF-MFI and SDI indicated that PAC adsorption was better pre-treatment than flocculation. The UF-MFI and SDI_{15} of raw seawater were 4,531 s/L^2 and 5.8, respectively. After the pre-treatment of flocculation with $FeCl_3$ (of 1 mg/L), the UF-MFI was decreased to 2,305 s/L^2 and 4.5. On the other hand, a pre-treatment with PAC adsorption with 0.05 g/L gave rise to higher organic removal and also resulted in a very low UF-MFI and SDI_{15} value of 1,329 s/L^2 and 3 respectively.
3. Detailed molecular weight distribution (MWD) of seawater organic matter was examined after different pre-treatments. MWD of the initial seawater mainly ranged from 1,510 Da to 130 Da. It is observed that $FeCl_3$ flocculation and PAC adsorption as a pre-treatment is partially removed the organic matter corresponding to 1,510 Da and 530 Da respectively.

2.5 SUBMERGED HOLLOW FIBER SYSTEM AS PRE-TREATMENT FOR THE SEA WATER REVERSE OSMOSIS: EFFECT OF THE OPERATION CONDITIONS AND THE CHARACTERIZATIONS OF THE PERMEATE QUALITY

2.5.1 Introduction

The application of membrane system as pre-treatment for the RO plant in the water industry is now being considered as a viable solution due to its several advantages compared with the conventional pre-treatment methods (Pearce, 2007; Chua *et al.*, 2003 and Fritzmann *et al.*, 2007). However, the fouling of MF/UF membrane for seawater RO pretreatment is still an key issue influencing the operation cost of pre-treatment system. As a result, there is still more research needed to improve membrane configurations and operation conditions in order to reduce energy consumption and fouling control. Relaxation and backwashing as physical cleaning techniques have been widely used as standard operating strategies to limit fouling. The effect of backwash duration, backwash flow rate and backwash frequencies on membrane fouling were investigated previously by some researchers (Chua *et al.*, 2003 and Pervov *et al.*, 2003). However, they are mostly based on the spiral wound system and the pressurized hollow fibre system with the aid of small dose of coagulant. Not many reports has been seen by using the submerged hollow fibre system for the SWRO pre-treatment even though submerged hollow fibre system have been widely used for the surface water and wastewater treatment such as MBR system. It is believed that submerged hollow fibre might have the chance of produce higher permeate with lower energy cost (Fane *et al.*, 2005).

In this study, the effects of operation conditions on submerged hollow fibre system without coagulant for both seawater from Sydney area and synthetic sea water were studied. Three different MF/UF membranes were investigated based on the fouling performance. The effects of operation parameters (filtration duration, backwash duration, backwash flux etc.) on the TMP and fouling mechanisms are investigated. Based on the findings, the system operation optimization would be available. Furthermore, the composition of the feed, permeate and the waste water collected after backwash were analysed for the better understanding of the fouling mechanisms.

2.5.2 Method and materials

The filtration was carried out under constant flux mode using a bench scale submerged hollow fibre system. Figure 2.30 shows the diagram of the experimental system.

A full system operation cycle consists of filtration, backwash/relaxation, system drain and the refill. During the filtration, the permeate was sucked out by the pump head 2 of peristaltic pump A. In the same time the same amount of feed solution was feed into the acrylic tube (9 cm in diameter and 30 cm in length) to maintain the feed water level constant by the pump head 1 of the peristaltic pump A. During the backwash, the rotation direction of permeate pump was reversed. The permeate was used to backwash the membrane. After the backwash, the solenoid valve in the bottom of filtration tube was opened and the waste solution after backwash was discharged. After discharging, the solenoid valve is closed and pump B is activated to refill the filtration tube. Then next filtration cycle starts. The operation of the system is automatically controlled by the computer. The TMP and the flux are also recorded by computer with time.

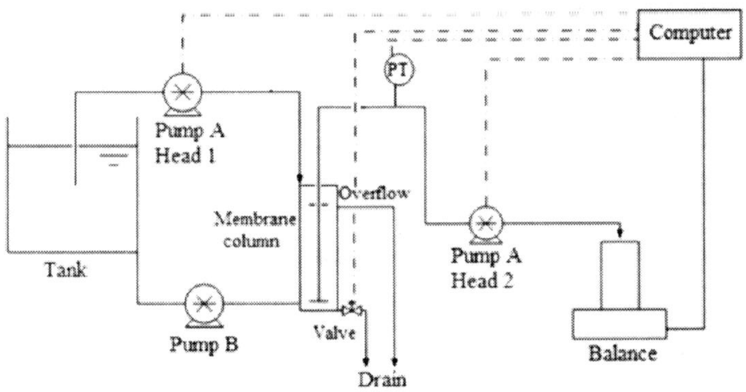

Figure 2.30 Experimental diagram of submerged hollow fibre system

The hollow fibre module used in all experiments was potted in the lab. Three different types of membranes were used in this study. They are polypropylenen membrane (PP) (0.22 μm, od/id: 650/390 μm), Polyvinylidene fluoride membrane (PVDF) (0.04 μm, od/id: 800/500 μm) and PVDF 0.04 μm (od/id 1,300/650 μm) (from Memcor, Australia).

Both real seawater from Sydney area and the synthetic seawater were used in this study. The compositions of the synthetic sea water are 33 g/L red sea salt with 5 mg/L humic acid (Sigma Aldrich), 10 mg/L 22 nm silicia colloidal (LUDOX® TM-50, Sigma Aldrich) and 10 mg/L 70–100 nm mixture size silica colloidal solutions (SNOWTEX®-ZL, Nissan Chemical Industries Ltd).

The compositions of the seawater water, the permeate water and the water collected after backwash were also studied using liquid Chromatography-Organic Carbon Detection (LC-OCD).

Morphology study of the membrane surface was conducted using scanning electron microscopy (SEM) with the model Hitachi S-900. The membrane samples are dried in a vacuum oven for at least 20 hours before coating with chromium for SEM analysis.

2.5.3 Data analysis

Three parameters, which are used to evaluate the effects of backwash in controlling membrane fouling, are shown as follow.

ΔTMP_f represents the TMP increase between the beginning and the end of the filtration.

$dTMP/dt_i$ represents the TMP increase rate or the fouling rate in the filtration cycle i. The value indicates the fouling cake property in the corresponding filtration cycle.

Reversible fouling (%) represents percentage of the fouling which can be removed by the backwash in the corresponding cycle.

2.5.4 Results and discussions

1. **Effect of different membrane and operation modes**

   In this section, the fouling profile of the real seawater filtration with periodical backwash was compared with the filtration with periodical relaxation and the continuous filtration. Three different membranes are used for different modes of filtration for the comparisons.

   For the PVDF Membrane (0.04 um, OD/ID: 1,300/650), the filtration with periodical backwash has less fouling compared with other two operation modes. It seems that backwash not only removed the reversible fouling cake but also changed the structure of the fouling layer resulting in the low fouling rate during the filtration cycle. SEM pictures of the fouled membrane also indicate that the backwash flux penetrated through the weakest parts of the fouling cake, leaving some open structure in the fouling cake.

For the PVDF Membrane (0.04 um, OD/ID: 800/500), the filtration with periodical backwash has less fouling compared with other two operation modes. However, for this membrane, the backwash results in higher fouling rate during the filtration cycle than the continuous filtration.

For the microfiltration membrane (PP membrane, 0.2 um), no benefit was observed for the backwash mode of filtration. For the MF membrane, the major fouling mechanisms is the pore blocking rather than the cake filtration, which results in the less percentage of reversible fouling removed by backwash.

The effects of operation modes on the membrane fouling for three different hollow fibres are summarized in Table 2.16. Even though the MF membrane has the higher permeability than those of UF membranes, which is mainly due to the high permeability of clean MF membrane, the MF membrane does show higher TMP rise and less reversible fouling removed than UF membranes (PVDF 0.04 μm (OD/ID:1,300/650). PVDF 0.04 μm (1,300/650) has the best performance of the three membranes considering membrane fouling limitation. So this membrane was chosen for the further investigation of backwash operation condition, including the filtration duration, backwash strength, backwash duration and the aid of aeration during backwash on the fouling limitation.

Table 2.16 The effect of operation modes on the membrane fouling for three hollow fibres

Filtration mode	PVDF (0.04 μm, OD/ID: 1300/650)			PVDF (0.04 μm, OD/ID: 800/500)			PP (0.2 μm OD/ID: 650/390)	
	Backwash mode	Relaxation mode	Continuous mode	Backwash mode	Relaxation mode	Continuous mode	Backwash mode	Continuous mode
TMPf (16 hr)	16.2	26.3	27.6 (5 hr)	37.3	46.8	51.7	19.7	21.1
Percentage of reversible fouling(%)	87.2	76.1	–	80.2	74.3	–	41	–
dTMP/dt	2.0	4.3	5.3	7.2	7.1	3.0	1.3	1.1

2. **The investigation of operation conditions of the filtration with periodical backwash on membrane fouling by using the real sea water**
   - **The effect of filtration duration on membrane fouling**
     It was found that the final TMP did not increase when filtration duration increased up to 3,600 s. Further increment in filtration duration appears to promote the fouling. Prolonged filtration duration might cause more compact cake layer, which was more difficult to be removed by backwash. Prolonged filtration duration also results in the higher fouling rate during the filtration cycle.
   - **The effect of backwash strength/flux on membrane fouling**
     It appears the existence of optimum backwash strength for fouling control. It might promote the membrane fouling if the backwash flux is either too low or too high. In our experiments for the real sea water, the increase of backwash flow rate up to one and half times of the permeate flow rate leads to the process improvement. Further increment in the backwash flow rate shows worse fouling. When the backwash flux is too high, the membrane and the residual fouling cake might also be fouled by residuals/impurities in the backwash solution, which results in less reversible fouling and higher fouling rate in the filtration cycle.
   - **The effect of backwash duration on membrane fouling**
     It was found that ΔTMP_f decreased from 13.1 to 6.3 kPa with the increase of backwash time up to 30 s. Further increase of backwash duration did not show obvious benefits. ΔTMP_f only drops to 5.7 when the backwash duration increase to 60 s. The percentage of fouling removed by backwash increases slightly with the increase of backwash duration up to 30 s. However it drops slightly when the backwash time further increase. This indicates that the excess backwash volume might also foul the membrane or the residual fouling cake due to the impurities in the backwash flux. Again, similar trend was observed for the fouling rate. It decreased from 2.6 to 1.1 kPa/hr when backwash duration increased up to backwash duration 30 s. No obvious improvement was observed with further increment of backwash duration.
   - **The effect of backwash flux and backwash duration based on the same backwash volume on membrane fouling**
     For the same amount of backwash volume used, the final TMP increase did not vary a lot due to the change of backwash strength and backwash duration even though the high backwash strength with short duration has the lowest TMP increase after 16 hrs filtration. The moderate backwash strength showed slightly lower fouling rate (dTMP/dt) during the filtration cycle. It seems that too high backwash strength/flux might also cause some pore blocking of the cake or membrane by the impurities in the backwash solution. This is consistent with what was found in the effect of backwash strength alone that the excess backwash flux actually results in higher fouling rate in the filtration cycle.

- **The aid of aeration during backwash on membrane fouling limitation**

  It seems in Figure 2.31 that, in the low air flow rate, the backwash with aeration did show lower final TMP (ΔTMPf 9.0 kPa for 15 ml/min the air flow rate) than that of backwash without air bubble (ΔTMPf 14.4 kPa). However, it is unexpected that the further increases of the air flow rate actually limit the benefit, with ΔTMPf increasing to 15.5 kPa for the 198 ml/min air flow rate. It appears that the percentage of reversible fouling is quite constant (84.7 ± 2%) with the change of air flow rate from 0 to 198 ml/min. This indicated the existence of bubble did not improve reversibility. However, it was found that the aeration during the backwash did influence the structure of the cake (Figure 2.31b). The values of dTMP/dt for air flow rate of 15, 30 ml/min are only 1.6 and 1.4 kPa/hr respectively, while the value of dTMP/dt increased to 2.7 kPa/hr for the air flow rate 198 ml/min. Moderate flow rate did slow down the fouling rate during the filtration, however, with further increment in air flow rate, the fouling rate increases again. It was also observed that, with the increase of air flow rate, the shape of the bubble changed from small bubble to ellipsoidal bubbles and to spherical cap-vortex bubble. The bubble shape difference might contribute to this unexpected high fouling rate observed in the high air flow rate.

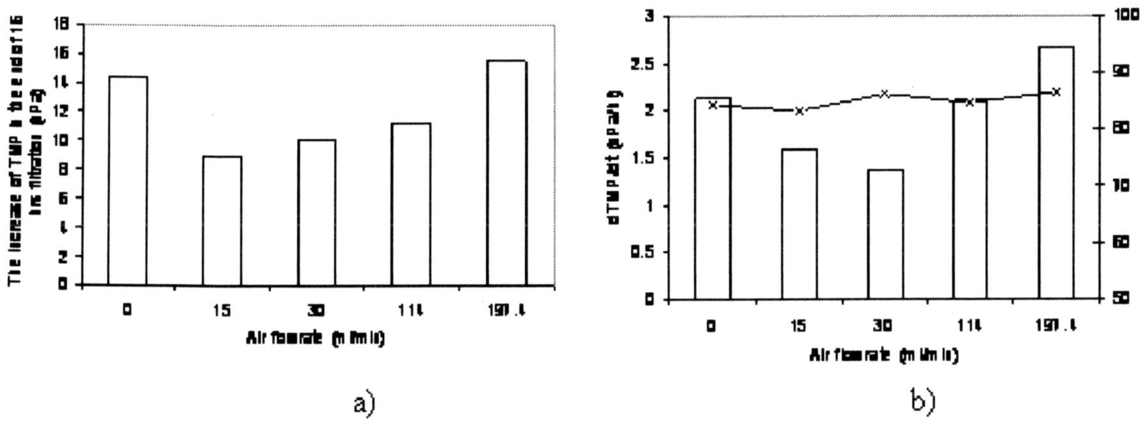

Figure 2.31 The effect of air bubble during backwash on the membrane fouling. The filtration flux 73 LMH, filtration duration 3,600 s, backwash flux is 1.5 times of filtration flux, backwash duration 30 s, PVDF 0.04 mm (od/id 1,300/650) hollow fibre

3. **The Characterization of the seawater**

   The compositions of sea water, the permeate and the wastewater collected after backwash were also studied using LC-OCD. It was found that the DOC removed by UF membrane is around 10%. It was also found that around 60% biopolymers were removed by UF membrane. This is significant reduction in organic matter with size larger than 20 kDa. There are more biopolymers found in the wastewater collected after backwash, indicating the fouling layer formed by the biopolymers was removed by the backwash (Figure 2.32).

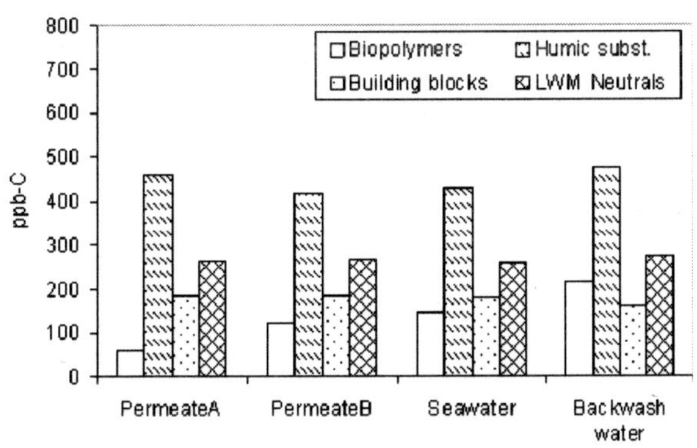

Figure 2.32 The characterization of seawater, the permeate and waste water collected after backwash by LC-OCD

4. **Investigation of effect of backwash operation conditions using the synthetic seawater**
   - **Effect of individual operating conditions**

     The effect of operation condition such as filtration duration, backwash duration and backwash strength was investigated individually by using the synthetic seawater as the feed solution. The results are consistent with those found using the real seawater which was shown in the previous section.

- **Effect of operation flux on membrane fouling**
  The effect of filtration flux was investigated on the flux of 70 and 80 LMH. Other operation conditions such as filtration duration (1 hr), backwash duration (60 s) and backwash strength (same as the filtration flux) were kept constant. As expected, higher filtration flux results in higher final TMP. With higher flux, the accumulation of cake layer is much faster resulting in higher fouling rate. However, it is also found that the percentage of foulant removed by backwash increased with higher filtration flux. The elevated filtration flux accelerates the accumulation of larger foulants lead to a quick cake formation. This cake would act as a secondary membrane entrapping the smaller particles that contributes to the irreversible fouling. Also, the high filtration flux promotes the growth of the cake in height rather than the density. As a result, it might be able to be removed easily by backwash.
- **Investigation of the interaction within different operation conditions by experimental design**
  2^3 experimental design was applied to investigate if there is any interaction among operation conditions (filtration duration, backwash duration and backwash strength) and which is the most important effect. The filtration flux of 100 LMH was used in the experimental design test in order to distinguish the effects more clearly. The values of operation conditions and three responses which are used to analyse the effects of operation conditions are listed in Table 2.17.

Table 2.17 Operation conditions and results of the 2^3 experimental design

Label	A-Backwash duration (s)	B-Backwash strength (times of filtration flux)	C-Filtration duration (hr)	ΔTMPf (kPa)	Reversible fouling (%)	dTMP/dt (kPa/hr)
1	10	0.5	0.5	12.6	81.6	2.2
a	90	0.5	0.5	7.4	91.1	2.5
b	10	2	0.5	9.1	91.5	3.0
ab	90	2	0.5	8.4	91.4	2.6
c	10	0.5	1.5	25.9	69.9	3.6
ac	90	0.5	1.5	17.8	89.8	4.0
bc	10	2	1.5	17.6	87.1	2.7
abc	90	2	1.5	16.0	86.2	3.5

Via regression analysis and only taken the major effects into model, each response can be expressed by the equations as follow.

$$\Delta TMP_f = 14.3 + \left(\frac{9.95}{2}\right)\left(\frac{C-1}{0.5}\right) + \left(\frac{-3.9}{2}\right)\left(\frac{A-50}{40}\right) + \left(\frac{-3.15}{2}\right)\left(\frac{B-1.25}{0.75}\right) + \left(\frac{-2.74}{2}\right)\left(\frac{A-50}{40}\right)\left(\frac{B-1.25}{0.75}\right) \quad (2.1)$$

$$\frac{dTMP}{dt} = 3.0 + \left(\frac{0.85}{2}\right)\left(\frac{C-1}{0.5}\right) + \left(\frac{-0.53}{2}\right)\left(\frac{B-1.25}{0.75}\right)\left(\frac{C-1}{0.5}\right) \quad (2.2)$$

$$Re\,versibility = 0.86 + \left(\frac{-0.076}{2}\right)\left(\frac{A-50}{40}\right)\left(\frac{B-1.25}{0.75}\right) + \left(\frac{0.071}{2}\right)\left(\frac{A-50}{40}\right) + \left(\frac{0.06}{2}\right)\left(\frac{B-1.25}{0.75}\right) + \left(\frac{-0.057}{2}\right)\left(\frac{C-1}{0.5}\right) \quad (2.3)$$

As seen in Equation 2.1, the effect C (filtration duration) is the most important effect in the ΔTMP_f, followed by effect A (backwash duration) and effect B (backwash strength). No important interaction of these three factors on the ΔTMP_f was observed. The change of filtration duration from low to high level causes an increase in the ΔTMP_f. This confirms the previous finding that prolonged filtration duration results in the higher ΔTMP_f. The increase of backwash duration results in a decrease in the ΔTMP_f. The trend is much more obvious than presented in previous section. This may be due to the higher flux used (100 LMH compared to 70 LMH) for the experimental design, which may amplify the effect of backwash duration on the ΔTMP_f. It was also found that the increase of the backwash flux decreases the ΔTMP_f, which also supports the findings in the previous sections.

As seen in Equation 2.2, for the response of fouling rate, the factor C (filtration duration) is the most important effect followed by the interaction of effect B and C (backwash strength and filtration duration). Longer filtration duration leads to increasing fouling rate which consequently results in higher ΔTMP_f. The interaction of backwash strength and filtration duration on dTMP/dt was examined. For the experiments with shorter filtration cycle, where the membrane is not significantly fouled, the higher backwash strength actually results in higher fouling rate comparing with the corresponding values using lower backwash strength. While the membrane was fouled more significantly during the longer filtration duration, the higher backwash strength actually decreases the fouling rate comparing with those under lower backwash strength. The interaction of backwash strength and filtration duration highlight the importance of optimum backwash strength related to the fouling layer. These results again support our finding in previous section, mentioning the existence of optimum backwash strength. Excess backwash might not only waste the energy but also promote the fouling rate.

As seen in Equation 2.3, for the response of foulant reversibility which can be removed by backwash, the significant effects are the interaction of factors A and B (backwash duration and backwash strength) followed by factor A (backwash duration). The interaction of backwash duration and backwash strength was investigated. When applying the low backwash duration, the increase of backwash duration removed more percentage of reversible fouling. However, in the cases of high backwash strength, the increase of backwash duration did not show any benefit. This again highlights the existence of optimum backwash volume. Excess backwash volume shows no benefit or even caused less percentage of reversible fouling. This might be due to the fouling of membrane or existing cake by excess backwash solution.

2.5.5 Conclusions

For the seawater in Sydney area, the PVDF 0.04 um (ID/OD: 1,300/650) hollow fibre has the best performance during the filtration with periodical backwash when compared with the other two hollow fibre membranes. For this type of UF membrane, the mechanisms of backwash in fouling limitation are the removal of part of fouling cake and the change of fouling cake structure which in turn changes the fouling rate during the filtration cycle. The LC-OCD analysis shows that the UF membrane mainly removed the biopolymer in the feed solution. The biopolymer cake can be partially removed by backwash.

In the submerged hollow fibre ultrafiltration process, the performance of the system can be improved by optimizing the operation conditions. Experiments using the real seawater and the synthetic seater showed consistent results. The increase of filtration duration up to a limit does not promote the membrane fouling. However, prolonged filtration duration would significantly foul the membrane. Increase of backwash strength also limits membrane fouling within a certain range. Outside this range, further increment in backwash strength lead to sever fouling. The effect of backwash duration on fouling limitation is very much depending on the backwash strength. For low backwash strength, longer backwash duration shows some benefit. While at high backwash strength, the prolonged backwash duration shows no benefit or even makes things worse. This indicated the existence of optimum backwash volume. For the same backwash volume used, the moderate backwash strength is preferred rather than a long time weak backwash or a quick, extremely strong backwash.

The existence of the aeration during backwash did not remove more percentage of foulant from the cake. However, the bubble does influence the structure of the remaining cake and then resulting in the different fouling rate during the filtration cycle.

2.6 ULTRAFILTRATION-BASED HYBRID PROCESSES FOR PRE-TREATMENT TO SEAWATER REVERSE OSMOSIS DESALINATION

2.6.1 Introduction

Seawater reverse osmosis (SWRO) desalination has been widely used in the last decades for drinking water production. However, severe fouling of RO membranes has been inevitably encountered. The main types of RO fouling are scaling, colloidal fouling, biofouling and organic fouling. The biofouling and organic compounds are responsible for more than 60% of SWRO fouling (Khedr, 2000).

Conventional pre-treatments – i.e. coagulation, sedimentation and filtration – were initially used as RO pretreatment. But because of unpredictable variation and also of constant degradation of seawater quality, particularly concerning open intake sources, low pressure membrane filtration as microfiltration (MF) and ultrafiltration (UF) has gradually gained acceptance as the preferred pre-treatment to SWRO. In recent years, pilot-scale testing and installations of MF and to a larger extent UF have grown up (Bu-Rashid and Czolkoss, 2007; Pearce, 2007; Vedavyasan, 2007 and Xu *et al.*, 2007) due to the reliability of these processes for producing superior quality feed to RO regardless of raw seawater quality. UF membranes represent absolute barrier to particulate, colloids, macromolecules, algae and bacteria. However, approximately 99% of organic carbon in seawater is dissolved organic matter (<100 nm): about 24% of the marine organic carbon is high molecular weight dissolved organic carbon (1–100 nm) and about 75% low molecular weight dissolved organic carbon (<1 nm) (Benner *et al.*, 1997). As a consequence, the marine organic carbon cannot be totally removed by MF or even UF which present pore size in range of 0.1–1 μm and 0.01–0.1 μm or less, respectively. Thus, Shon *et al.* (2008) reported that UF membrane removed only an insignificant (20% removal) quantity of marine organics of 1,200, 950 and 90 Da contained in surface seawater. Moreover another drawback of membrane pretreatment lays in the inherent property of the membrane to retain foulants from raw water and then become fouled itself.

Therefore, the aim of INSA in the frame of MEDINA project (WP2) is to develop and study at lab-scale hybrid processes based on ultrafiltration to take advantage of the membrane process performances – such as the total rejection of suspended solids- and also to overcome its drawbacks. Thus the idea is to couple ultrafiltration with:

1. a coagulation step
2. an adsorption step

in order to enhance organic matter removal and also to reduce the low-pressure membrane fouling. The different pre-treatments studied are summarized in Figure 2.33.

An important point is that all experiments are performed with real seawater from Mediterranean sea, provided by Veolia Water (partner 2).

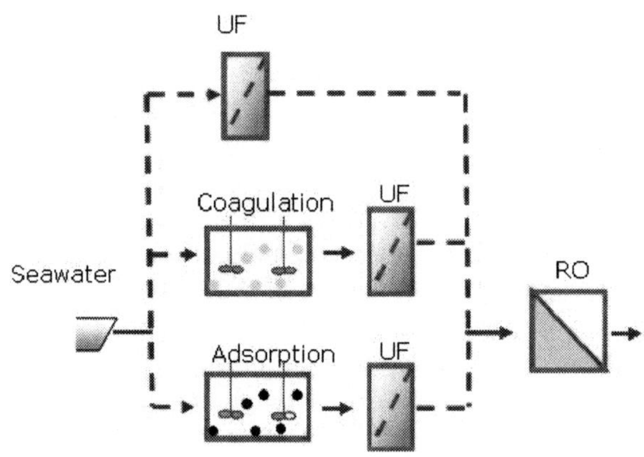

Figure 2.33 Schematic of the different studied pretreatment

2.6.1.1 Preliminary tests with synthetic solutions

INSA first focused activity on preliminary ultrafiltration tests with synthetic solutions, on the choice of equipments and protocols to be able to operate with real sea water solutions.

During the first period of the project, some experiments with a preliminary synthetic seawater (composed of NaCl, $CaCl_2$ and commercial humics) were carried out.

Dead end filtration tests were performed at constant pressure (0.4 bar) with two polyethersulfone flat sheet UF membranes (30 and 100 kDa MWCO – Orelis) in order to characterise permeate flux and humic acid retention rate. Submersion tests of clean membranes (without filtration) were also performed in 10 mg/L HA solutions at constant pH with and without salts, in order to estimate adsorption fouling resistance.

Results clearly showed an influence of salts on interactions between the membrane and organic matter.

- permeate flux decline during filtration is more significant in presence of salts than with humic acid only. Permeate flux decline seems to be still more significant in presence of calcium chloride, what is probably due to an aggregation and a precipitation of humics
- the maximal retention of humics is 98% when only humics are in solution and decreases when salts were added, due to electrostatic interactions.

Fouling is mainly due to adsorption, but also to particulate and colloidal deposit formation at membrane surface, and/or to pore blocking. Total fouling resistance appears to increase in presence of calcium chloride for the same HA an Na content.

INSA performed, then, tests on real seawater to adapt protocols and equipments as necessary. In collaboration with Véolia, INSA organised to sample, receive and to store samples of sea water.

2.6.1.2 Experiments with real seawater

2.6.1.2.1 Materials and methods

Seawater, adsorbants and coagulants

Seawater from Mediterranean Sea with a salinity of 39 g/L and an average total organic carbon concentration (TOC) of 1.3 mg/L was used in all experiments.

Two types of adsorbent particles were studied: bentonite and powdered activated carbon (PAC). Bentonite used is in dry powdered form (SÜD-CHEMIE) with mean diameter of 5–6 μm and specific surface of $3.5 \times 10^6\,m^2/m^3$. Two PAC types, named PAC A and PAC B were supplied by PICA. It has to be noted that PAC B presents a volume of micropores twice higher than PAC A.

Three concentrations: 50, 100 and 200 mg/L were studied for both PAC types. The PAC was added into seawater and then UF was immediately done.

Ferric chloride (41% $FeCl_3$, Brenntag) and aluminium sulphate (99.5% purity, Prolabo, France) were used as coagulants. According to previous Jar test experiments, the optimum concentration and pH is 20 mg/L and 6.5, respectively for $FeCl_3$ and 10 mg/L and 7 for $Al_2(SO_4)_3$.

In order to reduce chemical consummation, smaller dose of $FeCl_3$ was experimented in the range 2–20 mgL^{-1} at pH 6 ± 0.5.

Membrane and filtration devices

A flat sheet regenerated cellulose membrane (YM30 from Millipore) with a molecular weight cut-off (MWCO) of 30 kDa was used in all the experiments. Pure water permeability of the membrane at 20ºC was determined between 220–250 $Lh^{-1}m^{-2}bar^{-1}$. Characteristics of the membrane are shown in Table 2.18. Contact angle measurements were performed with a goniometer (GBX, Digidrop) and confirmed the hydrophilic nature of the membrane.

Table 2.18 Characteristics of the UF membrane

Characteristics	Values
Water permeability @ 20ºC ($Lh^{-1}m^{-2}bar^{-1}$)	220–250
Membrane resistance (m^{-1}) ($\times 10^{12}$)	1.4–1.6
Contact angle (º)	15.9
Zeta potential @ pH 7 (mV) (Kim *et al.*, 1996)	−2

The filtration device is schematised in Figure 2.34. The dead-end stirred cell (Amicon, Millipore) has a maximum capacity of 400 mL. The membrane surface is 39.6 cm^2. A 5 L stainless steel solution reservoir was connected to a circuit of compressed air allowing fixing transmembrane pressure (TMP) in the range 0.4–2.4 bar. Permeate mass was measured by an electronic balance connected to a personal computer. Temperature during all filtration tests was 20ºC.

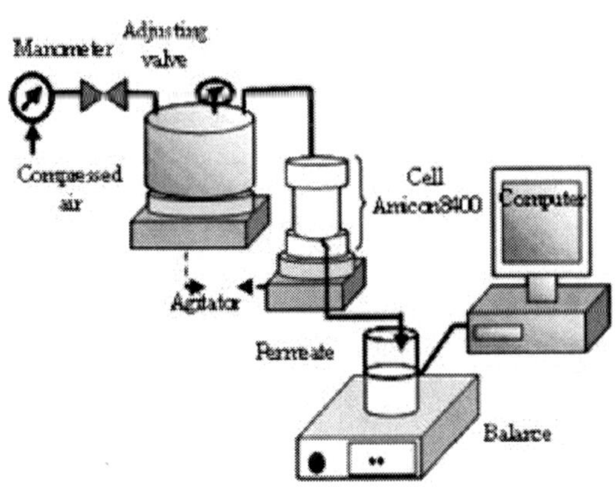

Figure 2.34 Schematic of dead end filtration set up

Filtration methods

Two operating modes of filtration were investigated. The first one is the pressure step method. This method consists in filtering feed water with constant TMP during the production of a fixed permeate volume and then increasing the TMP to another value. Each constant TMP was applied for a 100 mL permeate volume, which corresponds to duration between

5 and 15 min. The TMP step was equal to 0.4 bar. The range of studied TMP was between 0.4–2.4 bars. According to the filtration model, for each constant TMP, a decrease of permeation flux versus time is observed due to membrane fouling. This flux decline can be described as an increase of fouling resistance versus time.

$$J_{(20^\circ C)} = \frac{\Delta P}{\mu_{(20\ C)}(R_m + R_f)} \tag{2.4}$$

where $J_{(20^\circ C)}$ is the flux, DP is the TMP, $\mu_{(20^\circ C)}$ is the viscosity of the permeate and R_m and R_f are the resistance of the membrane and fouling, respectively.

Thus, the fouling rate (dR_f/dt) is the slope from the graph of fouling resistance versus time at each constant TMP and is calculated by:

$$\frac{dR_f}{dt} = \frac{R_{f^2} - R_{f^1}}{t_2 - t_1} \tag{2.5}$$

The second filtration mode is a constant pressure mode. Permeate flux values are measured during time and plotted versus filtered volume per unit area.

Analytical methods

The total organic carbon (TOC) concentration was measured by TOC-meter (TOC-Vcsh, Shimadzu, France). The non-purgeable organic carbon (NPOC) method was used. The detection of TOC-meter is limited to 0.1 mg/L TOC in presence of high salt concentration as in the case of seawater.

Molecular weight (MW) distributions of different samples were determined by high-pressure size exclusion chromatography (HP-SEC, AKTA) method with a fluorescence detector at 350–445 nm. SEC column used in this study can measure MW in range of 10–100 kDa.

The apparent retention rate (R) of NOM is defined by

$$R = \left(1 - \frac{C_{permeate}}{C_{feed}}\right) \times 100 \tag{2.6}$$

where $C_{permeate}$ and C_{feed} are TOC concentration in permeate and feed water, respectively.

SDI_3 measurement

The measurement of SDI_3 was carried out by ASTM D-4189 standard with a time interval of 3 minutes. Membrane filters (HAWPO, Millipore) with 47 mm diameter and average pore size 0.45 μm were used under constant pressure at 2.07 bar. The filtrates were collected and recorded for determining SDI_3 values by:

$$SDI_3 = \frac{\left(1 - \frac{t_i}{t_f}\right)}{3} \times 100 \tag{2.7}$$

where t_i is initial time required to filter the first 500 mL after pressure has been set to 2.07 bar and t_f is time required to filter 500 mL at 3 minutes after the start of measurement.

2.6.1.3 Results and discussion

1. Coupling of coagulation and UF

Figure 2.35 shows the variation of retention rate for the coupling coagulation/UF at different $FeCl_3$ and $Al_2(SO_4)_3$ concentrations. $FeCl_3$ and $Al_2(SO_4)_3$ concentration varied between 2–20 mg/L and 5–10 mg/L, respectively. First of all, UF alone performs retention rate about 10%.

Concerning the coupling $FeCl_3$ coagulation/UF, experiments carried out at sub-optimum dose perform retention rates nearly equal to that obtained without $FeCl_3$; i.e. varying between 13–16%. This retention rate slightly increases up to 23% when $FeCl_3$ at optimum dose is added. Likewise, sub optimum dose of $Al_2(SO_4)_3$ shows lower retention rate compared with experiment done at optimum concentration this is approximately 15%. However, the retention rate by coagulation/UF – even when the optimum dose of coagulants is used – is always lower than that obtained by PAC adsorption/UF at sufficient PAC concentration. It suggests that combining of $FeCl_3$ or $Al_2(SO_4)_3$ coagulation/UF is not efficient enough to remove marine dissolved organic molecules compared with PAC adsorption/UF.

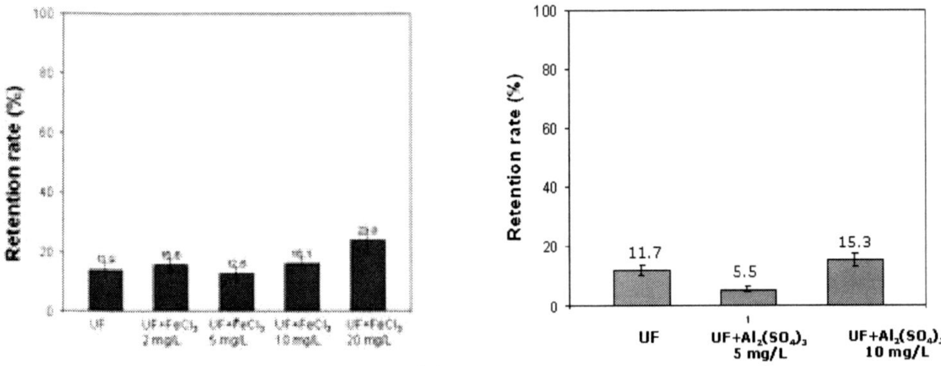

(a) (b)

Figure 2.35 Variation of retention rate for UF coupled with (A) $FeCl_3$ and (B) $Al_2(SO_4)_3$

Fouling velocity increases with pressure for all $FeCl_3$ concentrations and no significant deviation was observed for the different $FeCl_3$ concentrations.

In conclusion, it appears that combining coagulation and ultrafiltration for this application and for this kind of membrane cannot improve significantly UF performances. As a consequence, the coupling adsorption/UF was more deeply studied.

2. Coupling of bentonite adsorption and UF

The effect of bentonite addition on fouling rate as well as NOM retention rate was determined for different bentonite concentrations. Figure 2.36 shows the variation of the fouling rate (A) and the retention rate (B) at bentonite concentrations of 200, 500 and 1,000 mg/L. Average values of retention rate are presented for each dose of bentonite since this rate remains constant during the experiment, for all the experiments. As shown in the figure, the fouling rate increases with TMP for all bentonite doses. An addition of bentonite with a concentration between 200 and 500 mg/L seems to have no effect on the fouling rate which is very similar to that of test done without bentonite. However, the addition of 1,000 mg/L bentonite allows a slightly lower increase of the fouling rate. In that case of highly concentrated bentonite, more adsorption sites are available to fix organic matter. It can be assumed that the membrane is less adsorbed and thus less fouled by NOM, what is traduced by a lower fouling rate. The average retention rate for the different doses of bentonite was measured. This retention rate of seawater by UF alone is approximately 5%. The addition of bentonite at different doses can not improve significantly the retention rate that stays between 3 and 10%. Consequently, the hybrid process coupling bentonite adsorption and UF is not efficient enough to remove the NOM in seawater.

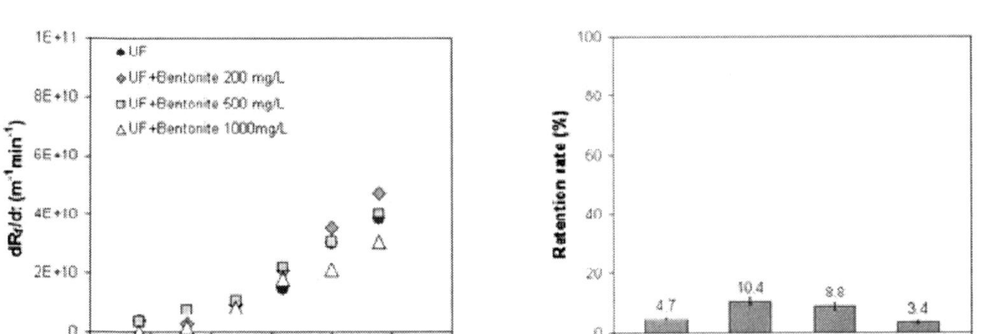

(a) (b)

Figure 2.36 Effect of bentonite concentration on (A) the fouling rate and (B) the TOC retention rate

The effect of precoating membrane by bentonite particles was also studied. The idea of these experiments was that the bentonite particles precoated on membrane surface might improve the NOM retention by trapping organic molecules in their cake deepness. To carry out this experiment, 500 mg/L of bentonite suspension was filtered for 1 hour at constant pressure of 2 bar. The mass of deposited bentonite on membrane surface is approximately 40 g/m^2. Then seawater without bentonite was filtered by pressure step method. The results show a considerable increase of fouling rate compared to experiments performed with no precoated membranes. A contrary result was obtained for a dynamic MF membrane precoated by mineral bentonite in the case of a membrane bioreactor for wastewater treatment (Ye *et al.*, 2006). The authors showed that precoating membrane presents many advantages including stable permeate flow rate. Of course, the characteristics of the feed solution were very different in their study. Moreover the particle size distribution of the bentonite that precoats the membrane plays an important role in structuring the deposit. The size distribution of bentonite particles used in our experiments may be too small (5–6 μm); therefore, this bentonite deposit is too dense and compact and cause pore blocking on membrane surface. The retention rate of membrane precoated by bentonite is nearly similar to that of filtering suspended bentonite through the membrane. So, there is no interest of precoating this membrane with that kind of particles.

3. Coupling of adsorption with Powdered Activated Carbon (PAC) and Ultrafiltration

The variation of retention rate for PAC adsorption/UF at different PAC concentrations was investigated. PAC concentration varied between 50–100 mg/L for both PAC types. Here, UF alone performs retention rate between 5%–10%. The notable discrepancy of retention rate values for UF alone may be first due to TOC measurement uncertainty especially in permeate that presents low organic concentration, close to limit of TOC detection. The second point is the variation of seawater characteristics according to season and to samples received from Mediterranean Sea.

Concerning the coupling adsorption/UF, the retention rate appears to increase with PAC dose for both PAC types. It is thanks to an increase of available attractive site to adsorb organic molecules. The retention rate can reach about 50% and 40% when 200 mg/L of PAC B and PAC A was used, respectively. This confirms the important role of PAC surface area, a higher BET surface area resulting in higher organic matter retention rate.

Figure 2.37 presents fouling velocity versus TMP for PAC B (A) and PAC A (B). It can be noticed that fouling rate increases exponentially with TMP whatever experiments and it decreases when increasing PAC dose for both PAC types. Regarding Figure 2.37a, it is observed almost the same fouling rate when 50 mg/L PAC B was added compared with test without PAC. However, the fouling rate is reduced when 100 mg/L PAC B was used and 200 mg/L PAC B presents the lowest fouling rate. Comparing between PAC B and PAC A (Figure 2.37b), at the same PAC concentration, PAC B presents lower fouling velocity than PAC A for all PAC doses. This confirms previous results concerning the influence of PAC surface area: PAC B which presents the highest surface area allows adsorbing a higher amount of marine organic molecules and thus reducing membrane fouling by these molecules. In addition, the pore size distribution can play a role on membrane fouling: PAC with larger microporous volume (PAC B) performs lower fouling velocity compared with the one which has a smaller volume of micropores (PAC A). It could confirm the previous hypothesis i.e. the presence of low molecular weight dissolved organic carbon (<10 kDa which is not seen in HP-SEC analysis) that can easily penetrate and be adsorbed into PAC micropores and then cause less UF fouling.

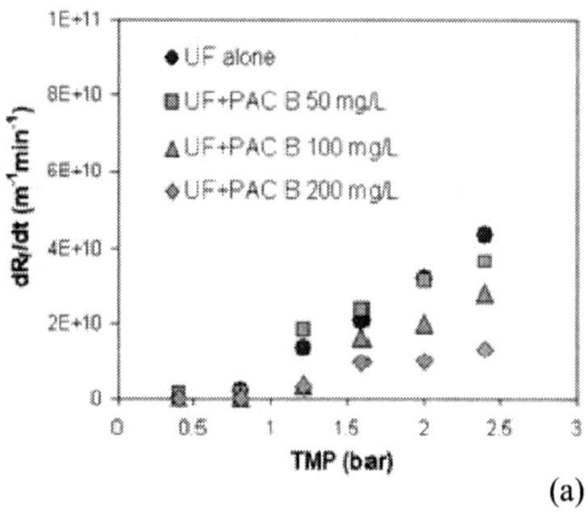

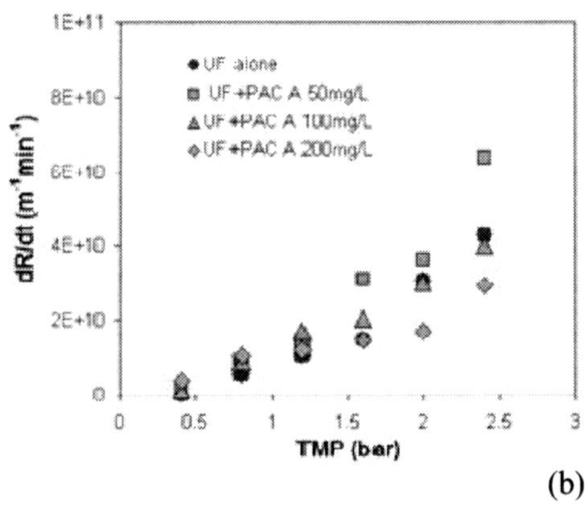

Figure 2.37 Effect of PAC dose on fouling velocity in function of TMP for (A) PAC B and (B) PAC A

Molecular weight (MW) distribution of raw seawater and of different permeate samples was achieved by high-pressure size exclusion chromatography (HP-SEC) analysis. As results from TOC meter analysis, the UF can slightly remove marine organic matter with MW roughly 10 kDa (considering area under the graphs by integrating method). PAC addition can improve organic removal for both PAC types. However, it is remarkable that PAC A presents area under the graph less than PAC B; this result is contrary to the one from TOC measurement that shows nearly similar retention rate. It must be noted that TOC meter measures all types of organic carbon present in samples while HP-SEC method used in this work can only measure organics in range 10–100 kDa. Actually, PAC B could be able to remove organics with MW about 10 kDa less efficiently than PAC A but it may effectively eliminate lower MW organic matter that is majority of marine organic compounds as reported by Benner *et al.* (1997).

4. Comparison of flux decline for the different hybrid processes

Flux decline of the different hybrid processes is presented in terms of normalized flux (J/J_0) conducted at temperature 20°C as a function of filtered volume. UF alone presents a severe flux decline reaching a value (J/J_0) of 0.48 for 1 L filtered volume. The UF experiments with PAC A and PAC B present higher flux decline at the beginning of filtration, then the PAC B performs less severe flux decline than UF alone with a normalized flux of 0.60 for 1 L filtered volume when 100 mgL^{-1} PAC B was used. On contrary, UF combined with PAC A is responsible for a flux decline almost the same as that without PAC. This indicates that PAC can reduce the flux decline by trapping and adsorbing dissolved organic carbon, especially when PAC with higher BET surface area and microporous volume i.e. PAC B was used.

5. Evaluation of fouling potential (SDI_3)

Fouling potential for the different studied pre-treatments was determined in terms of silt density index (SDI_3). This is a parameter commonly used in desalination plants to evaluate water quality before feeding to RO. Thus the SDI_3 of permeate obtained with the different pretreatments (UF alone, UF + PAC adsorption, UF + coagulation) was measured. Real seawater used in these experiments presents an average SDI_3 of 26. Permeate filtered by UF membrane alone shows a lower SDI_3 of 9. Furthermore, permeate of UF coupled with PAC adsorption and $FeCl_3$ coagulation presents SDI_3 approximately 4–6 that is fairly lower than that of UF alone. Therefore, it should be noted that pre-treatment by UF can greatly reduce fouling potential in term of SDI_3 and an addition of PAC or $FeCl_3$ allows a further reduction of the fouling potential compared with that by UF alone. These results seem to go against results on organic removal which demonstrated lower organic retention by $FeCl_3$ coagulation/UF compared with PAC adsorption/UF. This suggests and confirms mostly the fact that SDI is not fully adequate to predict organic fouling potential.

2.6.1.4 Conclusions

Combining of PAC adsorption/UF can be an effective pretreatment before SWRO desalination compared with coagulation/UF that shows only about 20% organic removal even when optimum condition of $FeCl_3$ coagulation was accomplished. PAC adsorption/UF reduces fouling velocity and increases retention rate from 10% for test without PAC up to 50% when 200 mg/L PAC B was used. These results are confirmed by molecular weight distribution of organic matter carried out by HP-SEC method: PAC addition enhances performances of UF membrane by removing organics with MW approximately 10 kDa. In addition, PAC B can enhance UF permeate flux compared with that done without PAC. PAC A nonetheless is unable to reduce flux decline. It appears that PAC with higher BET surface area and larger volume of micropores allows improving UF performances not only in terms of fouling velocity and retention rate but also permeate productivity. On the other hand, fouling potential as SDI_3 decreases from 26 for seawater solution to 9 when UF was used as pretreatment. The PAC and $FeCl_3$ addition to UF allows further reduction of SDI_3 from 9 to 4–6.

REFERENCES

Al-Ahmad M., Abdul Aleem F. A. and Ubaisy A. (2000). Biofouling in RO membrane systems part 1: fundamentals and control. *Desalination*, **132**, 173–179.

American Society for Testing and Materials (ASTM). (1995). *Standard Test Method for Silt Density Index (SDI) of Water*. pp. D4189–4195.

Bacchin P., Aimar P. and Field R. W. (2006). *J. Membr. Sci.*, **281**, 42–69.

Benner R., Biddanda B., Black B. and McCarthy M. (1997). Abundance, size distribution, and stable carbon and nitrogen isotopic compositions of marine organic matter isolated by tangential-flow ultrafiltration. *Marine Chemistry*, **57**, 243–263.

Bouhabila E. H., Ben Aim R. and Buisson H. (1998). *Desalination*, **118**, 315–322.

Bu-Rashid K. A. and Czolkoss W. (2007). Pilot tests of multibore UF membrane at Addur SWRO desalination plant, Bahrain. *Desalination*, **203**, 229–242.

Chen V., Fane A. G., Madaeni S. S. and Wenten I. G. (1997). *J. Membr. Sci.*, **125**, 109–122.

Chua K. T., Hawlader M. N. A. and Malek A. (2003). Pretreatment of seawater: results of pilot trials in Singapore. *Desalination*, **159**(3), 225–243.

Courties C., Vaquer A., Troussellier M., Lautier L., Chretiennot-Dinet M. J., Neveux J., Machado C. and Claustre H. (1994). Smallest eukaryotic organism. *Nature*, **370**, 255.

Defrance L. and Jaffrin M. Y. (1999). *J. Membr. Sci.*, **157**, 73–84.

Defrance L. and Jaffrin M. Y. (1999). *J. Membr. Sci.*, **152**, 203–210.

Drioli E., Criscuoli A. and Curcio E. (2002). *Desalination*, **147**, 77–81.

Drioli E., Curcio E., Criscuoli A. and Di Profio G. (2004). *J. Membr. Sci.*, **239**, 27–38.

Drioli E., Curcio E., Di Profio G., Macedonio F. and Criscuoli A. (2006). *Chem. Eng. Res. Des.*, **84**, 209–220.

Espinasse B., *et al.* (2002). *Desalination*, **146**, 91–96.
Fane A. G., Yeo A., Law A., Parameshwaran K., Wicaksana F. and Chen V. (2005). Low pressure membrane processes ~ doing more with less energy. *Desalination*, **185**(1–3), 159–165.
Field R. W., Wu D., Howell J. A. and Gupta B. B. (1995). *J. Membr. Sci.*, **100**, 259–272.
Flemming H. C. (2002). Biofouling in water system – cases, causes and countermeasures. *Applied Microbiology and Biotechnology*, **59**, 629–640.
Frebrero I. L. (1984). *A Study of Seawater Quality in Laem Chabang and the Evaluation of Impacts of Deep Sea Port on Marine Environment*. AIT thesis no. EV-85-6, Thailand.
Fritzmann C., Löwenberg J., Wintgens T. and Melin T. (2007). State-of-the-art of reverse osmosis desalination. *Desalination*, **216** (1–3), 1–76.
Hamoda M. F. and Al-Attar I. M. S. (1995). *Water Sci. Technol.*, **31**, 61–72.
Hermans P. H. and Bredee H. L. (1936). Principles of the mathematical treatment of constant pressure filtration. *Jounal of the Society of Chemical Industry*, **55**, 1–4.
Hermia J. (1982). Constant pressure blocking filtration laws: application to power-law nonnewtonian fluids. *Institut of Chemical Engineers*, **60**, 183–187.
Huber S. A. and Frimmel F. H. (1994). Characterization and quantification of marine dissolved organic carbon with a direct chromatographic method. *Environ. Sci. Technol.*, **28**, 1194–1197.
Judd S., Judd C. (2006). *The MBR Book: Principles and Applications of Membrane Bioreactors for Water and Wastewater Treatment*. Elsevier.
Kaakinen J. W. and Moody C. D. (1985). *Characteristics of Reverse-Osmosis Membrane Fouling at the Yuma Desalting Test Facility*. ACS Symposium Series, pp. 359–382.
Khedr M. G. (2000). Membrane fouling problems in reverse osmosis desalination applications. *Desalination and Water Reuse*, **10**(3), 8–17.
Le Clech P., *et al.* (2003). *J. Membr. Sci.*, **227**, 81–93.
Lebaron P., Servais P., Agogue H., Courties C. and Joux F. (2001). Does the high nucleic acid content of individual bacterial cells allow us to discriminate between active cells and inactive cells in aquatic systems? *Appl. Environ. Microbiol.*, **67**, 1775–1782.
Neveux J. and Lantoine F. (1993). Spectrofluorometric assay of chlorophylls and pheophytins using the least squares approximation technique. *Deep Sea Research*, **40**, 1747–1765.
Ord J., Kinser K. J. and Cassanova R. A. *IDA World Congress – Atlantis*, The Palm – Dubai, UAE November 7–12, 2009.
Pearce G. K. (2007). The case for UF/MF pretreatment to RO in seawater applications. *Desalination*, **203**(1–3), 286–295.
Pearce G. K. *IDA World Congress – Atlantis*, The Palm – Dubai, UAE November 7–12, 2009.
Pervov A. G., Andrianov A. P., Efremov R. V., Desyatov A. V. and Baranov A. E. (2003). A new solution for the Caspian Sea desalination: low-pressure membranes. *Desalination*, **157**(1–3), 377–384.
Pierce G. K. (2007). The case for UF/MF pretreatment to RO in seawater applications. *Desalination*, **203** 286–295.
Potts D. E., Ahlert R. C. and Wang S. S. (1981). A critical review of fouling of reverse osmosis membranes. *Desalination*, **36**, 235–264.
Rapenne S., Pontié M., Thekkedath A., Duchesne J., Jacquemet V., Leparc J. and Suty H. (2005). Tools for membrane autopsies and antifouling strategies in seawater feeds: a review. *Desalination*, **181**, 75–90.
Ridgway H. F. and Flemming H. C. (1996). Membrane biofouling. In: *Water Treatment Membrane Processes*, J. Mallevialle, P. E. Odendaal and M. R. Wiesner (eds.), McGraw-Hill, pp. 6.1–6.62.
Rittschof D. (1999). Fouling and natural products as antifouling. In: *Recent Advances in Marine Biotechnology*, vol. 3, M. Fingerman, R. Nagabhushanam and M.-F. Thompson (eds.), Science Publishers, Inc., U.S.A. pp. 244–249.
Rosenberger S., Laabs C., Lesjean B., Gnirss R., Amy G., Jekel M. and Schrotter J. C. (2006). *Water Res.*, **40**, 710–720.
Schippers J. C. and Verdouw J. (1980). The modified fouling index, a method of determining the fouling characteristics of water. *Desalination*, **32**, 137–148.
Schloss P. D. and Handelsman J. (2005). Introducing DOTUR, a computer program for defining operational taxonomic units and estimating species richness. *Appl Environ Microbiol*, **71**, 1501–1506.
Shon H. K., Vigneswaran S. and Cho J. (2008). Comparison of physico-chemical pretreatment methods to seawater reverse osmosis: detailed analyses of molecular weight distribution of organic matter initial stage. *Journal of membrane science*, **320**, 151–158.
Tansakul C., Laborie S. and Cabassud C. (2009). Membrane hybrid processes for pretreatment before seawater reverse osmosis desalination. *Desalination and Water Treatment*, **9**, 279–286.
Uygur A. and Kargi F. (2004). *Enzyme Microb. Technol.*, **34**, 313–318.
Vedavyasan C. V. (2007). Pretreatment trends – an overview. *Desalination*, **203**, 296–299.
Vincelet C. and Jacquemet V. (2008). MEDINA deliverable D 1.1.4 – documentation of modified protocols of analytical tools applied to seawater/brackish water sources: LC-OCD protocol modification.
Visvanathan C., Boonthanon N., Sathasivan A. and Jegatheesan V. (2002). Pretreatment of seawater for biodegradable organic content removal using membrane bioreactor. *Desalination*, **153**, 133–140.
Visvanathan C., Boonthanon N., Sathasivan A. and Jegatheesan V. (2002). *Desalination*, **153**, 133–110.
Vrouwenvelder J. S. and Van der Kooij D. (2001). Diagnosis, prediction and prevention of biofouling of NF and RO membrane. *Desalination*, **139**, 65–71.
Wu D., Howell J. A. and Field R. W. (1999). *J. Membr. Sci.*, **152**, 89–98.
Xu J., Ruan G., Chu X., Yao Y., Su B. and Gao C. (2007). A pilot study of UF pretreatment without any Chemicals for SWRO desalination in China. *Desalination*, **207**, 216–226.
Ye M., Zhang H., Wie Q., Lei H., Yang F. and Zhang X. (2006). Study on the suitable thickness of a PAC-precoated dynamic membrane coupled with a bioreactor for municipal wastewater treatment. *Desalination*, **194**, 108–120.

Chapter 3

Development of tools for RO fouling characterization and understanding

S. Rapenne[1], C. Barbe[1], G. Schaule[2], M. Strathmann[2], S. Grobe[2], J.-P. Croué[3], L. Mondamert[3], W. Hijnen[4], D. van der Kooij[4], C.-L. de O. Manes[5,6], N. J. West[7,8], P. Lebaron[5,6,7,8] and M. Herzberg[9]

[1]Anjou-Recherche, Veolia Environment (France)
[2]IWW (Germany)
[3]Laboratoire de Chimie et Microbiologie de l'Eau (France)
[4]KWR (The Netherlands)
[5]UPMC, UMR 7621, LOMIC, Laboratoire Océanologique (France)
[6]CNRS, UMR 7621, LOMIC, Observatoire Océanologique (France)
[7]UPMC, UMS 2348, Observatoire Océanologique (France)
[8]CNRS, UMRS 2348, Observatoire Océanologique (France)
[9]ZIWR (Israel)

3.1. MEMBRANE AUTOPSIES

3.1.1 Introduction

Membrane fouling is still a major issue in RO desalination plants, since a decline of membrane performances can occur, leading to a higher exploitation cost. To prevent and/or discard fouling phenomena, foulant material must be identified precisely by performing autopsy. This identification will lead to a better operation of SWRO plants by adjusting chemical cleaning strategies and/or pretreatments.

Autopsy is a standard procedure to characterize deposits at the membrane surface. In the autopsy process, after optical inspection of the module, the membranes are cut open and their surfaces as well as any other inner surface (spacer, glued part) are examined with a suite of analytical methods.

Chemical, physical and microbiological analyses identify the type of foulants components inferring about their effect on membrane performances. Scaling, organic fouling, particulate fouling, and biofouling are the four major types of fouling that can occur on RO membrane surfaces. In most cases, the results allow a clear discrimination between different types of mineral components but there are still difficulties in defining more precisely the nature of organic

 Membrane-Based Desalination: An Integrated Approach (MEDINA). Edited by Enrico Drioli, Alessandra Criscuoli and Francesca Macedonio. ISBN: 9781843393214. Published by IWA Publishing, London, UK.

foulants, organic colloidal matter or established biofilms. Therefore advanced tools are necessary to move towards a better characterization and understanding of membrane fouling especially for the organic and biofouling types.

In the framework of MEDINA project, membrane autopsy results provide information on the choice of new pre-treatments (subject of WP2) as well as on the assessment of the efficacy of cleaning strategies (subject of WP4) as an integrated approach.

Comprehensive autopsies were performed on membranes from different full-scale and pilot plants.

3.1.2 Presentation of studied sites

Spiral wound RO modules were collected at different sites. The characteristics of the studied sites, such as raw water type and pre-treatments applied, are presented in Table 3.1.

Table 3.1 Main characteristics of the studied sites

Site	Water type	Pretreatment type
Site A	Open intake seawater	Conventional – coagulation + granular filtration
Site D	Open intake seawater	Conventional – coagulation + granular filtration
Site H	Open intake seawater	Conventional – coagulation + granular filtration
Site G	Seawater from beachwells	Granular fitlration
Site E	Open intake seawater/seawater from rock-wells	Conventional – coagulation + granular filtration
Site C	Estuarine brackish water	1. Conventional – coagulation + granular filtration 2. Ultrafiltration

Each sampled module was located in front position of a pressure vessel from a RO train. The plants were composed of several stages and passes and the membranes were sampled from the first stage/ first pass of one of the RO units. Moreover, each sampled module differed in usage and exploitation time and chemical cleaning frequency and nature (Table 3.2).

Table 3.2 Exploitation time and chemical cleaning frequency for sites A, H, D, G and E

Membrane origin	Number of sampled modules	Exploitation time (years)	Usage time after the last cleaning/preservation (days)
Site A	1 (November 2008)	0.5	10
Site H	1 (September 2008)	5	17
Site D	1 (November 2007)	2	60
Site G	1 (November 2008)	3	120
Site E	1 (February 2009)	4	255
Site C(1)	1 (June 2008)	7	Not documented
Site C(2)	1 (September 2008)	7	0

For each site, only one module was autopsied, except for site D and C. Five modules were autopsied for site D and the results are presented elsewhere (Rapenne *et al.*, 2009 and de O. Manes *et al.*, 2011). Here is presented a comparison from the module of November 2007 of Site D with those from the other studied sites. For site C, two autopsies were performed before and after chemical cleaning.

3.1.3 Membrane sampling protocol and analytical tools

General information about appearance of the membrane element before opening was obtained and documented. During opening of the module element, visual inspection and odour was documented. It is recommended to analyze layers in the first step directly on the membrane and/or spacer surface. Therefore the total surface of several membrane sheets was inspected for the type of distribution of layers on the membrane surface. If the fouling layer is homogenously distributed over the whole membrane surfaces, one sample from a defined area has to be chosen for further investigations. Any decision to be taken (otherwise the reason for a decision why) more than one sample has to be documented (Schaule *et al.*, 2009).

As the analysis requires opening of all membrane sheets, drying out of the layer could occur. Precaution has to be taken and it might be useful to screen using light microscopy if fungi are present in the layer. If screening is positive for fungi health precaution has to be taken.

The inspection of the inner parts of the membrane element included

1. Visual inspection of outer and inner parts (e.g. colour)
2. Notice of odour
3. Consistency and texture of the layer
4. Distribution of the layer on the membrane sheet + spacers
5. Further observations

More general results obtained by global analysis are the basis and completion to the results obtained with detailed analysis. All results will be referred to one square centimeter of membrane surface. For special purposes results will be referred to volume of layer respectively weight of layer e.g. Na/gram wet layer.

All information obtained by 1–5 will be documented in a protocol. Colour of the layer will give a hint of the presence of e.g. humic substances, iron precipitations or pigments of microbial origin. The consistency of a layer can be determined qualitatively by touching the mucus layer softly.

General information about portion of organic and inorganic components will be obtained by determining wet weight, water content, and ignition loss of the layer. This will classify the type of layer. As the characterisation and classification of sludge are performed based on EN and ISO standards, most of the procedure has been modified for layers which are scrapped from membranes sheets. The layers may be considered as sludge in the moment they are removed from the surface. Then applying the same standardised analytical tools as for sludge is an acceptable way.

To quantify the water content, layer from a defined area has to be scrapped gently (Schaule *et al.*, 2000). The weight of this sample before and after evaporation will define the water content. Water which will be evaporated with 105°C until constant weight is achieved is defined as water content. High water content in the amount of 90% or more is an indication for a strong biological, especially microbiological, fraction. Extracellular polymeric substances produced by bacteria have a strong tendency to build hydrogels with a very high water content (Costerton *et al.*, 1981 and Strathmann, 2003). On the other hand, inorganic deposits have much lower water content.

In addition to this usual protocol, several advanced analytical tools were investigated in the frame of the Medina project. The autopsy analyses include chemical, microbiological and microscopic examination, as well as the use of cross-flow test cells to characterize membrane performances. The autopsy tools are classified into four groups: microscopic examination (membrane surface analysis), global analysis, organic analysis and microbiological identification. The results of these analyses enabled to determine what type of fouling occurred on membrane surfaces: organic fouling, scaling or biofouling or, as often, a combination of them.

Table 3.3 presents the different analytical tools used by the partners involved, as well as the expected results.

3.1.4 Results

A general overview of the autopsy results are found in Table 3.4 below for each membrane.

Table 3.3 Analytical tools used to perform autopsies

Analytical tools	Partners	Results	Analyses performed on
Hydraulic Characterization			
RO Bench Scale	LCME	Permeability tests	Membrane samples
Microscopic Observation			
Light Microscopy	IWW	Microscopic observations of foulant material	Scrapped deposit slightly suspended in deionised water
Atomic Force Microscopy (AFM)	BGU	Topography of membrane surface	Membrane samples
Confocal Laser Scanning Microscopy (CLSM)	BGU	Images of thin deposits after staining (EPS and dead microorganisms detection)	Membrane samples
Scanning Electron Microscopy (SEM) – Energy Dispersive X-ray spectroscopy (EDX)	IWW, LCEE, BGU, AR-VW	Microscopic observation and elements concentration	Membrane samples
Inorganic Matter Analysis			

(*Continued*)

Table 3.3 Analytical tools used to perform autopsies (Continued)

Analytical tools	Partners	Results	Analyses performed on
Inducted Coupled Plasma – Optical Emission Spectrometry (ICP-OES)	AR-VW, IWW	Ions and metallic elements quantification	AR-VW: Suspended deposit in deionised water + Digestion in aqua regia IWW: Digestion of scrapped deposit in aqua regia LCME: Lyophilised deposit after acid mineralisation
Organic Matter Analysis			
Elemental analysis	LCME	C, H, N, P and S elements quantification	Lyophilised deposit
Drying and ignition loss	AR-VW, IWW, LCME	Wet weight, dry weight and volatile matter (organic/inorganic proportion)	AR-VW, IWW: Scrapped deposit LCME: Lyophilised deposit
Total Organic Carbon meter – Catalytic combustion	AR-VW, IWW	Total Organic Carbon (TOC) determination	Scrapped deposit (if sufficient) Otherwise, suspended deposit in deionised water
Liquid Chromatography – Organic Carbon Detection (LC-OCD)	AR-VW, IHE	Dissolved Organic Carbon and NOM fractions	Suspended deposit in deionised water
Attenuated Total Reflectance – Fourier transformed Infrared Spectroscopy (ATR-FTIR)	AR-VW, LCME, BGU	Functional groups of membrane foulant determination	Membrane samples
KBr pellet – FTIR	LCME	Determination of the functional groups of scrapped deposit	Lyophilised deposit
^{13}C-NMR	LCME	Identification of carbon atoms in organic molecules	Lyophilised deposit
Gas Chromatography – Mass Spectrometry (GC/MS)	LCME	Sugars and Phospholipids determination	Lyophilised deposit
Flash Pyrolysis – GC/MS	LCME	Biopolymers determination	Lyophilised deposit
Thermochemolysis – GC/MS	LCME	Fatty acids and Aromatics identification	Lyophilised deposit
Fluorescence Excitation-Emission Matrix (F-EEM)	IHE	Nature of organic matter (humic-like/protein-like)	Suspended deposit in deionised water
Spectrometry at 490 nm (Dubois method)	IWW, KWR[a], AR-VW	Carbohydrates quantification	Suspended deposit in deionised water
Spectrometry (Lowry method)	IWW	Protein quantification	Suspended deposit in deionised water
Spectrometry (Filisetti-Cozzi and Carpita method)	IWW	Uronic acid quantification	Suspended deposit in deionised water
Microbiological Analysis			
DAPI staining – Microscopy	AR-VW, IWW, KWR[a]	Total bacteria count	Suspended deposit in deionised water
Culture on media	IWW	Colony count, determination of specific species (manganese and iron oxidising bacteria)	IWW: Suspended deposit in deionised water
Single Strand Conformational Polymorphism (SSCP)	LOB	Characterization of total and active bacterial diversity by rDNA and rRNA profiles	Membrane samples
Clone Librairies	LOB	Identification of bacterial groups/ species	Membrane samples
Fluorescent In Situ Hybridization (FISH)	LOB	Detection of bacteria species using specific probes	Membrane samples
Adenosine TriPhosphate (ATP) quantification by luminometer	KWR[a]	Measurement of active biomass (living cells)	Suspended deposit in deionised water
CTC staining – Microscopy	AR-VW	Actively respiring cells quantification	Suspended deposit in deionised water

[a] *These parameters were determined in membrane samples after isolation of the fouling layer in sterilized tap water.*

Table 3.4 Characteristics of the deposits observed or collected after autopsy of RO membranes originating from different desalination sites

	Site D	Site H	Site E	Site A	Site G	Site C(1)	Site C(2)
Visual observation of deposits	Orange-brownish deposit – Not slimy	"Apple sauce" aspect – Slimy	Grey – not slimy	Orange – Not slimy	Absence of deposit	Brownish	Brownish
RO bench scale Confirmation of hydraulic performances at full-scale?	OK	NO	OK	NO	n.a.	n.a.	n.a.
Light microscopy on deposits	Amorphous particles and bacteria in coccoid, rod-shape and filamentous form but different proportions among modules						
	Amorphous, non-crystalline nature	Bacterial numbers very high	Amorphous, non-crystalline nature	Bacterial numbers very high	Amorphous, non-crystalline nature	Amorphous with some bacteria	–
CLSM	Total cells: + EPS: +	Total cells: +++ EPS: +++	Total cells: ++ EPS: +++	n.a.	Total cells: + EPS: ++	n.a.	Total cells: + EPS: +++
SEM-EDX	C, O, S => membrane material + organic matter and carbonates Na, Cl, Mg => seawater						
	Fe		Al, Si	Fe	Fe, Al ,Si	Fe, Al, Si	Fe
Dry matter (μg/cm²)	110	40 (AR-VW) and 96 (IWW)	150	Deposit not sufficient	Absence of deposit	1090	660
Loss on ignition 550°C: organic/inorganic (%)	60/40	80/20	60/40	Deposit not sufficient	Absence of deposit	30/70	30/70
Loss on ignition 900°C: carbonates/inorganic residual	–	85/15	75/25	–	–	–	30/70
ICP	Fe, Ca, Mg, S, Na	Ca, Fe, Mg, S, Na	Ca, Fe, Mg, S, Na, Al, Si	Fe, Na	Ca, Fe, Mg, S, Na, Al, Si	Fe, Al	Fe, Al
Elemental analysis	Confirms loss on ignition analyses at 900°C						
FTIR-ATR/KBr pellet	Proteins and polysaccharides	Proteins and polysaccharides	Proteins and polysaccharides + Aluminosilicates	Proteins and polysaccharides	Proteins and polysaccharides + Aluminosilicates	Aluminosilicates	Aluminosilicates
^{13}C-NMR	Predominance of polysaccharides	Predominance of polysaccharides				Predominance of polysaccharides	
Flash pyrolysis GC/MS (most abundant organic structures)	Proteins + Polysaccharides	Proteins + Polysaccharides	Proteins	Proteins	Proteins	Proteins + Polysaccharides	Proteins

(*Continued*)

Table 3.4 Characteristics of the deposits observed or collected after autopsy of RO membranes originating from different desalination sites (Continued)

		Site D	Site H	Site E	Site A	Site G	Site C(1)	Site C(2)
Thermochemolysis GC/MS (origin of the deposit)		Bacterial and terrestrial (and/or algal)	Bacterial and terrestrial (and/or algal)	Bacterial	Bacterial	Bacterial	Bacterial and terrestrial (and/or algal)	Bacterial and terrestrial (and/or algal)
EEM-fluorescence + FI		Protein-like FI: Microbially-derived	Protein-like FI: Microbially-derived	Protein-like FI: Terrestrially-derived	Protein-like FI: Microbially-derived	Protein-like FI: Terrestrially-derived	Protein-like and humic-like FI: Terrestrially-derived	Protein-like and humic-like FI: Terrestrially-derived
LC-OCD (peak)		Biopolymers	Biopolymers + Neutrals	Biopolymers	Biopolymers	Biopolymers	Biopolymers	Biopolymers
EPS (μg/cm^2)	Carbohydrates	25.7	4.6	5.1	–	–	5.2	99.5
	Proteins	0.1	1.3	4.4	–	–	1.0	0.06
	Uronic acids	–	11.9	–	–	–	0.1	0.06
EPS by GC/MS		Bacterial and terrestrial origin	Bacterial origin	Bacterial origin	n.a.	n.a.	Bacterial and terrestrial origin	Bacterial and terrestrial origin
Total cells count (log) – AR-VW		6	5.5	6	4	4	7	6
Clones libraries (type of species)		Proteobacteria (Alpha, delta, gamma) + Bacteroidetes + Plactomycetes	–	Proteobacteria (Alpha, delta, gamma) + Bacteroidetes + Plactomycetes	–	–	–	–
ATP (pg/cm^2)		2,444	1,117	2,489	12,243	256	8,590	–
Actively respiring cells (log) – Bacterial activity (%)		5 log – 4%	5 log – 67%	5 log – 3%	3 log – 2%	2 log – 0.5%	6 log – 32%	5.5 log – 13%

All the analyses showed that the phenomenon of biofouling was prevalent on all the sites under study even if hydraulic performances proved that this phenomenon was controlled (not shown). Light microscopy of slightly suspended deposit was performed for modules from site D (Nov. 07), site H (Sept. 08) and site A (Nov. 08) (Figures above) as well as for site G (Nov. 08), site E (March 09) and site D (March 09) (no pictures). A typical image of light microscopy of membrane layer is given in Figure 3.1 and Figure 3.2.

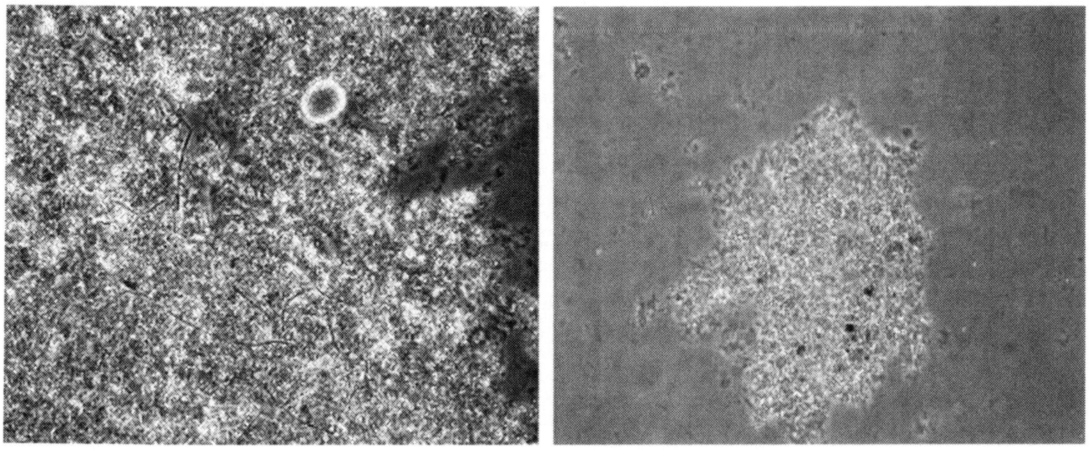

Figure 3.1 Light microscopic images (phase contrast) of fouling deposit obtained from RO membranes from Site A (left image) November 2008 and Site D (right image) September 2007; magnification: 400 fold

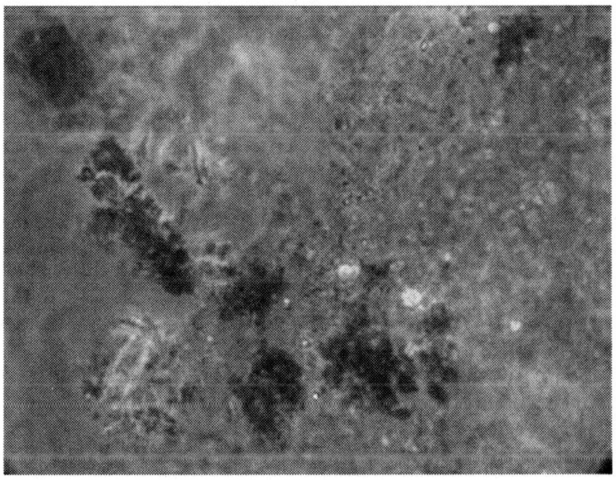

Figure 3.2 Light microscopic image (phase contrast) with staining for iron deposits, Side C June 2008; magnification: 400 fold

In all cases, the deposits were dominated by amorphous particles and bacteria in coccoid, rod-shape and filamentous form. Proportions of both, particles and bacteria, were varying between the different modules. In case of modules from site A and site H, bacterial numbers were very high, thus dominating the microscopic picture of the deposit. Especially in case of site H, bacteria in a slimy matrix could be observed. This was a hint for the presence of EPS and biofilms on the membrane surface. In case of modules from site D, site E and site G many particles were visible. Most of the particles were of amorphous, non-crystalline nature and had a brownish color. Some algae were only observed in the deposit from site A. Protozoae were only observed in low numbers in the deposit of site E.

Organic matter was dominant in all membrane samples even if inorganics, such as aluminosilicates and iron precipitates, could be put into in evidence on several sites.

FTIR revealed the presence of polysaccharides and proteins for all the samples, typical of a biofouling phenomenon.

More details on organics and microorganisms speciation can be found the subsequent sections. A general discussion on the relevancy of each tool will end this chapter.

3.2 SPECIFIC ORGANIC COMPOUNDS ANALYSES

Foulants were recovered from membrane surface using gentle scrapping and lyophilized before being subjected to a large variety of characterization tools to look at both the inorganic and organic composition of the material. Foulants were found to be predominantly organic in nature while inorganics were often dominated by Fe and Ca. The complexity of the organic matrix that incorporates a mixture of biopolymers requires the use of analytical techniques proposed for the characterization of polymeric structures. Among the various tools used for this study, this Chapter presents the results obtained with two important ones, solid state ^{13}C-NM and pyrolysis/thermochemolysis GC/MS. These tools are commonly used for NOM characterization.

3.2.1 ^{13}C-NMR

^{13}C-NMR spectroscopy was performed on lyophilised deposits from the autopsied modules. Figure 3.3 depicts the proportion of the different carbon atoms present in the organic molecules constituting the analysed deposits from sites D, C(1) and H. The associated NMR profiles are available on Figure 3.4. The proportion of the different carbon atoms follows the same trend whatever the considered module, meaning that organic molecules constituting analysed deposits, are similar for all deposits. Since the first results showed no significant differences between sites D, C(1) and H, it was decided not to run ^{13}C-NMR analysis for sites A and E deposits (G deposit was not obtained in sufficient quantity to perform ^{13}C-NMR analysis).

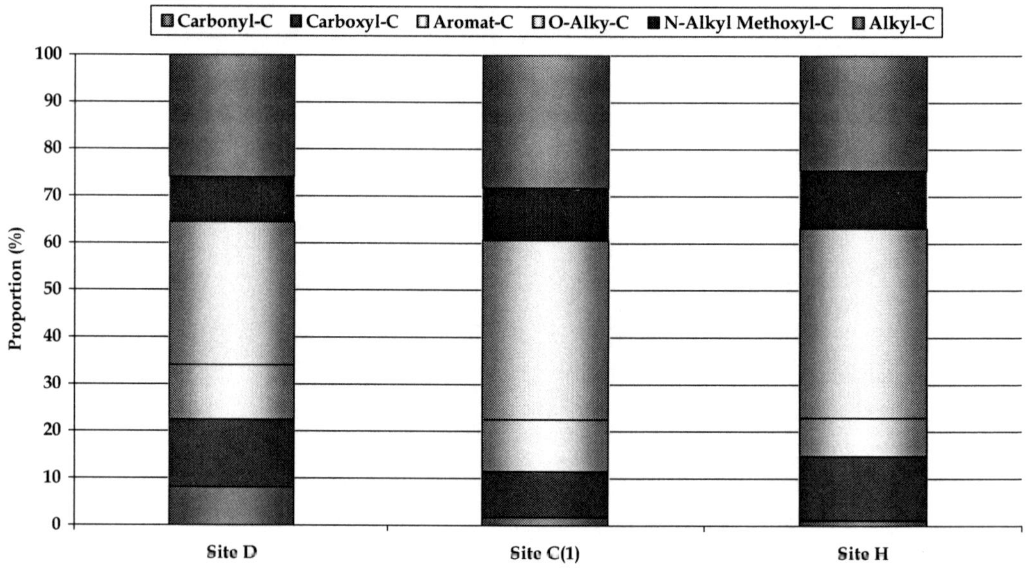

Figure 3.3 ^{13}C-NMR results for deposits from sites D, C(1) and H. This figure is available in colour at http://www.iwapublishing.com/template.cfm?name=isbn9781843393214

For all the spectra, the dominant bands between 60 and 90 ppm can be assigned to CO bounds typical of aliphatic hydroxyl groups in sugars, the peak around 90–110 ppm corresponds to anomeric carbon (i.e. carbohydrates). The peak at 20 ppm corresponds to the methyl carbon of the N acetyl group. The major band centered on 175 ppm (160–190 ppm band) corresponds to the presence of both carbonyl and amide groups present in N-acetyl secondary amide. It can be noticed that the relative abundance of this band is less important for site C (brackish water feed site) as compared to the two other sites (seawater feed site).

For both site D and site H deposits, the band centered on 130 ppm (115–140 ppm) should be related to the presence of aromatic structures. Note that the presence of aromatic structures may come from protein-type structures (i.e. presence of aromatic amino acids), possibly polyphenols from algae but also from humic-type structures that can be present in marine environment (i.e. marine-like humic substances). Surprisingly this band is nearly absent in site C spectrum while site C is fed by an estuarine brackish water expected to be more enriched in humic-type material. The absence of the aromatic band might be related to the lower abundance of the carbonyl/amide band indicating the lower abundance of protein structures in the site C foulant. This hypothesis agrees with the pyrolysis results that showed a major presence of polysaccharides in site C foulant.

The three spectra confirm the structural similarities of the deposit at different locations due to the predominance of polysaccharides in biofilms. It is interesting to note that these spectra show strong similarities with spectra (Figure 3.5) produced from deposits isolated from NF membranes isolated from modules harvested from a river water treatment plant

(Croué *et al.*, 2003) and also similar from colloids (high molecular weight structures) isolated from surface waters (Leenheer, 2009; Leenheer et Croué, 2003; Croué, 2004 and Leenheer *et al.*, 2000). Leenheer *et al.* (2000) considered this material as N-acetyl aminosugars type structure resulting from microbial activities (microbial residues). Note that the ^{13}C NMR spectrum of the foulant of site C exerts stronger similarities with river water colloids and NF foulants than foulants from site D and site H. The origin of the feed water, brackish/river waters versus seawaters might be responsible for the difference in the structural composition of the foulant (carbonate precipitation with seawater, larger proteins abundance in seawater deposits).

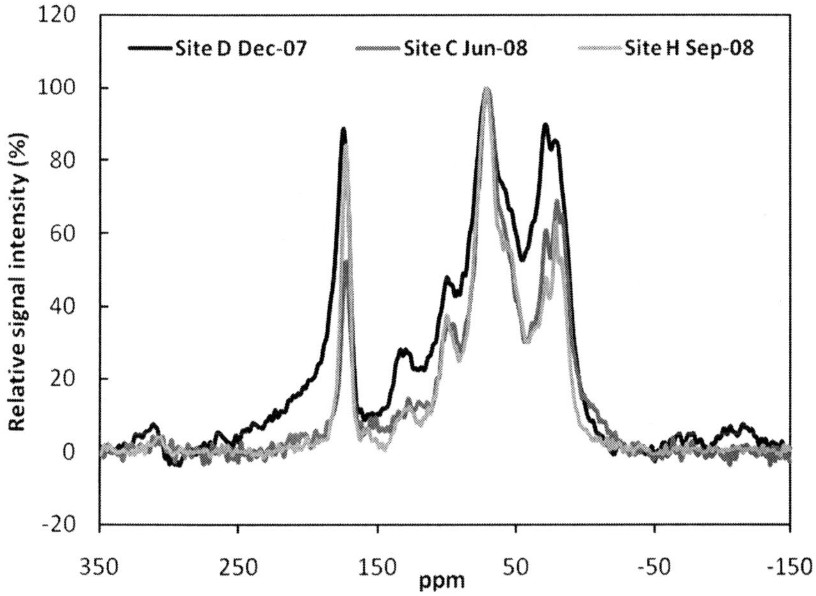

Figure 3.4 ^{13}C-NMR profiles of autopsied modules from sites D, C(1), and H. This figure is available in colour at http://www.iwapublishing.com/template.cfm?name=isbn9781843393214

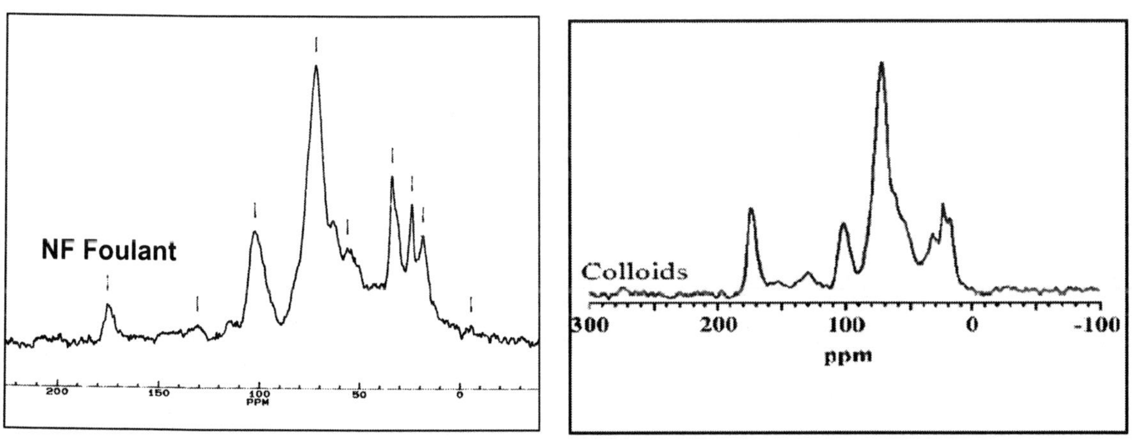

Figure 3.5 ^{13}C-NMR of NF foulant (surface water treatment plant) and colloids (high molecular weight structures isolated from river water)

Results from ^{13}C-NMR spectra are in agreement with the existence of a biofilm at the membrane surface, biofilm that showed minor differences in its overall composition from site to site. It is however questionable to attribute the existence of the foulant layer to microbial activities at the membrane surface only, colloidal structures present in the feed water may go through the pre-treatment steps and accumulate at the surface of RO membranes.

3.2.2 Flash pyrolysis – GC/MS analysis

Flash pyrolysis performed on lyophilised deposits generates by-products which are separated by gas chromatography and identified by mass spectrometry. Pyrolysis GC/MS can be used to compare the structural characteristics of aquatic organic matter and humic substances from various origins (Widrig *et al.*, 1996 and Christy *et al.*, 1998). Pyrolysis by-products generated by the thermal degradation of organic molecules are classified in different biopolymers classes (proteins, polysaccharides...) as described in Bruchet *et al.* (1990). Figure 3.6 presents an example of pyrochromatograms obtained by Pyrolysis GC/MS with the main fragments and biopolymers classes reported.

All pyrochromatograms obtained for the isolated membranes foulants showed specific fingerprints. However, peaks of acetic acid and acetamide remain the most intense for all the samples. The Toluene peak appears to be more intense for sites D, E and A suggesting a stronger abundance of proteins. For both site C and site H, the intense peaks of methylfurancarboxaldehyde, acetic acid and acetamide suggest the dominance of polysaccharides and aminosugars. Site G pyrochromatogram reveals the dominance of styrene and phenolic by-products. In order to provide possible comparison of the biopolymer distribution of the various deposits, a semi-quantitative approach was used. The area of chromatographic peaks that originate from known organic fragments was used to determine the relative proportion of the main biopolymers of bulk organic matter (Bruchet *et al.*, 1990). Figure 3.7 and Figure 3.8 present the relative abundance of biopolymers in the membrane deposits for each studied site.

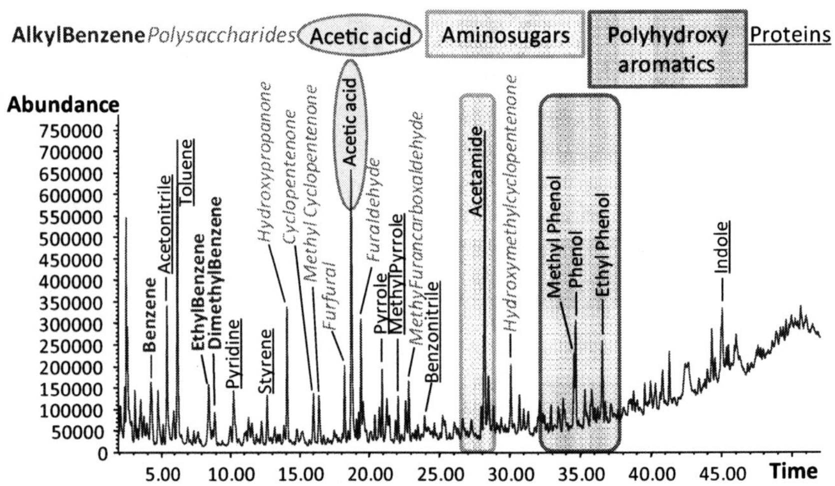

Figure 3.6 Example of a pyrochromatogram obtained by Pyrolysis GC/MS (site D)

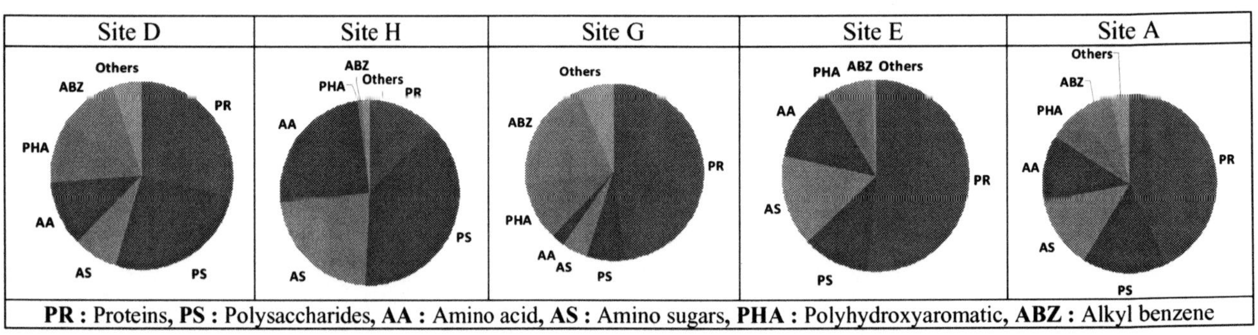

Figure 3.7 Biopolymers distribution within deposits of the autopsied modules from sites D, H, G, E and A

The foulants isolated from the different RO modules exert specific biopolymer distribution. The main dominant biopolymer class is proteins for site D, site G, site E and site A whereas polysaccharides are dominant for site C and H. Another interesting point is that for all sites, the surface area of the peaks attributed to proteins and polysaccharides always represents more than 50% of the identified peaks. Proteins and polysaccharides can be related to microbial origin whereas polyhydroxyaromatics also observed in foulant program are more generally linked to a terrestrial origin, i.e. humic substances. However, aromatic aminoacids produce phenolic derivatives upon pyrolysis. The predominance of proteins and polysaccharides indicates a strong microbial origin of the deposits. The strong abundance of polysaccharides and proteins, confirmed by the presence of amino sugar type structures, is characteristic of biofilms and organic colloids isolated from natural waters (microbial residues). Moreover, high relative proportion of proteinaceous material associated with a small proportion of polysaccharide fragments suggest that the material is recent (Widrig *et al.*, 1996), which was observed for site A, site G and site E.

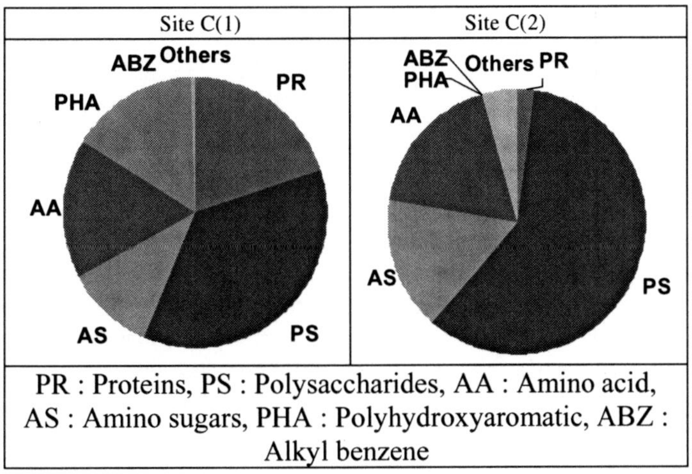

Figure 3.8 Biopolymers distribution within deposits of the autopsied modules from site C, before and after cleaning

A comparison between biopolymers distribution within deposits from site C put into evidence that proteins are impacted by chemical cleaning, since their proportion within deposits from site C(2), after the performed chemical cleaning, decrease radically.

3.2.3 Thermochemolysis TMAH analysis: fatty acids analysis

The thermochemolysis TMAH followed by GC/MS detection, a complementary technique that gives access to a better identification of the fatty acids and phenolic structures, was used for each membrane deposit. Thermochemolysis GC/MS analysis of carbohydrate monomers and aminoacids showed the production of a large variety of methoxy phenol derivatives. As a consequence this approach was not used to specifically address the polyhydroxyaromatic composition of foulants knowing the abundance of both polysaccharides and proteins in the matrix.

The fragmentogram m/z 74 (fatty acids) of the deposit from site D is presented on Figure 3.9, as an example.

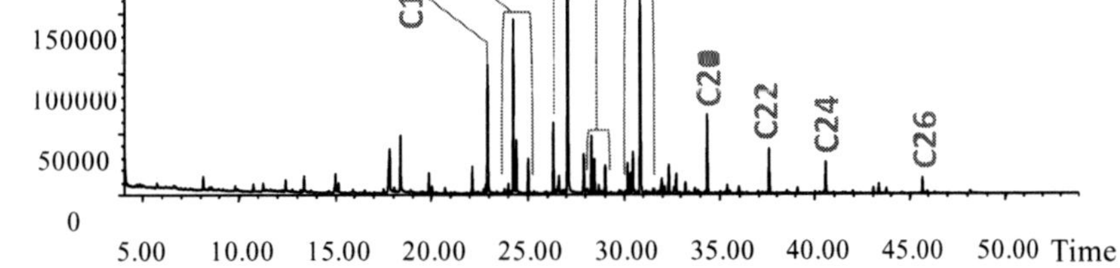

Figure 3.9 Fragmentogram m/z 74 (i.e. +.CH3–COOCH3) of site D foulant

Fatty acid methyl esters (FAMEs) were identified in the C12–C26 range with a predominance of the C16 for all samples. Most of them are saturated FAMEs; unsaturated fatty acids (C16unsaturated and C18unsaturated) specific of more recent material were also detected. The presence of branched C15 and C17 (iso and anteiso), always observed in foulant chromatograms is linked to bacterial origin. Small chain fatty acids are attributed to microbial origin while the presence of long chain FAMEs (C22–C26) is generally attributed to higher plant input (Durand *et al.*, 2005), but they can also find their origin in algae. Differences were observed in FAMEs distribution (m/z 74) between all the studied sites. Site D sample shows more intense peaks of short chain FAMEs (<C20) as compared to long chain FAMEs (≥C20). For sites G, E and A, only C14 to C20 FAMEs were detected. Site C samples presents intense peaks of long chain FAMEs. Relative proportions (determined based on relative surface area) of short chain and long chain FAMEs were calculated and results are given in Table 3.5, for each analysed module.

Table 3.5 Distribution of fatty acids in membranes foulants

%	Site D	Site H	Site G	Site E	Site A	Site C(1)	Site C(2)
Short carbon chains (C12–C20)	84	62	99	99	99	54	50
Long carbon chains (C22–C26)	16	38	1	1	1	46	50

All deposits from seawater desalination plants (sites D, H, G, E and A) present an abundance of FAMEs with short carbon chains, meaning that foulants at membrane surface is more from bacterial origin. This origin is more pronounced for deposits from sites G, E and A since FAMEs with short carbon chains represent 99% of the total fatty acids within the studied deposits. However, regarding the brackish desalination plant, the contribution of short carbon chains is similar to the long carbon chains within the deposits of site C. It means that foulant at membrane surface is from bacterial origin and terrestrial/algal origin (50/50). This analysis does not put into evidence a real change in the distribution of FAMEs within deposit from site C(2), after the chemical cleaning. Indeed, the same trend regarding the distribution before and after cleaning is observed.

3.3 QUANTITATIVE BIOMASS PARAMETERS

A number of analytical tools presented in Table 3.3 have been explored in the autopsy study to quantify the amount of biomass in the sampled modules of full-scale installations. Two of these parameters were related to major constituents of bacteria in biofilms, ATP in active bacterial cells and carbohydrates in the extracellular material of bacteria (CH; Dubois *et al.* (1956)). The ATP analysis is quick, simple and cheap with a detection limit of around 0.01 ng.cm^{-2}. The CH-analysis needs more care but can be done also quickly at low costs with a level of detection of 5–10 μg.cm^{-2}. These two parameters were significantly correlated in biofilms ($r^2 = 0.92$; $P < 0.001$) as demonstrated in Chapter 4.2 (development of a laboratory cleaning test Page 101). The proportional correlation between total bacterial cell count Hobbie *et al.* (1977) and ATP for biofilms in water treatment has been demonstrated in literature (Magic-Knezev and Van der Kooij (2004)).

Elucidation of the role of biomass in membrane fouling requires the assessment of a quantitative relationship between the amount of biomass in the feed channel and the occurring operational problems. In a separate laboratory study by Hijnen *et al.* (2011) the relationship was assessed between the Normal Pressure drop (NPD; kPa) in a model membrane feed channel (Membrane Fouling Simulator (MFS); Vrouwenvelder *et al.* (2006)) and the accumulated biomass determined with ATP and CH. The pressure drop was increased as a result of biofouling initiated by the dosing of acetate as a nutrient. ATP and CH concentrations in the fouled membrane sheets were positively and significantly correlated with the observed NPD increase (ATP, $r^2 = 0.52$, $P < 0.001$ and CH, $r^2 = 0.82$, $P < 0.001$) as presented in Figure 3.10. This was not observed for the total bacterial cell count (TDC). Moreover, the ATP and CH concentrations in these biofilm samples were significantly correlated ($r^2 = 0.62$; $P < 0.001$). Biofilm was visually enriched with iron deposits which were measured in the same biomass samples. No correlation was observed between the iron concentrations and NPD (Figure 3.10).

Membrane sheets (20/20 cm; feed spacer and two membrane sheets) from the modules of the different locations were cut from the feed, middle and outlet sites, fixated between plates and cooled transported within days to the laboratory. From these sheets samples of 2/2 cm were cut to isolate and quantify the biomass. There were distinct differences observed in the ATP concentrations of the locations presented in declining order (Figure 3.11). The concentration ranged from 0.34 to 14.6 ng.cm^{-2} and at most locations the highest concentration was observed in the feed site of the module. The CH concentrations in the same samples depicted in the same order as presented for ATP in Figure 3.11 showed also distinct differences. No declining order was observed in this figure, however, thus showing that there was no correlation between ATP and CH in these membrane samples.

The ATP concentrations in these membrane samples from full-scale modules were in the lower concentration range of the ATP concentrations observed in the laboratory study (Figure 3.10). This may lead to the conclusion of a minor role of biofilms in the fouling problems of these modules, despite the results presented in chapter 3.2. This conclusion, however, must be considered with care because of the following three reasons: (i) the sample transportation and analysis delay (not within the usually 24 hours after sampling) may lead to ATP leakage from the biofilm. Secondly, (ii) except samples from location C all samples came from desalination plants supplied with seawater. The laboratory study was conducted with freshwater (no seawater available at the laboratory). And finally (iii) the full-scale samples came from modules with prolonged operational time with varying chemical cleaning histories. Previously published ATP concentrations in fouled field modules (increased NPD) operated with fresh water and determined shortly after sampling Hijnen *et al.* (2009) were in the high range of concentrations presented in Figure 3.10, thus showing the predictive value of ATP.

The CH concentrations in the samples from the field modules ranged from 18–172 μg. cm^{-2} and were in the higher range of the CH concentrations observed in the laboratory biofouling study (Figure 3.10). This indicates a prominent role of biofilm formation in the fouling of those membranes as also demonstrated by the other studies (chapter 3.3). Thus, it remains undefined what the reason was for the low ATP concentrations in these modules with clear indications for the occurrence of biofouling.

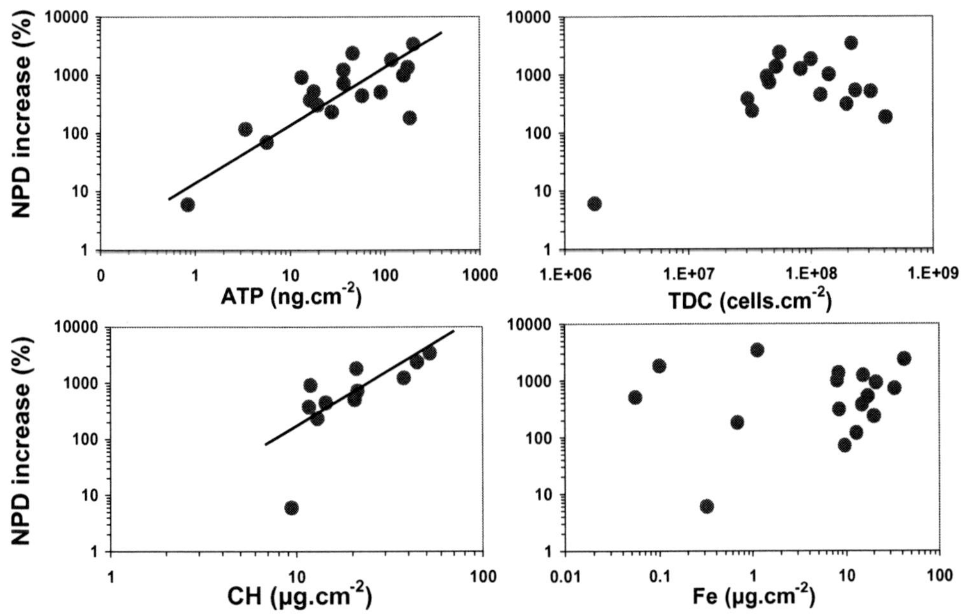

Figure 3.10 The biomass concentration determined with ATP and CH and Fe in the fouling layer of a biofouled model membrane system (MFS; membrane and feed spacer) correlated with the NPD at the time of autopsy

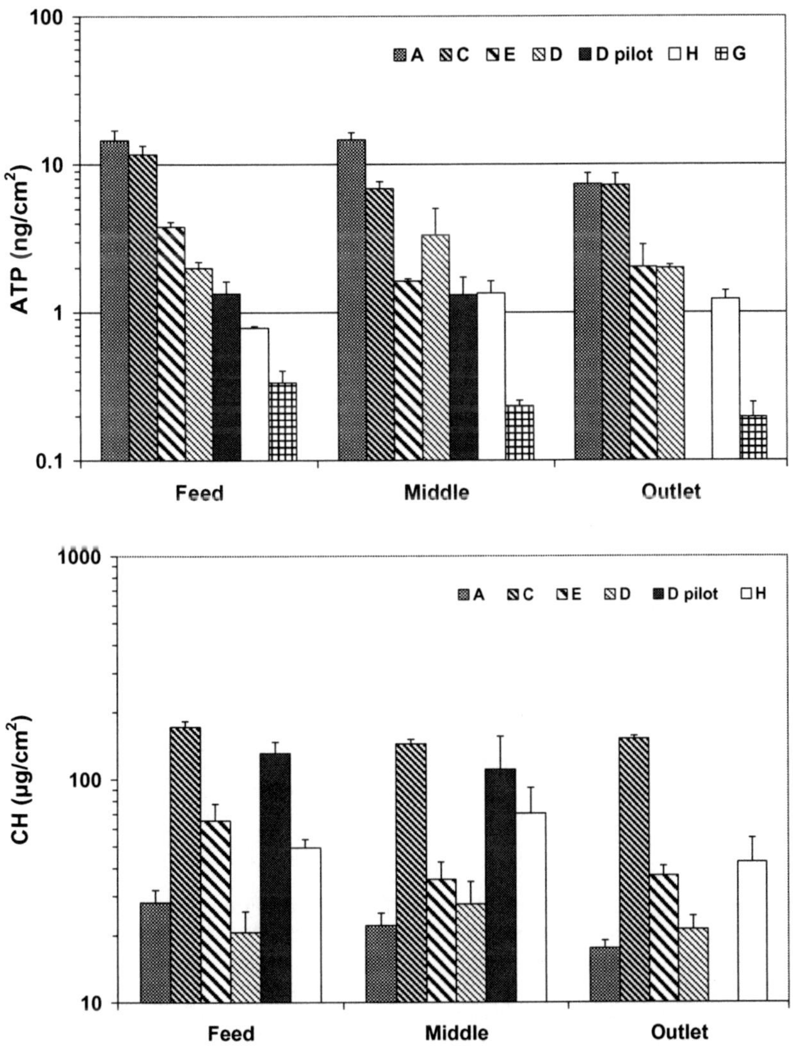

Figure 3.11 The ATP and CH concentrations in the sampled membrane modules of the different full-scale sites (Table 3.1) presented in declining order of ATP concentrations

3.4 MOLECULAR ANALYSIS

Molecular tools were applied to characterize changes in bacterial community structure of SWRO membranes from different desalination plants and to identify the (major) bacterial groups that colonize SWRO membranes. The analysis comprised a total of nine SWRO membrane samples from pilot and full-scale desalination plants Sites A (Nov08), D (five samples: Sep/2007, Nov/2007, Mar/2008, Jun/2008 and Feb/2008), E (Mar09), G (Nov08) and H (Sep08). The community structure of different membrane samples was analysed by 16 S ribosomal gene fingerprinting using the capillary electrophoresis single-stranded conformational polymorphism CE-SSCP technique. The analysis of 16 S ribosomal RNA gene clone libraries from one membrane sample from Site D (Nov/2007), which was in use for the shortest time allowed getting insights on the bacterial community composition and identifying the potential SWRO membrane primary colonizers.

Membrane autopsies were performed by using molecular tools accordingly:

1. Genetic fingerprinting (CE-SSCP); for bacterial community structure and diversity
2. 16 S ribosomal RNA gene sequencing and phylogenetic affiliation of clones; for bacterial community composition of the major (dominant) species.

3.4.1 Adaptation of molecular tools to analyze microbial community structure: homogeneity and reproducibility

The membrane samples of two modules from Site D Nov/2007 comprising coupons from the membrane sheet sampled at three positions corresponding to the feed end (inflow) of the module, the middle of the module, and the brine exit end (outflow) of the module, designated as feed-, middle- and brine- were analysed at different levels. First it was analysed the community diversity within feed, middle and brine locations (Figure 3.12A). Reproducibility tests were performed within the same location at distinct extraction dates (Figure 3.12B). Other parameter tested was the comparison between the communities present on the membranes to the active ones by the means of 16S cDNA fingerprinting (Figure 3.13).

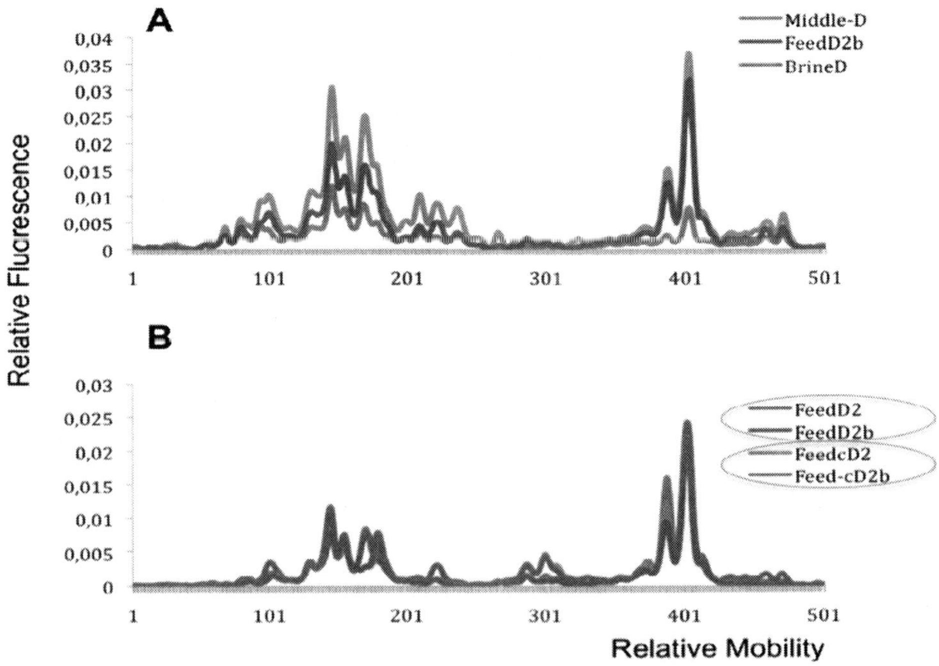

Figure 3.12 Bacterial community structure in a full-scale desalination plant (FSDP) as revealed by capillary electrophoresis single-stranded conformational polymorphism (CE-SSCP). (A) Superimposed normalised DNA profiles for samples taken from different positions in the seawater reverse osmosis (SWRO) membrane module relative to the water flow flux. (B) Superimposed normalised DNA and the corresponding cDNA profiles for seawater reverse osmosis (SWRO) membrane samples analysed at two independent dates (2 and 2b). This figure is available in colour at http://www.iwapublishing.com/template.cfm?name=isbn9781843393214

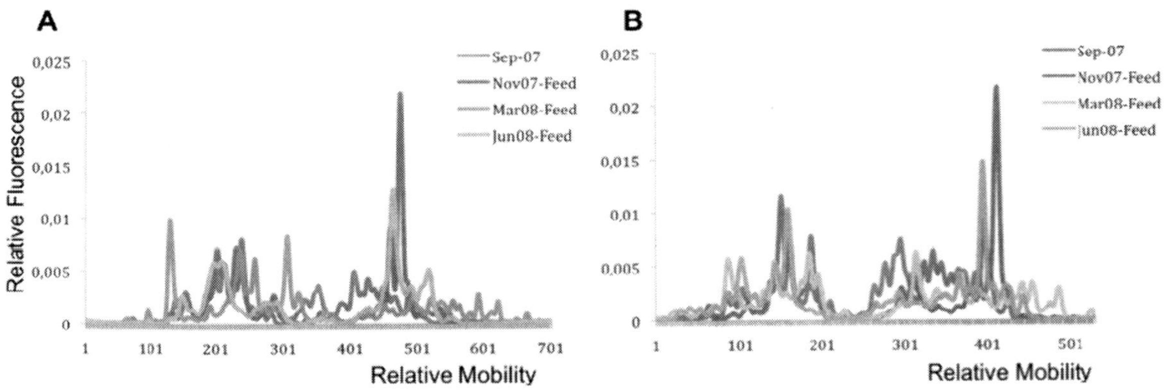

Figure 3.13 Bacterial community structure in a full-scale desalination plant (FSDP) as revealed by capillary electrophoresis single-stranded conformational polymorphism (CE-SSCP). (A and B) Superimposed normalised DNA and cDNA profiles showing different peak profiles for SWRO membrane samples respectively. This figure is available in colour at http://www.iwapublishing.com/template.cfm?name=isbn9781843393214

Similar profiles were obtained for the SWRO membranes from the three different module positions suggesting no changes in bacterial diversity only the relative abundances of some groups in the brine sample (Figure 3.12A). The reproducibility test confirmed the efficacy of the nucleic acids extraction procedures, adapted to the SWRO membrane samples (Figure 3.13B). Differences in the16 S genetic fingerprints of both DNA and cDNA derived samples indicates that total and active community structure changed according to the sampling date.

3.4.2 Bacterial diversity at different desalination plants

The bacterial communities from the active fractions of SWRO membrane samples with different usage times from pilot and full-scale desalination plants was compared by means of CE-SSCP. Table 3.6 summarizes the usage time after cleaning (AC) from the autopsied membrane samples.

Table 3.6 Pilot (Site A) and full-scale desalination plants (Sites D, E, G and H) used in this study and the period they were sampled followed by the number of days that the module was in use after a cleaning procedure

Membrane origin	Usage time (days*)
Site A (Nov08)	10
Site H (Sep08)	17
Site D (Feb09)	60
Site G (Nov08)	120
Site E (Mar09)	255

**After cleaning.*

16S rRNA genetic fingerprinting (cDNA) of the different samples revealed that the bacterial community structure was different at each facility (Figure 3.14).

3.4.3 Evolution of bacterial communities in SWRO membranes from one full-scale desalination plant (Site D) according to module usage time

Bacterial communities from SWRO membranes obtained from Site D (Sep/2007, Nov/2007, Mar/2008 and Jun/2008) were compared according to progressive module usage time. The four samplings periods are ordered herein as 1, 2, 3 and 4 according to the increasing usage time of the RO modules (10, 60, 120 and 330 days respectively). Genetic fingerprinting and 16S rRNA gene clone libraries analysis revealed that the bacterial communities present on seawater reverse osmosis (SWRO) membranes varied between different sampling times (Figure 3.15). Clone libraries analysis showed that *Proteobacteria, bacteroidetes* and *Planctomycetes* were dominant in the membrane libraries (Figure 3.16). Moreover it is shown that bacterial diversity increases with membrane usage time.

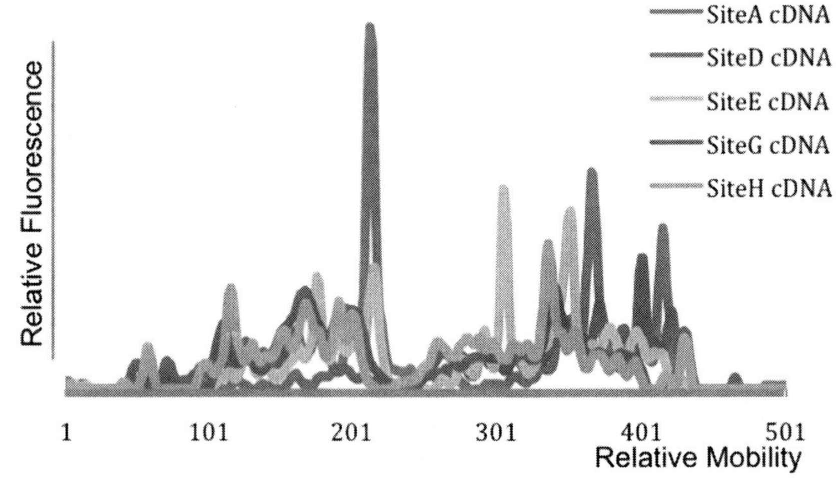

Figure 3.14 Bacterial community structure in SWRO membranes from desalination plants as revealed by capillary electrophoresis single-stranded conformational polymorphism (CE-SSCP). Superimposed cDNA profiles from Sites A, D, E, G and H. This figure is available in colour at http://www.iwapublishing.com/template.cfm?name=isbn9781843393214

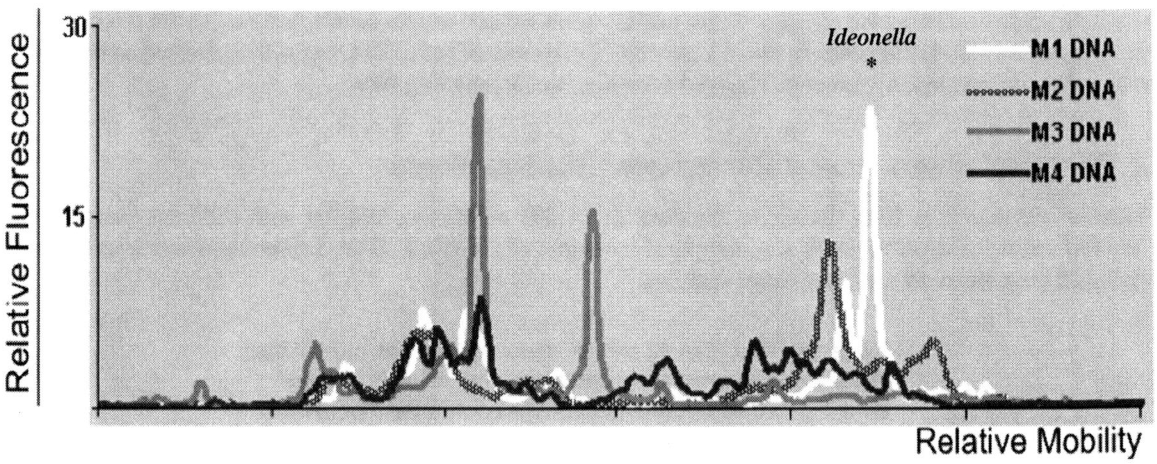

Figure 3.15 Bacterial community structure in SWRO membranes from a full-scale desalination plant (FSDP) as revealed by capillary electrophoresis single-stranded conformational polymorphism (CE-SSCP). Superimposed normalised DNA profiles obtained for SWRO membranes (M) with different usage time. Asterisks marks the position of *Ideonella* peak assigned by the clone library (named in bold). Numbers correspond to the ascending membrane usage time: 10, 30, 120 and 330 days respectively

The clustering of sequences at the 97% OTU level revealed that five OTUs were unique to the membrane used for the shortest period comprising one *Betaproteobacteria* genus, *Ideonella*, and three different *Bacteroidetes* OTUs. Interestingly, the typical freshwater genus *Ideonella* was the dominant OTU for this membrane but was absent from those in use for longer periods (indicated in Figure 3.16). This result suggests that bacteria affiliated to the *Ideonella* genus could be one of the SWRO membrane primary colonizers.

3.4.4 Conclusions

The assessment of bacterial community diversity and composition in the SWRO membranes by molecular tools was accomplished with different methods each giving different levels of resolution. Genetic fingerprinting, CE-SSCP, was used to follow bacterial community structure over different Sites and sampling dates and within one Site, membrane samples in use for different time-periods. Our results showed that bacterial community structure not only changed according to the Site studied but also according to the SWRO membrane usage time. 16S DNA clone libraries and sequence data set analysis were performed for some selected samples in order to identify the major bacterial groups/species present on the membranes and their relative abundance. We show that, for one particular desalination plant (Site D), *Proteobacteria* and *Bacteroidetes* dominated the SWRO membranes. However the occurrence and abundance of some bacterial groups seemed to be different amongst the membranes. Some of them were exclusive for the membranes with short usage time, like *Ideonella spp.*, and others for those with longer usage time like *Planctomyces*. Together, our data indicate a dynamic colonization process of SWRO. We have validated the experimental approach used in this study by statistical analysis of the data and by comparing literature data on freshwater systems (de O. Manes *et al.*, 2011). The molecular tools applied to this study are specifically useful for the understanding of bacterial community populations colonizing the SWRO membranes in order to address functional properties of certain genera/species in biofilm formation.

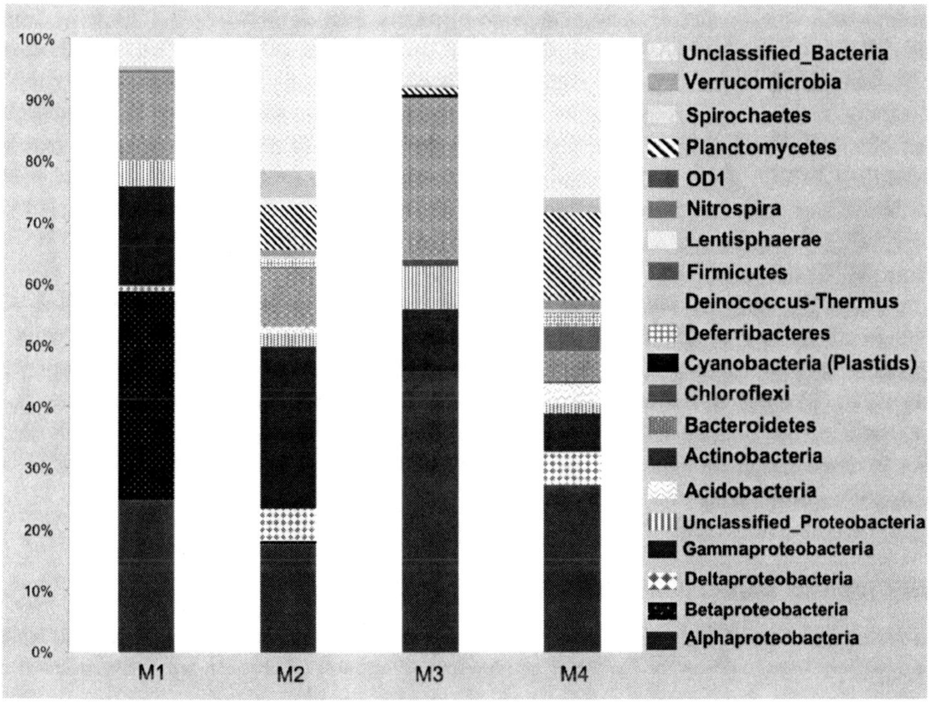

Figure 3.16 Schematic representation of the hierarchical classification and distribution of the 16S ribosomal RNA gene sequences into major bacterial groups from clone libraries obtained from SWRO membrane (M) samples. Numbers correspond to the different sampling periods

3.5 GENERAL CONCLUSION ON AUTOPSY TOOLS RELEVANCY

3.5.1 Tools for microscopic observation

Starting with physical removal of the fouling layer and applying light microscopy can provide the operators with general idea if large amount of microorganisms are present and if there might be a problem of biofouling. Also general amorphous structure with deposits implied that EPS and mineral deposits were also part of the fouling layer. Following light microscopy, confocal microscopy showed that large amount of EPS and bacteria existed on the membranes, but it was hard to find a correlation between the confocal data and the degree of membrane fouling that was checked in the RO bench scale units as well as with the type of process from which the autopsies were taken from. But generally, confocal microscopy is a coherent and efficient tool that can be used to explain the decline in membrane performance. We should keep in mind that biofilm structure should have a major impact on the fouling behaviour and still more fundamental studies in this area should be done. SEM-EDX analyses, usually performed for autopsy, also confirm that all biofilm layers contained both biological material and inorganic composites: iron, magnesium, aluminium, and silica containing deposits were part of the fouling layers analysed. It should be mentioned that the sulphur peak in the EDX analysis is likely to be originated from the polysulfone RO membrane support layer. The contribution of AFM to the fouling layer characteristics is doubtful, since the layers analysed are too rough compared to the fine resolution of AFM that is generally used for flat and smooth surfaces. It should be mentioned that the use of AFM could be extended to analyse the adhesive properties of the extracted components of the fouling layers such as the EPS. Force measurements and force-distance curves before and after cleaning of the autopsies may have enlighten the biofouling propensities of pristine, cleaned, and fouled membranes.

3.5.2 Inorganic matter characterization

For the more detailed characterization of the elemental composition of the fouling layer, the inorganic part could be analyzed by ICP-OES (Mondamert *et al.*, 2009). Quantification of 16 different elements was possible in all investigated cases in the MEDINA project. Depending on the used sample digestion (e.g. by aqua regia) it is not possible to quantify all elements. E.g. silicium is underestimated by digestion with aqua regia, since not all silicates will be solubilized.

In case of Site E, 15 of 16 different elements were detected, whereas in case of Site H only 11 of 16 different elements were present. In general, results were similar for both samples in quality. The dominant elements of Site H and Site E were Al, Ca, Fe and Na. The origin of the elements seems to be from sea water.

It was common to both investigated deposits that more elements were detected by ICP-OES in comparison to SEM with EDX. With ICP-OES, deposits will be analysed overall, independent of the position they have on the autopsied membranes. In contrast, SEM with EDX only shows the chemical composition of the surface of the deposits (upper 5–10 μm). Additionally, in some cases single spots, like particles, could be identified by EDX and could contain high concentrations of elements like Fe. SEM with EDX offers the possibility to analyze specific area of the autopsied membranes, whereas ICP-OES gives a opportunity for an overall characterization of deposits without any areal-depended aim. A biofouling study in a model system (MFS) showed that Fe accumulated in biofilms was not correlated with the observed NPD. Nevertheless, further fundamental studies are lacking on the role of Fe deposits in other operation problems (MTC; salt rejection).

In conclusion, ICP-OES analyses were found to be very useful for an overall characterization of the membrane deposits independent of surface structure and inhomogeneous distribution of different components. By use of this method it was possible to gain a quantitative impression about the amount of elements in the deposit. Further, it was possible to differentiate between different inorganic deposits. In addition, by ICP-OES analyses it was possible to identify the main chemical components of the deposit. By all these results from ICP-OES it is possible to have a whole set of information to characterize the deposit. From our point of view these technique is suggested to be (together with SEM/EDX) one major method for membrane autopsies.

3.5.3 Organic matter characterization

Reverse osmosis membranes exert high rejection efficiency for organics and inorganics. As a consequence RO foulant is expected to incorporate a large diversity (in terms of size and structure) of organic and inorganic structures according to the fed water characteristics, the bacterial activity occurring at the membrane surface and the conditions of operation (recovery, cleaning strategies). The fouling layer accumulating at the membrane surface (concentration-polarization phenomena, precipitation, adsorption/sorption, bacterial growth i.e. biofilm formation) corresponds to a complex mixture of organics i.e. from low molecular weight compounds to biopolymers (originally present in the fed water or produced at the membrane surface by microorganisms), residual of chemical agents used along the treatment train (i.e. coagulant, antiscalant) and inorganic species (complexation, adsorption, precipitation phenomena).

The characterization of the organic constituents of foulant can be addressed by the determination of both global and specific parameters developed directly on the membrane surface or on the material recovered from the membrane surface by scraping. Because the water content of RO foulant is high, it is fundamental to carefully address membrane conditioning operations performed prior to shipment or prior to analysis i.e. module preserved in seawater, module rinsed with permeate water, module preserved in SBS.

FT-IR was an informative technique; however the organic signature of all foulants were found similar showing abundant amide (e.g proteins) and sugar contents. The use of TOF SIMS (Time of Flight Single Ion Monitoring Scanning) or TEM (Transmission Electron Microscopy) with specific staining should be explored for functional group identification and in-depth structural configuration.

Most of the organic characterization tools are developed on foulant scrapped from the membrane surface. The fouling layer is expected to be mainly insoluble. Autopsies conducted during the MEDINA project also showed that chemical cleaning did not generally remove a significant proportion of organic foulant. As a consequence only a minor organic fraction (few per cents) of foulant can be dissolved in ultrapure water or base. It is important to address this limitation when applying analytical techniques based on direct DOC characterization, i.e. fluorescence, LC-OCD, DOC and TN contents, of the soluble phase of foulant. Module/membrane conditioning can also have a major influence on the nature of the soluble fraction recovered since it may incorporate a high proportion of feed water. However one can also argue that relationships may exist between the soluble and the insoluble phases of the foulant, results that might be worth to demonstrate.

It is clear that analytical tools developed on dried/lyophilised foulant material were the most informative. Elemental analysis, loss of ignition at 550 or 900°C provide essential information regarding the organic carbon content of the foulant and the relative proportion of organic nitrogen. However it is important to notice that the presence of carbonates and nitrates will interfere with the organic composition analysis. FT-IR using KBr pellet or ATR and solid state

^{13}C-NMR are complementary approaches. Because they are only semi-quantitative analytical tools, the large abundance of EPS i.e. polysaccharides and proteins (biofilm constituents), mask the presence of other constituents that can be more specific of a given foulant. Pyrolysis GC/MS and thermochemolysis (TMAH) GC/MS were found to better discriminate foulants from different origins. The relative distribution of the major biopolymers can be evaluated providing valuable information on the nature of the deposit accumulated at the membrane surface. Thermochemolysis TMAH GC/MS gave insight in the aromatic and lipid distribution, results that can be used to precise the microbial and/or terrestrial origin of the foulant. During thermochemolysis, sugars were found to generate aromatic compounds that can also be attributed to marine humic-like structures, interferences that must carefully examined. However one can also argue that pyrolysis and thermochemolysis GC/MS are only looking at a minor fraction of the organic matter (few per cents). Specific GC/MS analyses, sugar monomers and lipids quantified after hydrolysis and liquid/liquid extraction plus derivatization are other approaches that can be used to tentatively determine the origin of the foulant. All membranes showed abundant polysaccharides and proteins content, biopolymers that characterize the biofilm nature of RO foulant. Specific analyses are needed to discriminate foulants from different origins. Pyrolysis GC/MS and thermochemolysis GC/MS are interesting analytical approaches to address the complexity of the biofilm. Other specific and quantitative analyses, such as lipids, amino acids, sugars and pigments should be more deeply explored to address the expected diversity of the organic composition of RO foulant.

3.5.4 Microbial characterization

Biofouling is considered to play a major role in membrane fouling of desalination plants. Nevertheless no single microbiological parameter could be linked to biofouling until now. The experience has demonstrated that the combination of several parameters will give in many cases of membrane autopsies a "picture" about microbiology involved in the development of the deposit.

General microbiological analytical parameters will include Light Microscopy, Total Cell Number and extracellular polymeric substances (EPS). Supplementary special microbiological tests as for e.g. iron oxidising bacteria may be informative. Furthermore the molecular characterization of the bacterial population attached and growing on membranes aimed at identifying the major species/groups, might shed light on the fouling layer bacterial composition and their involvement in biofilm formation.

Light microscopy is an important parameter because all the information about the different components will give an impression about the type of layer and how microorganisms are distributed. High microscopic visible viscosity of the deposit indicates biology involved. Light microscopy coupled with special staining e.g. for iron compounds reveals a semi quantitative image of the microbiological produced iron precipitates. The observations from light microscopy will be the basis together with water content, loss of ignition for the decision on further analytical tools to be performed such as FTIR or EPS analyses which may be useful.

Total Cell Number – and this is mostly focused on bacteria – showed a high variety in the results although the same method was applied. This could be due to complex deposits, matrix effects due to particles, precipitates which are sometimes very difficult to differentiate from bacteria. This experience is known from other deposits like the layers e.g. from drinking water distribution systems. Accurate numbers may be obtained when the deposit is dominated by microorganism (biofilm type) and Total Cell Numbers are $> 10^8$ bacteria/cm^2 (Flemming *et al.*, 1993). It is known that staining with Acridin Orange instead of DAPI may lead to higher cell numbers because many other organic "particles" and EPS are stained and Total Cell Number may be overestimated. The parameter Total Cell Number is useful when the cell density is higher than $>10^8$ bacteria/cm^2 because then per volume of 1 cm^3 a multilayer biofilm is given. Total Cell Numbers in the range of 10^5 to 10^7 bacteria/cm^2 are standard. Total cell number on other microorganisms than bacteria are difficult to estimate. Special staining or methods like FISH are necessary. However when FISH was attempted in MEDINA on SWRO membranes, it was very difficult to have an accurate cell count due to the high background interference generated by autofluorescent organic substances present in the fouling layer.

In summary, light microscopy and Total Cell Number may point out if bacteria and/or fungi may play a significant role for development of the deposit. These results may be completed by qualitative and quantitative analysis of EPS. With regard to biofilm, many different methods have been published and will give useful information e.g. about biofilm structure. CLSM with special staining may give a good semi quantitative image but not all type of EPS compounds may be stainable. Chemical methods like e.g. protein analysis according to Lowry are well established for pure EPS. But sample preparation for complex deposits is necessary in order not to get overestimation of EPS. All chemical analyses with "purified" EPS give numbers in the range of μg/cm^2. Experience with results is necessary (data base). Because EPS are gel like substances and can bind 80 to >99% water, it is not possible to figure out exactly how much volume will be formed. In summary, CLSM may be informative if all EPS are stainable, the results of the chemical methods differ very much, no standard sample preparation is existing and experience with values are necessary. Therefore comparison of results is difficult until now. If so called carbohydrate analysis is performed with original sample (deposit), no information about EPS is obtained.

Carbohydrate analysis with the Dubois method (CH) together with Adenosinetriphosphate (ATP) to quantify the amount of biomass has been evaluated as relatively quick, simple and cheap quantitative biomass parameters for membrane autopsy to verify the role of biofouling. In a model biofouling study under well defined conditions a positive correlation of CH and ATP concentrations with the pressure drop increase in a model membrane system (Membrane Fouling Simulator; MFS) was observed. The fouling conditions were stimulated by dosing acetate. ATP and CH concentrations were mutually correlated for these undisturbed active biofilms. A positive and proportional correlation between ATP and carbohydrates was also observed in the cleaning study (chapter 4.2). In disturbed biofilms after chemical cleaning, however, the CH/ATP ratio clearly increased (chapter 4.2) indicating the different character of both parameters (total biomass CH versus active biomass ATP).

Variable CH and ATP concentrations were measured in the field SWRO elements from the different desalination plants presented in chapter 3.1. The CH concentrations were in the same order of magnitude as observed in a model biofouling study which is a clear indication of the role of biofilm formation in the fouling process as indicated by other parameters. The ATP concentrations, however, were relatively low compared to the ATP concentrations in the model biofouling study. This was attributed to three possible causes: (i) prolonged transport and storage time before analysis (ATP has to be measured in fresh membrane samples), (ii) the use of seawater instead of fresh water and (ii) a history of membrane cleaning of the field modules. From the results, none of these causes could be identified as the most likely cause.

The assessment of bacterial community diversity and composition in the SWRO membranes by molecular tools was accomplished with different methods each giving different levels of resolution. Genetic fingerprinting, CE-SSCP, was used to follow community structure over different sampling dates. This method gave the diversity degree of a given sample and was suited to compare membrane from different sites and, within one site, membrane samples in use for different time-periods. 16 S DNA clone libraries and sequence data set analysis were performed in order to identify the major bacterial groups/species present on the membranes, for some selected samples, and their relative abundance. This molecular tool is specifically useful for the understanding of bacterial community populations colonizing the RO membranes in order to address functional properties of certain genera/species in biofilm formation.

REFERENCES

Bruchet A., Rousseau C. and Mallevialle J. (1990). Pyrolysis-GC-MS for Investigating High-Molecular-Weight THM Precursors and Other Refractory Organics. *Journal AWWA*, **82**(9), 66–74.

Christy A. A., Bruchet A. and Rybacki D. (1998). Characterization of natural organic matter by Pyrolysis/GC-MS. *Environment International*, **25**, 181–189.

Costerton J. W., Irvin R. T. and Cheng K.-J. (1981). The bacterial glycocalyx in nature and disease. *Annu. Rev. Microbiol.*, **35**, 299–324.

Croué J.-P. (2004). Isolation of humic and non-humic fractions: structural characterization. *Journal of Environmental Monitoring and Assessment*, **92**(1–3), 193–207.

Croué J.-P., Grasset L., Bacle S. and Jacquemet V. (2003). Nanofiltration Membrane Autopsy of Full Scale Unit: Characterization of the Organic and Inorganic Constituents of the Fouling Membrane. *AWWA Membrane Conference*, Atlanta Georgia (USA) May 2–5, pp 9.

de O. Manes C-L., West N. J., Rapenne S. and Lebaron P. (2011). Dynamic bacterial communities on reverse-osmosis membranes in a full-scale desalination plant. *Biofouling*, **27**, 47–58.

Dubois M., Gilles K. A., Hamilton J. K., Rebers P. A. and Smith F. (1956). Colorimetric method for determination of sugars and related substances. *Analytical Chemistry*, **23**(3), 350–356.

Durand C., Ruban V. and Amblès A. (2005). Characterisation of complex organic matter present in contaminated sediments from water retention ponds. *Journal of Analytical and Applied Pyrolysis*, **73**, 17–28.

Flemming H.-C., Schaule G. and McDonough R. (1993). How do performance parameters respond to initial biofilm formation on separation membranes. *Vom Wasser*, **80**, 177–186.

Ghiglione J. F., Larcher M. and Lebaron P. (2005). Spatial and temporal scales of variation in bacterioplankton community structure in the NW Mediterranean Sea. *Aquat Microb Ecol.*, **40**, 229–240.

Hijnen W. A. M., Cornelissen E. R. and van der Kooij D. (2009). Biofilm and pressure drop formation in spiral-wound membranes induced by low concentrations of easily assimilable organic carbon in the feed water, 4–6 September, Qingdao, China.

Hijnen W. A. M., Cornelissen E. R. and van der Kooij D. (2011). Threshold concentrations of biomass and iron for pressure drop increase in spiral-wound membrane elements. *Water Research*, **45**, 1607–1616.

Hobbie J. E., Daley R. J. and Jasper S. (1977). Use of nuclepore filters for counting bacteria by fluorescence microscopy. *Appl Environ Microbiol.*, **33**(5), 1225–1228.

Leenheer J. A. (2009). Systematic approaches to comprehensive analyses of natural organic matter. *Annals of Environmental Science*, **3**, 1–130.

Leenheer J. A. and Croué J.-P. (2003). Dissolved organic matter: recent discoveries and research challenges. *Environmental Sciences and Technology*, **37**, 19A–26A.

Leenheer J. A., Croué J.-P., Benjamin M., Korshin G. V., Hwang C. J., Bruchet A. and Aiken G. (2000). Comprehensive Isolation of Natural Organic Matter from Water for Spectral Characterizations and Reactivity Testing. In Natural Organic Matter and Disinfection By-Products: Characterization and Control in Drinking water, S.E. Barrett, S.W. Krasner and G.L. Amy Editors, ACS Symposium Series 761 American Chemical Society, 68–83.

Magic-Knezev A. and Van der Kooij D. (2004) Optimisation and significance of ATP analysis for measuring active biomass in granular activated carbon filters used in water treatment. *Water Res.*, **38**(18), 3971–3979.

Mondamert L., Labanowski J., Berjeaud J.-M., Rapenne S., Croué J.-P. and Rapenne S. (2009). Autopsy of RO desalination membrane: Part 2. Chemical characterisation of the foulant. *Desalination and Water treatment*, **9**, 73–81.

Rapenne S., Barbé C., Mondamert L., Manes C., Schaule G., Lebaron P. and Croué J.-P. (2009). Seawater Characterization and Membrane Autopsy Tools for Better Operation of SWRO Plants, Proc. IDA World Congress on Desalination and Water Reuse, Dubai.

Schaule G., Griebe T. and Flemming H.-C. (2000). Steps in biofilm sampling and characterization in biofouling cases. In: Biofilms, Investigative methods & applications (Eds.: Flemming H.-C., Szewzyk U., Griebe T.), Technomic Publishing Company, Inc., Lancaster.

Schaule G., Rapenne S., Strathmann M., Grobe S., Robert C. and Jacquemet V. (2009). Autopsy of RO desalination membranes: Part 1 microbial characterization of foulants. *Desalination and Water treatment*, **9**, 66–72.

Strathmann M. (2003). Visualisierung und Charakterisierung von extrazellulären polymeren Substanzen in Biofilmen (Visualisation and characterisation of extracellular polymeric substances in biofilms). PhD thesis, University Duisburg-Essen, Der andere Verlag, Osnabrück, Germany.

Vrouwenvelder J. S., Van Paassen J. A. M., Wessels L. P., Van Dam A. F. and Bakker S. M. (2006). The membrane fouling simulator: a practical tool for fouling prediction and control. *J. Mem. Sci.*, **281**, 316–324.

West N. J., Obernosterer I., Zemb O. and Lebaron P. (2008). Major differences of bacterial diversity and activity inside and outside of a natural iron-fertilized phytoplankton bloom in the Southern Ocean. *Environ Microbiol.*, **10**, 738–756.

Widrig D. L., Gray K. A. and McAuliffe K. S. (1996). Removal of algal-derived organic material by preozonation and coagulation: monitoring changes in organic quality by Pyrolysis-GC-MS. *Water Research*, **11**, 2621–2632.

Chapter 4

Development of cleaning strategies for RO membranes

D. van der Kooij[1], E. R. Cornelissen[1], W. A. M. Hijnen[1], M. Strathmann[2], D. Pütz[2], H.-C. Flemming[2], L. Mondamert[3], J.-P. Croué[3], S. Rapenne[4], C. Barbé[4], C. Robert[4], P. Brisson[4] and C. Machinal[4]

[1]KWR (The Netherlands)
[2]IWW (Germany)
[3]CNRS UMR 6008, LCME, Université de Poitiers (France)
[4]Anjou-Recherche, Veolia Environment (France)

4.1 CLEANING OF SPIRAL-WOUND MEMBRANES

4.1.1 Introduction

Fouling of spiral-wound reverse osmosis membranes used for desalination or water treatment is unavoidable and can cause operational problems including (i) an increase of the feed channel pressure drop (ii) and/or a decrease of the membrane flux and deterioration of the permeate quality. Therefore, an effective cleaning strategy is needed to ensure a reliable and cost-effective operation of the involved installation. A number of fouling processes can be distinguished, viz. (i) scaling, (ii) particulate and colloidal fouling, (iii) organic fouling and (iv) biofouling. Biofouling, which is caused by the accumulation and/or growth of microbial biomass in the feed channel, is a common type of fouling (Khedr, 2000 and Schippers *et al.*, 2004). Studies on the formation of biofilms on water-exposed surfaces have shown that microbial attachment to surfaces is very strong and removal of biofilms is difficult. Consequently, high concentrations of a variety of chemicals are used for membrane cleaning. Unfortunately, (i) cleaning effects of chemicals are not well-documented in the scientific literature, (ii) certain chemicals have undesired environmental effects and (iii) (frequent) use of certain chemicals e.g. oxidizing agents may cause performance reduction. Therefore, investigations on membrane cleaning were included in the MEDINA project to obtain quantitative information about effects and procedures for membrane cleaning with the use of chemicals.

4.1.1.1 Principles of biofouling processes and cleaning

Biomass accumulation is the result of the attachment of microorganisms to the membrane and the feed spacer, and subsequent growth by utilizing biodegradable compounds present in the feed water. Attached bacteria produce

 Membrane-Based Desalination: An Integrated Approach (MEDINA). Edited by Enrico Drioli, Alessandra Criscuoli and Francesca Macedonio. ISBN: 9781843393214. Published by IWA Publishing, London, UK.

extracellular polymeric substances (EPS) to anchor themselves to the surface, which results in the formation of a biofilm, a slimy and viscous layer in which also inorganic compounds accumulate (Flemming *et al.*, 2007). The attachment of bacteria to water-exposed surfaces has been recognized for over 60 years (Zobell, 1943) and the main physico-chemical processes governing attachment have been elucidated (Marshall *et al.*, 1971 and Bos *et al.*, 1999). Extracellular polymeric substances (EPS) with hydrophilic, hydrophobic and charged regions are excreted by bacteria and adsorb irreversibly to surfaces depending on charge, hydrophobicity and roughness of the surface. The rate and extent of biofilm formation (or biofouling) depends on the concentration and the nature of growth-promoting compounds in the feed water. A recent study has shown that concentrations of easily biodegradable compounds as low as a few micrograms per litre in the feed water can promote biofouling (Hijnen *et al.*, 2009). A far-reaching removal of growth-promoting compounds can be achieved by extended pre-treatment (van der Hoek *et al.*, 2000) but this approach requires large investments. Combining an effective cleaning procedure with adequate pre-treatment will lead to minimal costs but requires quantitative information about effects of pre-treatment processes and cleaning procedures.

Experiments with pure cultures have shown that removal of attached bacteria is difficult to achieve and requires high concentrations of chemicals. Corpe (1974) showed that alkaline treatment (0.1 N NaOH) was more effective in removing a pure culture of a marine bacterium from glass than acid treatment (0.1 N HCl) and that detergents and complexing agents enhanced biomass removal. Hydrolyzing enzymes had only limited effects on attached bacteria (Corpe, 1974), but are more effective in combination with certain other agents, e.g. detergents and or EDTA (Whittaker *et al.*, 1984). Overall, the use of mixtures of compounds and/or series of treatments with chemicals showed enhanced effects. However, the efficacy also depends on the concentration and composition of the attached biomass which is affected by the quality of the involved water, e.g. presence of metals. Chemical membrane cleaning generally includes the use of alkaline, acids, detergents, complexing agents, biocides, and denaturising agents. Apart from the selection of chemicals, conditions such as concentration, temperature, duration and shear rate during membrane cleaning affect the result of the cleaning procedure. A limitation in the application of chemicals is that membrane damage can occur, i.e. the applied chemicals and conditions should be compatible with the membrane. The biofouling problems experienced to date in membrane installations for desalination and water treatment demonstrate that improvement of pre-treatment and cleaning procedures is still needed. Only limited information about cleaning procedures and their effects in practice is available, in part because such procedures are tailor-made, based on trial and error and include the use of commercial products of unknown composition. Given the importance of cleaning procedures for maintaining proper operational conditions, investigations were conducted to elucidate the effects of cleaning procedures using chemicals.

4.1.2 Study objectives

The main objective of the study was to define optimized cleaning strategies for spiral-wound membranes used in seawater and brackish water desalination. Emphasis was on the removal of attached biomass. For achieving this objective, the following two approaches have been followed: (i) evaluation of applied cleaning strategies applied in practice, based on understanding the (lack of) success of cleaning and (ii) identification of the most effective cleaning procedure for removal of biomass and organic compounds from membranes used for desalination of seawater and brackish water in laboratory studies. The evaluation of the cleaning strategies applied in practice was based on a literature survey and collection of data from full-scale plants. The second part of the study included systematic studies in the laboratory and in pilot plants on the effects of cleaning strategies. These studies, which are primarily directed to the removal of biomass ('biofouling') from spiral-wound membranes, include (i) development of test methods for comparing the effects of a variety of chemicals under defined conditions in the laboratory, (ii) testing of selected chemicals in cleaning procedures on membrane performance (permeation, salt rejection) in the laboratory, (iii) testing of chemicals on the effects on deposits collected from membranes and (iv) testing of the effects of selected cleaning procedures on the performance characteristics of fouled membranes in a pilot plant. In these studies, attached biomass cultured under laboratory conditions and biomass on membranes from full-scale and pilot plants was used. The results of these investigations are reported in sections 4.2, 4.3 and 4.4 and are summarized below.

4.1.3 Summary of results

4.1.3.1 Evaluation of practical experiences

A review revealed that in the scientific literature only a few reports are available about cleaning of membranes with biofouling and that details about membrane cleaning procedures and effects of cleaning are not provided. The chemicals which are used are described in most cases, but the nature of the applied chemicals remains unclear when commercial products are used. Furthermore, most studies do not provide information on concentrations of chemicals, nor about duration and effects of cleaning. Such information may not have been registered properly, practical limitations in studying cleaning procedures in full-scale plants, or other reasons, e.g. commercial aspects, do not allow publishing details about applied procedures and obtained results. Collection of information about cleaning procedures directly from full-scale plants also did not lead to satisfying information, partly because of similar reasons. Furthermore, testing of cleaning procedures in full-scale plants is limited by a number of reasons, including the down time of the installation, the risk of damaging the installation and the disposal of the used chemicals. Consequently, the obtained information was not suited for drawing conclusions about successfulness of specific cleaning procedures and did not lead to conclusions about the efficacy of the applied cleaning strategies for desalination membranes. Recommendations to promote a more complete and effective registration of the applied procedures and their effects in full-scale plants or pilot plants have been made.

The limited availability of information about effects of membrane cleaning in full-scale plants emphasised the need for investigations on membrane cleaning under defined conditions in the laboratory, in bench-scale tests and in pilot plants.

4.1.3.2 Laboratory test for determining biomass removal efficacy

A laboratory test was developed for determining the effect of a cleaning procedure on the concentration of attached biomass grown on PVC-P and samples of membrane from full-scale plants (Section 4.2; Hijnen *et al.*, 2010). The two challenges in developing a reproducible and representative test were (i) accurate determination of the concentration of attached biomass and (ii) production of samples with a uniform biofilm concentration. Two analytical procedures for determining the concentration of attached biomass were selected and optimized, viz. analysis of the concentration of adenosine triphosphate (ATP) and the analysis of the concentration of carbohydrates (CH). ATP is a measure for the amount of active biomass and CH represents carbohydrates within the cells and the extracellular polymeric substances (EPS) excreted for attachment. Bacterial cells in contact with chemicals used for cleaning can lead to ATP release, thus limiting the use of this parameter as a measure for biomass. CH analysis is well suited and important because it includes EPS which may contribute to biofouling to a large extent. The production of samples with biofilm grown on PVC-P was standardized and yielded samples that could be used in the test, but a certain level of variation between samples produced in the laboratory seems inevitable. Such variation was also observed in samples from membrane elements. The test procedure was optimized to account for variations in surface distribution of biofilms and conditions in biomass analysis were identified for further optimization. A series of chemicals were used to determine effects on biomass removal from biofilm grown on PVC-P and biofilms grown on membranes. More than 1 log unit of reduction of the ATP concentration was observed with a number of chemicals, e.g. alkaline and detergents (SDS). The reduction of the CH concentration in most cases was less than 50%, and exceeded this value with alkaline in combination with detergents. Exposure of attached biomass to water at relatively high flow rates (>1m/s) and water/air mixtures showed maximum removals of about 60%. In general, the results demonstrated that more than 50% removal of attached biomass is difficult to achieve with commonly used chemicals. Similar results were obtained for membrane biofilms with both membrane samples in the laboratory test and cleaning in place (CIP) with intact membrane elements. Consequently, it was concluded that the test laboratory test procedure using biofilms grown on PVC-P is a promising tool to study and optimize membrane cleaning procedures. For more details see section 4.2.

4.1.3.3 Effects of chemicals on restoring membrane performance in laboratory tests and in a pilot plant

Various techniques were used to study the effect of cleaning on membrane performance under test conditions in the laboratory (cf. Section 4.3). These techniques included testing of membranes with artificial fouling layers, and fouled membranes from full-scale plants in a customized flat-sheet membrane test cell, with permeability and salts rejection as membrane performance parameters. Furthermore, lyophilized foulants scraped from full-scale membranes were investigated in batch tests to determine the effect of chemicals on the level of dissolution. In addition, a pilot plant enabling testing of cleaning procedures under full-scale conditions was developed and used.

The investigations revealed that artificial fouling layers, generated with constituents which are part of attached biomass were relatively easy to remove from the membrane surface. Most likely, the cohesion of the foulants in this layer was much less than in attached biomass grown on a surface. Chemicals generally had limited effects on the deteriorated permeability of fouled membranes collected from full-scale plants. The most pronounced effects were observed with the use of alkaline in combination with detergents. In a specific case, a reduction of the permeability was observed after the application of oxalic acid (pH 1.5) which was attributed to iron-oxalic acid complexes.

No effects were observed with enzymes, and the use of detergents in a number of cases caused a reduction of the permeability. Certain chemicals showed a clear removal of inorganic compounds, including iron, from the membrane surface or from lyophilized foulants. Oxalic acid (pH 2) was found to be highly effective for the removal of iron present on RO membranes from a selected full-scale desalination plant. However, the investigations also showed that the effect of a cleaning agent differs for different inorganic compounds, e.g. aluminum removal with sodium bisulfite (SBS, pH 2) was much more effective than oxalic acid and citric acid (pH 2) for the removal of Ca and Si. Consequently, (i) the choice of the most effective chemical for the removal of inorganic compounds depends on the composition of the foulants and (ii) complete removal of all inorganic constituents with one or a few chemicals will be difficult.

The limited effects on biofilms observed in the laboratory tests with membranes from full-scale plants were confirmed by applying different cleaning procedures on membrane elements from a full-scale plant in a pilot plant (cf. Section 4.4). A clear removal of inorganic constituents was observed, in particular when oxalic acid was used. However, the effect on feed channel pressure drop and membrane permeability was limited. Membrane autopsies revealed that the removal of organic foulants, mainly consisting of bacterial cells and biomass components (EPS) measured as carbohydrates was limited. The limited effects observed in the pilot plant were attributed to a relatively low level of fouling.

4.1.4 General discussion

4.1.4.1 Membrane-cleaning paradox

Controlling fouling of RO membranes is needed to reduce operational costs and ensure water production, and can be achieved by combining appropriate feed-water pre-treatment and membrane cleaning. The limited availability of quantitative information about membrane cleaning procedures and their effects in the scientific literature is in contrast with reports about the general occurrence of (bio)fouling and its impact on the operation of membrane installations (Khedr, 2000 and Schippers *et al.*, 2004). This paradox situation may have a number of reasons, including the trial and error approach applied in many installations, the tailor-made approach for each situation, but also commercial interests of owners of membrane installations and manufacturers of cleaning chemicals. Cooperation in (international) research projects supported by governments seems helpful in collecting the required information about membrane-cleaning. Collecting such information is needed to meet the growing world wide demands for freshwater for which desalination with membranes is the approach for the future.

4.1.4.2 Limited removal of attached biomass

The different studies described in sections 4.2, 4.3 and 4.4 demonstrate that it is very difficult to remove attached biomass from the surface of membranes and PVC-P. With the (combinations of) chemicals used, usually less than 50% (in comparison with the reference sample) of the attached biomass was removed. These observations are consistent with reports on the removal of pure cultures of bacteria from surfaces using chemicals and with reports about the attachment processes (Corpe, 1974; Zobell, 1943 and Bos *et al.*, 1999). Observations on the effect of cleaning on membrane permeability (in laboratory tests and in the pilot plant) and on the feed channel pressure drop (in the pilot plant) were consistent with the limited biomass removal experienced in the laboratory studies. Obviously, biomass was the main cause of the impaired membrane performance, despite effective removal of inorganic compounds. However, the limited removal which can be achieved may be sufficient in full-scale plants for restoring the desired level of production.

The results of the investigations described in this section indicate that of the tested compounds NaOH in combination with detergents at a high pH generally are most effective for biomass removal and certain acidic chemicals are effective in the removal of inorganic constituents present in the fouling layer. However, based on the obtained results it is not possible to define the most effective cleaning procedure, because of the differences between effects on inorganic constituents and the limited biomass removal.

4.1.4.3 Tools and tests

Progress was made with testing cleaning chemicals and procedures in laboratory experiments and in a pilot plant. A laboratory test method using biomass attached to plastic materials was optimized to obtain defined biomass concentrations, analytical procedures for biomass quantification and standardized conditions for testing (duration, temperature). Progress in cleaning procedures using chemicals, possibly in combination with making the membrane less vulnerable to (bio)fouling, may be made by using different combinations and different sequences. Testing of new approaches in full-scale plants on a trial and error basis is not attractive for reasons given above. However, the tools and tests described in the following sections can be used for the development and selection of more effective cleaning procedures, using chemicals. Tools and tests that can be applied for determining the effects of cleaning procedures include: (i) techniques for measuring the concentration of attached biomass and other foulants, (ii) a method for the production of representative samples of attached biomass, (iii) laboratory equipment to assess the (effects of cleaning procedures on) membrane performance characteristics, and (iv) protocols for pilot plant testing. Optimization and standardisation of the tests may be needed to promote application.

4.1.4.4 Restrictions in the use of chemicals

The use of certain chemicals, which may be effective for cleaning, is restricted by their damaging effects on the membrane and the impact on the environment. Permeability testing and measuring the salts rejection of the membrane, as done with test systems in the laboratory and in the pilot plant can be used for determining the compatibility of cleaning agents and procedures. The environmental aspects of the use of cleaning agents were not investigated but definitely have consequences for the use of certain chemicals. EDTA (at pH 12) was found to be highly effective for the removal of Si and iron, but certain countries have classified this compound and certain other products (e.g. specific detergents) as hazardous (Lattemann, 2010). Consequently, studies on the most effective cleaning agent should be done with compounds which do not damage the membrane or the ecosystem. Furthermore, other approaches, e.g. effect s of specific hydrodynamic conditions to induce high shear, with or without using air in combination with chemicals need further exploration (Cornelissen *et al.*, 2007). A final consideration related to the application of cleaning procedures is that cleaning should be balanced with pre-treatment. Consequently, optimisation of pre-treatment in combination with optimisation of cleaning may lead to the most cost-effective and reliable operation of the installation.

4.1.5 Conclusions and recommendations

A number of conclusions can be drawn from the information collected in literature review, full-scale plants and investigations in the laboratory:

- The nature and efficacy of cleaning procedures applied in full-scale plants is poorly documented in the scientific literature. Development of protocols for registration of cleaning procedures and efficacy in full-scale plants is needed to promote a more systematic evaluation of cleaning procedures in practice;
- Low levels of biomass removal were found in lab, pilot and full-scale installation ($<50\%$). The removal of biomass attached to the surface of membranes is hampered by the protective nature of EPS and other constituents (e.g. inorganic compounds) of the biofilm;
- Determination of the concentration of carbohydrates and ATP gives quantitative information about the total amount of attached biomass and the amount of active biomass, respectively;
- Laboratory tests can be used to determine and compare the effects of chemicals on the removal of attached biomass provided that biofilms are produced under standardised dynamic conditions;
- Assessment of the effects of chemicals on permeability and salt rejection in a test cell system is an attractive approach for determining the effects on membrane performance. Analysis of the amount and nature of the foulants on the membranes are helpful in assessing the main cause of performance loss;
- Compatibility and environmental impact limit the application of certain effective chemicals and balancing pre-treatment with cleaning is needed for the prevention and control of biofouling problems.

The following recommendations can be made based on the results of the study:

- Standardisation of laboratory tests for determining the effects chemical treatment on biofouling;
- Further testing of (combinations) of chemicals in standardised procedures to identify the most effective procedure for the removal of attached biomass. The tools developed in the investigations described in this report can be used for this purpose.

4.2 DEVELOPMENT OF A LABORATORY METHOD FOR TESTING MEMBRANE CLEANING PROCEDURES

4.2.1 Introduction

Periodic cleaning in place (CIP) is commonly applied for controlling biofouling of spiral-wound membranes (SWM) used for desalination and in water treatment to ensure water production. The collection of detailed quantitative information about effects of cleaning procedures at full-scale plants is an option to obtain the optimal cleaning conditions. Precondition is that the information is complete and collected systematically during daily practice, but inquiries show that this is not easy to establish and cleaning conditions are tailored to the local situation. An additional method to obtain the required information is described in this section. A laboratory test is developed which enables the quantification of the efficacy of membrane cleaning procedures to remove biomass from biofouled spiral-wound membrane elements.

4.2.2 Principle of the test

The major requirements for a robust and cost effective laboratory test which gives reproducible results on distinctive features with predictive value are: (i) standardized production of biofilm samples as surrogate for membrane biofilms, (ii) distinctive parameters to quantify the biofilm concentration; (iii) a simple test which simulates soaking and rinsing of a CIP procedure and (iv) the cleaning efficiencies (CE) calculated from this test using certain (combinations of) chemicals should give the same ranking as for the CE of the CIP procedures of spiral-wound membranes using the similar (combinations of) chemicals.

4.2.3 Production of biofilm samples

A biofilm production method was developed using segments of polymer tubing with attached biomass grown on the surface. The major tubing material used was polyvinylchloride (PVC-P) tubing (Ø 1.2 or 1.8 cm) because of its high biofilm formation potential (van der Kooij and Veenendaal, 2001). Polyamide (PA) and polypropylene (PP) tubing, which represent the materials of membrane and spacer, respectively, were used incidentally. Biofilm was grown on the inside surface of the polymer tubing by flowing non-chlorinated tap water (0.1–0.2 m/s) through these tubes (Figure 4.1). For PA and PP tubing biofilm formation was enhanced by supplementing acetate (100 µg C/l) to the water. The non-chlorinated tap water, produced from anaerobic groundwater, had a temperature range of 15–23°C and the average turbidity, pH and NPOC were 0.16 FTU, 8.2, and 2 mg C/l, respectively. To validate the results of the laboratory tests with biomass attached to polymer, additional tests were performed with membrane material from spiral-wound membranes (SWM), either in CIP tests with membrane elements or in a laboratory batch test with SWM samples obtained by sampling for membrane autopsies.

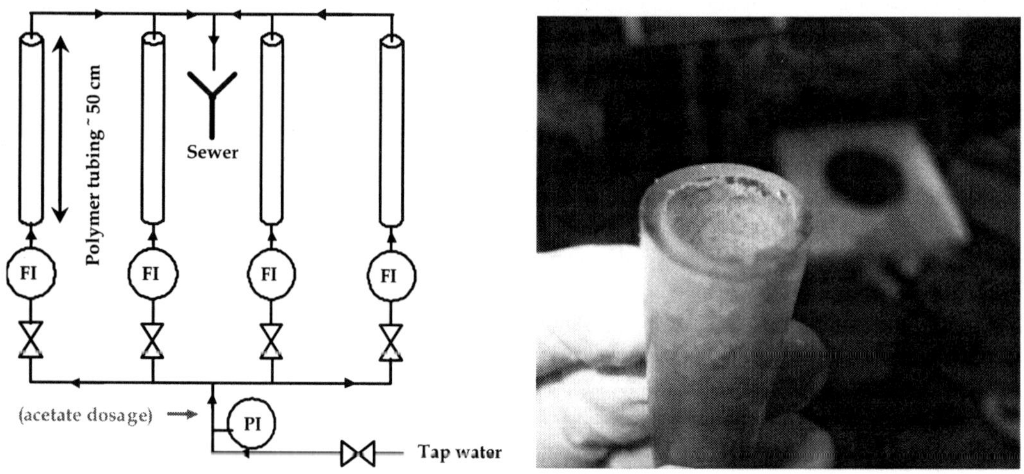

Figure 4.1 Biofilm production unit (BPU) with PVC-P tubes and an example of a biofilm on the surface of PVC-P tubing

4.2.4 Biofilm samples and biomass concentrations

A large variety of methods is available to quantify biomass concentrations. These methods include heterotrophic plate counts and total direct microscopic cell count (TDC), but also biochemical methods aiming at quantifying the constituents of bacterial cells and their excreted extra-cellular materials, e.g, adenosinetriphosphate (ATP), carbohydrates (CH), proteins, DNA, phospholipids, and muramic acids (Ridgway *et al.*, 1983; Flemming *et al.*, 1993; Ridgway and Flemming, 1996 and Vrouwenvelder *et al.*, 1998). CH measured with the method of Dubois (1956) and ATP were selected as the biomass parameters. The selection of these parameters was based on experimental data from a model membrane biofouling study showing correlation between these parameters and the normalized pressure drop in spiral-wound membranes (Hijnen *et al.*, 2010). ATP quantifies the active biomass in the biofilm and with the CH method the carbohydrates in the bacterial cells and in the extracellular material is determined. The set of parameters was supplemented with TDC as a microbiological parameter.

Microbial growth on the surface of the PVC-P tubes was clearly demonstrated by an increase of the ATP concentration from around 1 ng.cm^{-2} after two weeks to a more stable concentration of 20–60 ng.cm^{-2} after four weeks.

This parameter detected biofilm growth on the PVC-P after one month due to a lower level of detection CH content in the biofilm (5–10 μg.cm^{-2}) (Figure 4.2A).

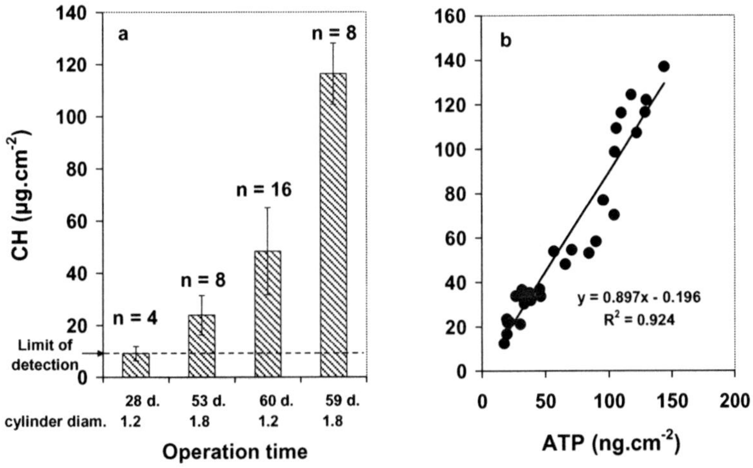

Figure 4.2 (A) The carbohydrate (CH) concentration of biofilms on PVC-P samples (Ø1.2 or 1.8 cm) after different periods of operation in the biomass production unit and (B) the relationship between ATP and CH concentrations of the PVC-P biofilm samples

The average ATP concentration in the biofilm samples used for cleaning tests ranged from 30–140 ng.cm^{-2}, depending on the operation time of the biomass production unit. Average CH concentrations ranged from 25–180 μg.cm^{-2} (>30 days of operation time). ATP and TDC were positively correlated but the relationship was not proportional, due to the varying ATP content in bacterial cells. Moreover, microscopic images of TDC showed clustered cells, especially in biofilm samples after cleaning which hampers an accurate quantification of biomass. From the relationship between ATP and CH in the biofilm, a ratio of 0.9 ± 0.05 μg CH per ng ATP was calculated (Figure 4.2b). Biofilm samples from spiral-wound membranes (SWM) used for validation tests had ATP concentrations ranging from <1–2 ng.cm^{-2} and CH concentrations ranged from <10–50 μg.cm^{-2}. Part of these samples were collected on site, transported and stored before use.

4.2.5 Cleaning test procedures

Three different tests were applied to determine the effects of cleaning procedures. The first set up consisted of flow set up using fixed PVC-P cylinders (Ø1.2 cm; 1.6–5.0 cm) with biofilm to test the effect of air-water cleaning (AWC). The cylinders were flushed with demineralized water at ambient temperature and different velocities (0.6–3.7 m/s) with or without air supply (0–7.4 m/s) to test effects of different air/water-ratios. Batch cleaning test (BCT) with biofilm samples (polymer tubing or SWM samples) were carried out in 200 ml solution agitated on a orbital shaker at 100 RPM during 1 hour and rinsed in 200 ml milliQ-water under the same conditions during 15 minutes, thus simulating soaking and mild rinsing. A column cleaning test (CCT) was used to verify the effect of flushing conditions after soaking (0.11 m/s; 15 minutes non chlorinated tap water). Prior to flushing the PVC-P biofilm cylinders in the column were soaked either in milliQ-water (reference) or in the cleaning solution during 20 minutes at room temperature.

The cleaning efficiency (CE) of the different tests was calculated from the reduction in biomass concentrations as (i) the log units reduction of the ATP concentration and (ii) the percentage reduction of CH concentration. The untreated or reference biofilm samples in the AWC and the first batch-cleaning tests (BCT) were randomly cut biofilm samples from the produced biofouled surface. Because of high variability in these data in the subsequent BCT and CCT protocols an adjoining biofilm sample next to the cleaned biofilm sample was treated in milliQ-water and used as reference biofilm sample for calculations of CE.

4.2.6 Cleaning efficiency for biofilms on polymer tubing

ATP was assumed to be a suited biomass parameter to assess cleaning efficiencies in the AWC experiments, because in the absence of chemicals no enhanced leakage of ATP from cells occurs. A maximum ATP reduction of 0.6 log (75% removal) was observed for PP samples with biofilm, whereas the removal from PVC-P did not exceeded 0.3 log (50%, Figure 4.3). Increasing the air supply (air/water A/W ratio) had only a positive effect for PVC-P at an A/W ratio of 1.5 and 2.0.

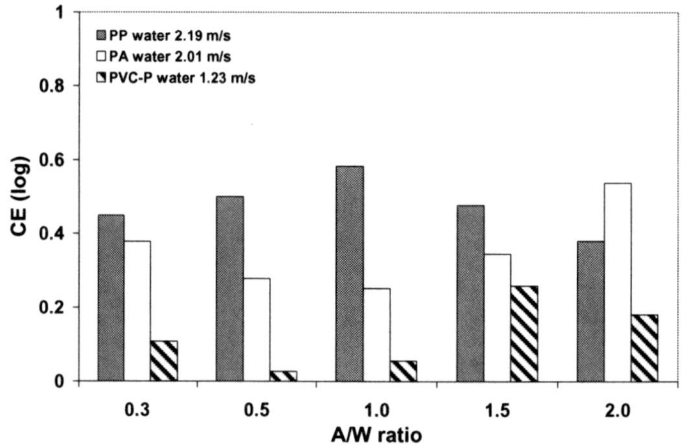

Figure 4.3 The effect of the air velocity at a constant water flow (see legend) on the reduction of the ATP concentration of biomass attached to PVC-P

In membrane cleaning it is common practice to apply multiple cleaning treatment using alkaline, detergent and acid steps in single or combined solutions with variable sequences. In a batch test with biofilm grown on PVC-P (60 days) the cleaning efficiencies were determined in milliQ water (control) and in MilliQ with NaOH, sodiumdodecylsulphate (SDS) and citric acid in one and in two steps with variable sequences (concentrations of 1%). The control showed some reduction of ATP and CH (Figure 4.4).

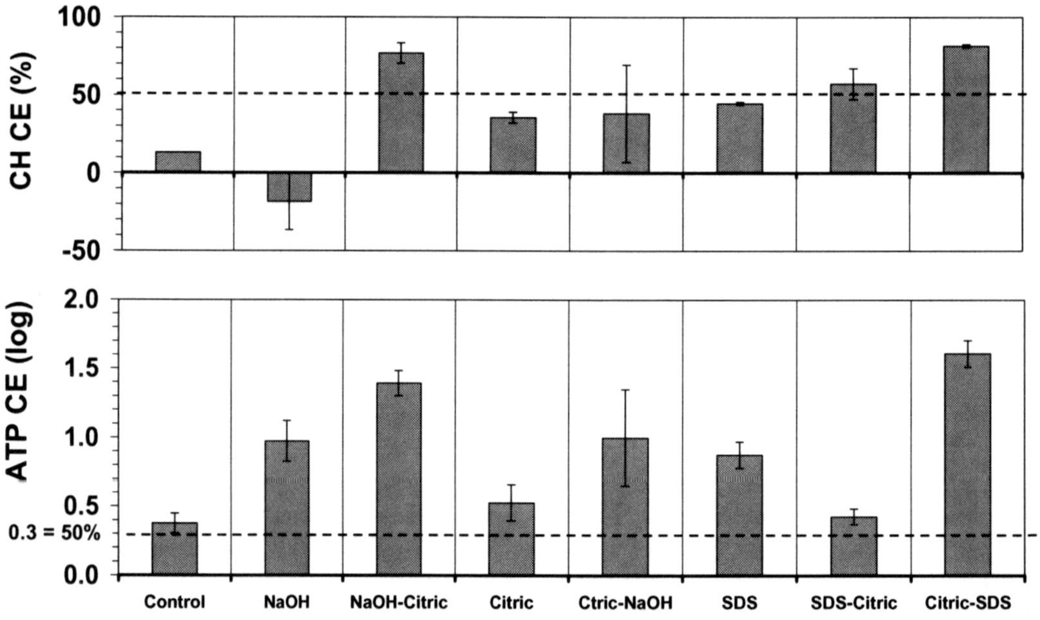

Figure 4.4 The cleaning efficiency for treatment of PVC-P biofilms with NaOH (1%, pH 12.7), SDS (1%, pH 7.1) and citric acid (1%, pH 2.2) in the batch test in one step or in two steps with different sequences (error bars show the variation in biofilm concentrations of samples of one test)

ATP reduction for all treatments, including the control, was >50% (0.3 log) and was highest after the two-step treatment with citric acid + SDS and by the treatment with NaOH + Citric. Except for the two-step treatment with SDS + citric acid all two-step treatments showed an increase in ATP reduction. The NaOH treatment caused an unexplained negative CE for CH. The other CE values ranged from 35.5% (citric acid) to the highest removal of about 80% for the two-step treatments with citric acid + SDS and NaOH + citric acid. These trends were also observed for ATP.

The same chemicals were tested in column tests with additional flushing (15 min) after soaking. No increase in cleaning efficiency was observed despite the additional flushing at velocities observed in SWM (0.1 m/s). In most of these column tests the reduction of the ATP concentration was less than 50% (0.3 log) and the reduction of the CH concentration was less than 20%.

4.2.7 Validation tests

A major requirement of the cleaning test is that the test is predictive for cleaning efficiency and biomass removal by CIP procedures in spiral-wound membranes. In the batch test flushing (shear) is mild compared to the flushing in the membrane elements at flow rates of 0.1–0.2 m/s. In the column test, however, the effect of the additional flushing on biomass removal was negligible as presented above. A pronounced difference between CIP in membrane elements and laboratory tests is the accessibility of the biofilm. In the laboratory test the biofilm is optimal accessible for chemical solutions compared to the accessibility of the surfaces of feed channel in SWM elements during a CIP. To assess the predictive value of the laboratory test validation is necessary. A few comparative tests were carried out.

The cleaning procedures (CP) presented in Table 4.1 were applied as CIP procedures with membrane elements in a pilot plant (CIP-1) and at bench scale (CIP-2) and compared with similar cleaning procedures in the laboratory batch test (BCT) and the laboratory column test (CCT), respectively. All CPs were two-step treatments. CP #1 is an alkaline + acid, #2 a detergent + acid, #3 a detergent/alkaline + acid and #4 an acid + detergent treatment.

Table 4.1 Cleaning procedures (CP) applied in the CIP tests with SWM elements and in the laboratory tests with SWM and PVC-P biofilms

CP	Tests[a]	First step		Second step	
		Agents	pH	Agents	pH
#1	CIP-1	NaOH 0.1%	12	HCl 0.2–0.5%	2.0
	BCT/CCT	NaOH 1%/NaOH 0.05%	12.7/12	Citric 1%	2.2
#2	CIP-1	STPP 2%+NaDBBS 0.025%	10	Citric 2% DBNPA 0.05%	2.0
	BCT	SDS 1%	7.1	Citric 1%	2.2
#3	CIP-1/CIP-2/BCT	NaOH 0.1%+SDS 0.03%	12	Oxalic 1%	1.5
#4	CIP-2/BCT	Citric acid 1%	2.2	SDS 1%	7.1

[a]*CIP = cleaning in place (elements); BCT = batch test and CCT = column test.*

The data presented in Table 4.2 show similarities but also differences in CE for the two parameters under different test conditions (tested biofilms and cleaning procedures) with no systematic pattern. Overall from this table and from the results described above it can be concluded that the cleaning efficiencies observed in the batch tests using PVC-P biofilms are in the same order of magnitude as observed for SWM material. Furthermore, the results show that in most cases ATP removal was more than 1 log unit and CH removal was generally as less than 50% (0.3 log unit). Chemical cleaning results in biomass removal and inactivation and therefore ATP reduction overestimates biofilm removal and CH reduction is a more appropriate parameter to quantify CE. Only in four out of eleven cleaning tests the CH removal was >50%, viz. three laboratory batch tests on PVC-P biofilm with CP #1, #2 and #4 and one CIP tests with SMW biofilm (CIP-2). The higher CH removal observed in the laboratory batch test may indicate the beneficial effect of a better accessibility of the biofilm, but this may also caused by higher biofilm densities on the PVC biofilm samples as deduced from the CH concentrations described before. Nevertheless, the results with SWM biofilms are consistent with those of the laboratory tests and demonstrate that biofilm removal of more than 50% is hard to achieve with CIP procedures in spiral-wound membranes. The wide range of CE values presented in Table 4.2 demonstrates a high variability and no data on reproducibility were collected. This hampers the drawing of decisive conclusions. Biomass quantification with CH analysis in inhomogeneous biofilm samples is identified as one of the possible causes for this variability.

Table 4.2 The cleaning efficiency (CE) of the different CIP procedures tested with SWM and PVC-P biofilms and determined in different tests based on ATP and CH; variation is presented as the range of 2–4 separate biofilm samples from one test

CP	Biofilm substratum	Test[a]	CE ATP (log) AVG (range)	CE CH (%) AVG (range)
#1 Alkaline – acid	SWM	CIP-1	0.3 (0.3–0.3)	4 (−22–31)
#1 Alkaline – acid	PVC	CCT	0.4 (0.4–0.4)	8 (2–15)
#1 Alkaline – acid	PVC	BCT	1.4 (1.3–1.5)	77 (70–83)
#2 Detergent – acid	SWM	CIP-1	2.2 (2.1–2.4)	28 (27–29)
#2 Detergent – acid	PVC	BCT	0.4 (0.4–0.5)	57 (48–67)
#3 Det./alkaline – acid	SWM	CIP-1	0.5 (0.3–0.7)	44 (31–56)
#3 Det./alkaline – acid	SWM	CIP-2	1.2 (1.1–1.3)	21 (7–36)
#3 Det./alkaline – acid	SWM	BCT	1.8 (1.8–1.8)	29 (26–31)
#4 Acid – detergent	SWM	CIP-2	1.2 (0.9–1.6)	57 (39–74)
#4 Acid – detergent	SWM	BCT	1.2 (1.2–1.2)	12 (−5–29)
#4 Acid – detergent	PVC	BCT	1.6 (1.5–1.7)	82 (81–83)

[a]*CIP with intact SWM elements and BCT/CCT laboratory test with separate biofilm samples.*

4.2.8 Discussion and conclusions

The results of the current study showed that:

- removal of attached biomass was rather limited, viz. CE <0.5 log ATP and <0.2 log CH (50%) for biofilms attached to PVC-P or SWM material in laboratory tests and CIP tests with intact SWM elements;
- chemical treatment and mild rinsing in a batch test were more effective in removing biofilms than intensive mechanical cleaning with air/water flushing;
- biofilm removal observed in CIP and laboratory protocols with SWM and PVC-P biofilms were in the same order of magnitude. This demonstrates that the laboratory test protocol can be used to evaluate biofilm removal efficiency of membrane cleaning procedures;
- different cleaning procedures did not show distinctive differences in biomass removal. However, the number of tests was too limited and the variability in the observed CE values was too high to draw decisive conclusions about effects of different chemical under different conditions (concentration, combination, time and temperature). Furthermore, no systematic data on the effect of individual chemicals were collected which hampers comparison with literature data.

The production of a reproducible and homogenous biofilm and the quantification of biomass concentrations with reproducible analytical methods are preconditions for a robust laboratory test with distinctive features. This study confirmed that ATP is a suitable measure for the amount of active biomass present on a surface and can be used to study mechanical cleaning procedures. However, when chemicals are used, the parameter is no longer proportional to the total amount of biomass, because damage of cellular membranes lead to release of ATP from the cells. CH analysis appears to be valuable, but the CH concentration may vary in biofilms depending on age. Consequently, this put demands on the biofilm production period. Attached biofilms with an age of about 2 months were found suitable for testing. Application of total direct cell count was hampered by conditions at which TDC values were not representative for the amount of biomass.

On the basis of the results the batch test in a rotary shaker was selected as a preferred laboratory test: PVC-P biofilm samples are mildly agitated in a cleaning solution on an orbital batch for one hour and rinsed for 15 minutes under the same conditions. The results can be verified with SWM biofilm samples. The cleaning efficiency is assessed with CH analysis. Since ATP is a cheap and easy to determine biomass parameter and ATP reduction showed similar trends in biomass removal this parameter is a valuable additional parameter. Future research must be focused on optimization of biofilm production and biomass quantification and on the assessment of the reproducibility of the laboratory test. More systematic studies with different cleaning procedures and comparative studies with CIP tests are required to select the most optimal cleaning procedures for biofilm removal and validate the results for the membrane cleaning practice.

4.3 EFFECTS OF CHEMICALS ON MEMBRANE PERMEABILITY AND FOULANTS IN LABORATORY TESTS

4.3.1 Introduction

Biofouling of spiral-wound membranes can result in an increase of the pressure drop (NPD) in the feed channel and/or in a reduction of the membrane flux. In both cases, cleaning of the membranes is needed to prevent a reduction of the production capacity. Assessment of the effects of cleaning in full-scale plants is difficult and detailed information about effects of various cleaning procedures is lacking.

To be efficient a cleaning agent should minimize the intermolecular bindings between the foulant constituents and inhibit the interactions between the foulant species and the membrane surface. The efficiency of the cleaning agent will then depend on the foulant composition (organic/inorganic constituents) and the chemical nature of the membrane surface. Foulant characterization showed that deposits isolated from all RO studied membranes mainly consists of organic foulants (the residue after combustion represents an average of 25 to 35% in mass of the deposit) with a composition in biopolymers corresponding to a biofilm matrix (polysaccharides and proteins) and presence of multivalent cations (Fe, Al, Ca). According to Hong and Elimelech (1997), multivalent cations like Fe^{3+} and Ca^{2+} are the promoters of membrane fouling by humic substances as they increase their adhesion force by minimizing their negative charges. Silica is also suspected of enhancing fouling due to its interaction with humic material and lower molecular weight acids or/and by polymerization. EPS, the major constituent of the RO fouling layer strongly interact with minerals. Referring to Omoike *et al.* (2004), Kwon *et al.* (2006) indicate that orthophosphates and phosphate groups in EPS are known to form covalent bonds to Fe-hydroxides. Results obtained by Kwon *et al.* (2006) also demonstrated high binding strength between major components of EPS and silica. These authors mentioned that once the phosphate-containing polymers bind to a silica surface, higher energy is needed to separate the polymers from silica surfaces due to the strong H-bonds as compared to other polymers. Zhu *et al.* (2009) indicate that higher protein content in EPS favours interactions with silica. Moreover Al and Si can also precipitate as aluminum silicates on RO membrane (Greenlee *et al.*, 2009). Strong interactions between inorganic species and organic materials (i.e. humic substances and EPS) facilitate the development of biofilm and its high stability. Hence, the need of breaking inorganic/organic bindings appears evident with the objective to remove fouling layers from RO desalination membrane surface.

Therefore laboratory tests were developed and applied to evaluate the impact of cleaning procedures. The laboratory test described in Section 4.2 was dealing with the removal of attached biomass from the surface, and analysis of the biomass concentration before and after cleaning. This section deals with test for assessing the effect of cleaning on the membrane permeability, which is directly related to the water flux through the membrane. In addition to determining the effects on permeability also the impact of the cleaning procedure on the foulants and on the salt retention was tested.

IWW performed cleaning test on mixed fouling layers and on fouled RO membranes from a full-scale plant using a bench-scale membrane test unit. LCME conducted batch experiments directly on deposits (scraped and lyophilized) to select the most efficient cleaning agents. Selected agents were then tested on fouled membranes received from RO plants using a RO bench-scale unit. The aim of the present work was to compare the efficiency of cleaners/cleaning sequences on fouled RO desalination membrane isolated from full-scale plant.

4.3.2 Methods and materials

4.3.2.1 Equipment and test procedures at IWW

The efficacy of certain cleaners and cleaner combinations towards different fouling layers was determined using (i) model fouling layers and (ii) natural fouling layers on RO membranes from full-scale plants using a flat-sheet membrane test cell. Four different cleaner combinations were tested which were known from literature and also recommended by membrane manufacturers.

4.3.2.1.1 Mixed fouling layers

Mixed model fouling layers consisting of bacterial cells, EPS model substances (e.g. alginate) latex beads and laponite clay were generated and tested (Table 4.3). For the generation of the mixed model fouling layer a mixture of all four components was prepared in tap water. A final concentration of 200 mg/l of each of the components was adjusted in an 80 l feed water volume. Bacterial cell mass was added at a final wet weight concentration of 125 mg/l. Before adding the bacteria into the feed water tank, the cell mass was suspended in a small amount of tap water by vigorous mixing. Cell separation was checked by light microscopy. The feed water mixture was filled to the feed water tank and was constantly mixed by a stirrer to avoid sedimentation of the particles. The soaked BW30 membrane sheets were assembled in the 3 channels together with the standard FILMTEC™ feed and permeate spacers. The operation of the system was started in cross flow mode with fully opened concentrate recycle valve. Subsequently, the concentrate recycle valve was closed until a feed pressure of 10 bars was reached. Permeation was run for 2 hours resulting in an increase of feed pressure up to 22 bars. Afterwards, the test cells were disassembled and the membrane sheets with the fixed model fouling layer were place in wet conditions together with the feed spacer into a plastic bag and were sealed for short time storage at 4°C. These membrane sheets were subsequently used for the cleaning experiments described below. For comparison, untreated pieces were analyzed with respect to the fouling layer composition.

Table 4.3 Components used for fixed model fouling layers

Component	Model for
Drinking water bacteria	Bacteria
Alginate	Extracellular polymeric substances (EPS)
Humic substances	Organic fouling
Calcium carbonate particles	Scaling
Clay particles	Inorganic particulate fouling

4.3.2.1.2 Test cell

For cleaning experiments a customized flat sheet membrane test cell system (MemCell, ABB Maschinenfabrik Meyer AG, Switzerland) was used. The feed channel dimensions of the test cell are 40 mm × 200 mm × 0.8 mm resulting in an active filtration area of 80 cm². A permeate collection plate made of sintered steel was installed in the cell. Standard feed spacers from FILMTEC™ were used in combination with the BW30 membranes (Dow Chemical Company). A stainless steel heat exchanger coupled to a thermostat (HAKE F3) was installed to the feed water circle to maintain a feed water temperature of 20ºC or to obtain a cleaning solution temperature of 35ºC during cleaning experiments. Thus, membrane characterizations and cleaning experiments were possible under defined and reproducible conditions.

4.3.2.2 Chemicals and cleaning sequences

The chemicals used for the cleaning are presented in Table 4.2. The same cleaning protocols were used on mixed model fouling layers and the fouled membranes from full-sale plants (site D) and in pilot studies. In a first step the five alkaline/detergent cleaners were tested separately according to steps 1 to 5 of the procedure (Table 4.4). Subsequently, the most effective alkaline cleaner was selected used in combination with the 7 acidic cleaners for the full cleaning strategy given in Table 4.5.

Table 4.4 Cleaning agents used in batch tests

Type of cleaning		Chemical agent	Abbreviation	Concentration (w/v %)	pH
Alkaline	a	Sodium hydroxide	NaOH	0.1	12
	b	Sodium hydroxide + Sodium Dodecyl Sulfate	SDS	0.1 0.03	12
	c	Ethylene di-ethyl amine tetra acetate	EDTA	1	12
	d	Sodium tripolyphosphate + Sodium dodecyl benzenesulfonate	STPP	2 0.025	10
	e*	Ultrasil-67 + Ultrasil-69	Ultrasil	0.5 1	9.5
Acidic	1	Hydrochloric acid	HCl	0.2–0.5	2
	2	Sodium bisulfite	SBS	1	2
	3	Citric acid	Citric	2	2
	4	Sodium hydrosulfite	SHS	1	2
	5	Sulfamic & citric acid	Acids	3	2
	6	Nitric & phosphoric acid	–	0.5	2
	7	Oxalic acid	Oxalic	1	2

**Ultrasil mixture contains enzymes i.e. protease.*

In the first run, membranes from site C and site D with "natural" fouling layers obtained from the autopsies in WP3 were investigated. Three cleaners were selected according to the guidelines from the membrane manufacturers. In detail, cleaning was performed with (1) 0.1% NaOH & 0.03% SDS, (2) 2% STPP & 0.8% EDTA, (3) 2% citric acid and (4) deionized water as control. The system was recirculated with the cleaning solution for 5 min followed by a soaking time of 60 min. Afterwards, the cleaning solution was flushed out using deionized water at maximum cross flow.

4.3.3 Analysis of inorganic compounds

ICP-OES analyses were performed for the determination of the inorganic fouling layer components. For this purpose, pieces of approx. 10 cm² were cut from the membrane samples before and/or after cleaning. These membrane pieces were directly used for the digestion procedure. At IWW sample mineralization is performed by microwave digestion using aqua regia (3 ml HCl and 1 ml HNO_3) (Hoenig, 2001). The obtained solution is completed with ultra pure water (MilliQ) to a fixed volume. Measurements were performed using an ICP-OES Vista Pro (Varian).

4.3.3.1 Procedures and equipment at LCME

4.3.3.1.1 Batch experiments performed with lyophilized foulants

Deposits accumulated on RO membrane harvested from site D modules were then scraped from the membrane surfaces with a plastic spatula in the presence of ultrapure water. The mixed solution ultrapure water/deposit was introduced in a round bottom flask and lyophilized. Lyophilized deposits were used to perform batch experiments after extensive characterization (inorganic and biopolymer composition).

Table 4.5 Cleaning sequences applied to fouled membranes

Step		Chemicals	Temp.	Mode	Flow rate*)	Duration
1		NaCl solution (1% w/v)	20°C	Recirculation & permeation	1.0 × Q	60 min
Measurement of permeate flux and salt retention						
2	Hot rinsing	Deionized water	35°C	Flushing	1.1 × Q	20 min
3	Alkaline/detergent step	Alkaline cleaning solution	35°C	Flushing	0.5 × Q	5 min
			35°C	Soaking	none	60 min
			35°C	Flushing	0.5 × Q	5 min
			First flush collected for cleaning solution analysis (optional)			
			35°C	Recirculation	0.5 × Q	30 min
			20°C	Soaking	none	18 h
			20°C	Recirculation	1.0 × Q	60 min
			First flush collected for cleaning solution analysis (optional)			
4	Rinsing	Deionized Water	20°C/30°C	Flushing	1.1 × Q	20 min
5		NaCl solution (1% w/v)	20°C	Recirculation & permeation	1.0 × Q	60 min
Measurement of permeate flux and salt retention						
6	Acid step	Acidic cleaning solution	20°C	Flushing	0.5 × Q	5 min
			20°C	Soaking	none	60 min
			20°C	Flushing	0.5 × Q	5 min
			First flush collected for cleaning solution analysis (optional)			
			20°C	Recirculation	0.5 × Q	30 min
			20°C	Soaking	none	90 min
			20°C	Recirculation	1.0 × Q	60 min
			First flush collected for cleaning solution analysis (optional)			
7	Rinsing	Deionized water	20°C	Flushing	1.1 × Q	20 min
8		NaCl solution (1% w/v)	20°C	Recirculation & permeation	1.0 × Q	60 min
Measurement of permeate flux and salt retention						

*) *Q corresponds to the regular flow rate of an 8 inch module resulting in a tangential flow of 0.2 m/s.*

Deposits were subjected to various cleaning tests in batch. 50 mg of lyophilised deposit were treated with 500 mL of cleaning solution in a 1,000 mL beaker under controlled temperature (35°C) and stirring (500 rpm). The cleaning contact time was arbitrarily fixed at 24 hours. It is understood that such long period of cleaning might not be appropriate with regard to industrial application however results have to be interpreted as a potential cleaning effect. At the end of cleaning, deposit and cleaning solution were separated by centrifugation (4,000 rpm, 10 minutes). Recovered deposit was severally washed: deposit was put in 200 mL of ultrapure water under stirring and the mixed solution was centrifuged again. This step was repeated several times (depending on the cleaning agent) to eliminate residual contents of cleaning agent. Completion of washing was confirmed by checking the conductivity of water after washing (<100 µs/cm). Deposits were then lyophilised prior to analyses.

Lyophilised deposits were analysed by ICP-OES after acid mineralization (Mondamert *et al.*, 2009) and also subjected to pyrolysis GC/MS analysis (Bruchet *et al.*, 1990)

4.3.3.1.2 Cleaning sequences operated at bench scale

The objective of this part of the work was to determine the efficiency of two cleaning sequences at bench scale using coupons of harvested fouled RO membrane (1st stage/1st pass of the RO unit) from site D (presented at chapter 3.1).

Two cleaning sequences were tested; Sequence-A performed with NaOH and HCl and Sequence-B combining STPP and oxalic acid in series. Solutions of cleaning agents were prepared as indicated in Table 4.2 and used on membrane coupons harvested from the module recovered from site D.

The first line (in black on Figure 4.5) was used to measure the permeability of the membrane. Permeability measurements were carried out at a constant cross-flow velocity of 0.2 m/s, corresponding to a concentrate flow Q of about 1 L/min with synthetic seawaters using a transmembrane pressure (TMP) of 60 bars. The second line (in grey on Figure 4.5) was used to recirculate the cleaning solution on the membrane surface. Cleaning tests were performed at low pressure in absence of permeation (no permeate production). The recirculation of the cleaning solution was operated with a low pressure pump connected to a 2 L tank containing the cleaning solution. The cleaning solution is continuously under stirring in the tank and the temperature is controlled.

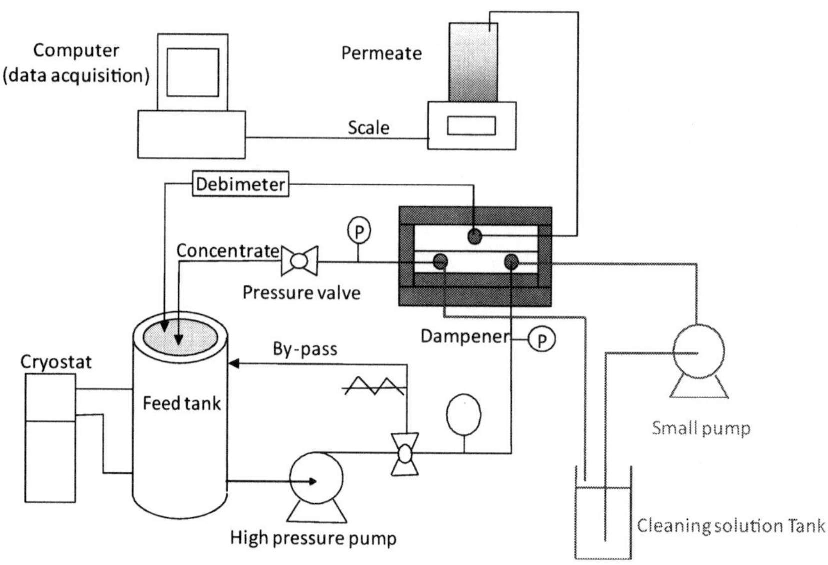

Figure 4.5 Modified setup of the RO membrane bench-scale unit used for cleaning experiments

In accordance with industrial practice, bench test experiments consisted of a two step cleaning sequence; an alkaline cleaning followed by an acidic one (Table 4.5) performed in a similar way with three successive operations. Before cleanings, a previous flushing was performed at 1.1 Q (1 L/min) with ultrapure water. Cleaning solutions were then recirculated on membrane surfaces at 0.5 Q (no permeation) before soaking (Q = 0). During alkaline cleaning experiments the temperature was maintained at 35°C while acidic cleanings were conducted at 20°C. It can be noticed that for both alkaline and acidic procedures two soaking steps were performed in order to reach the maximum efficiency of the cleaning procedure. The overall contact time in alkaline condition ranged from 17 to 20 hours. A shorter contact time i.e. 4 hours was used with acids.

Permeability and salt rejection measurements were performed before and after each cleaning step. Permeability measurements were carried out with synthetic seawater (adapted from Grasshoff, 1976), temperature was controlled at 20 ± 2°C. Salt rejections were determined based on conductivity measurements of both feed and permeate waters.

At the end of the whole cleaning sequence (alkaline + acidic cleaning), the membrane coupon was retrieved from the cell and deposit was gently scrapped and lyophilized to perform ICP-OES (Ca, Fe and Si essentially, after an acidic mineralization) and pyrolysis GC/MS analyses. During all the cleaning procedure it is important to notice that the membrane remains in the filtration cell to avoid any risk of deterioration.

4.3.4 Results and discussion

4.3.4.1 Impact of cleaning agents on foulant characteristics: batch experiments (LCME)

At full scale acidic cleaning is generally operated after alkaline cleaning. Experiments performed in batch mode on foulant suspensions (results not shown) showed that EDTA (alkaline cleaning) was the most efficient cleaner for Fe and Ca removal while STPP (Sodium tripolyphosphate) was also efficient to remove Ca (out-competes the performance of NaOH and SDS). None of the alkaline agent used had a significant impact on the relative abundance of Si. Because the use of EDTA at full scale might be prohibited the benefit of acidic cleaning was evaluated after STPP cleaning. For acidic cleaning oxalic acid exerted high Fe removal efficiency. If STPP did not significantly affect the biopolymer distribution of the foulant, almost all acidic cleaners, especially oxalic acid, led to preferential removal of lower molecular weight sugars and proteins.

4.3.4.2 Permeation tests with mixed fouling layers (IWW)

A mixed model fouling layer was generated on FILMTEC™ BW30 membrane sheets (Spettmann *et al.*, 2008). The extent on model fouling as well as the general composition as determined by dry weight content, loss on ignition, ICP-OES analyses and total cell count determination was within the range of fouling deposits observed on membranes autopsied from full scale RO plants performed in MEDINA WP 3 (Mondamert *et al.*, 2009 and Schaule *et al.*, 2009). The applied cleaning procedures removed inorganic deposits by approx. 93%–99% and bacteria were removed by around 99% (2 log units) as compared to the mixed model fouling layer before cleaning (Figure 4.6). This result was confirmed by SEM/EDX analyses. The most effective cleaning was achieved by using NaOH/SDS as alkaline/detergent cleaning step followed by oxalic acid as acidic cleaning step.

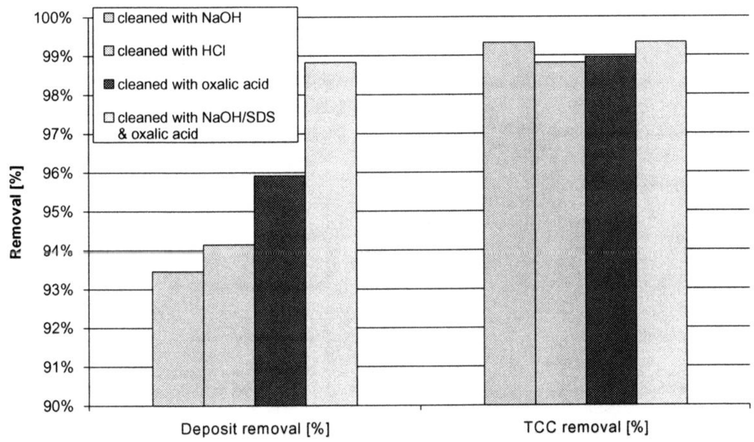

Figure 4.6 Removal of deposit and total cell counts (TCC) in the mixed fouling layers. This figure is available in colour at http://www.iwapublishing.com/template.cfm?name=isbn9781843393214

The acidic cleaning with hydrochloric acid resulted in a permeability increase of 17%. Alkaline cleaning resulted in a permeability increase of 56%. These cleanings resulted in a permeability of 5.5 l/m^{2}hbar and 5.7 l/m^{2}hbar, respectively. However, cleaning procedures with oxalic acid and NaOH/SDS & oxalic acid resulted in a permeability decrease of 6% and 4%, respectively, both resulting in 4.5 l/m^{2}hbar permeability.

Figure 4.7 shows that the salt retention was slightly improved after cleaning with HCl and NaOH. The achieved values of 99.0% and 99.3%, respectively, are in the range given by the manufacturer. The cleaning procedures with oxalic acid and NaOH/SDS & oxalic acid resulted in a loss of salt retention to a value of approx. 83%. This could be due to an alteration or damage of the membranes caused by the oxalic acid treatment and should be investigated further.

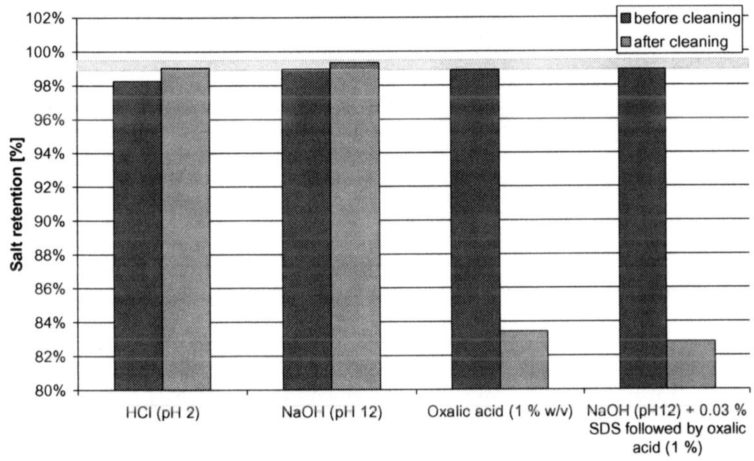

Figure 4.7 Salt retention of fouled membranes before and after cleaning. (The recommended salt retention as given by the manufacturer is 99.0%–99.5% and is marked as light blue bar in the Figure). This figure is available in colour at http://www.iwapublishing.com/template.cfm?name=isbn9781843393214

4.3.4.3 Effects on membranes from full-scale plants

4.3.4.3.1 IWW contribution

The performance of the fouled membranes as well as of the membranes after alkaline/detergent cleaning was measured with respect to the permeate flux and the salt retention. The increase in permeability compared to the membrane before cleaning and the salt rejections observed are compiled in Figures 4.8 and 4.9, respectively. The mean value for permeability before cleaning was found to be 1.17 ± 0.15 l/m^{2}hbar. The highest increase in permeability was caused by cleaning with NaOH in combination with SDS and by STPP/NaDDBS (Figure 4.8). The permeability increase was slightly lower with NaOH without SDS. Na_4EDTA and P3-Ultrasil showed only minor increases in permeability.

The effects of the alkaline treatment steps on the foulants on the membrane presented in Figure 4.8 show that treatment with NaOH/SDS gave the highest removal of the inorganic compound on the fouling layers.

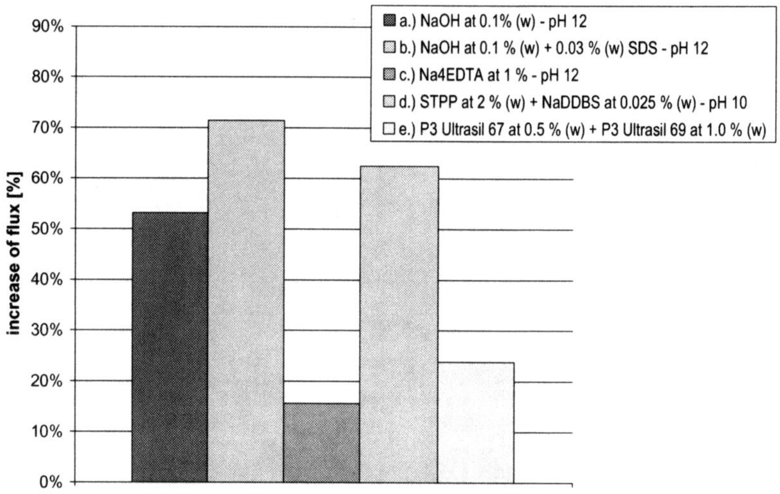

Figure 4.8 Increase in permeability after alkaline/detergent cleaning

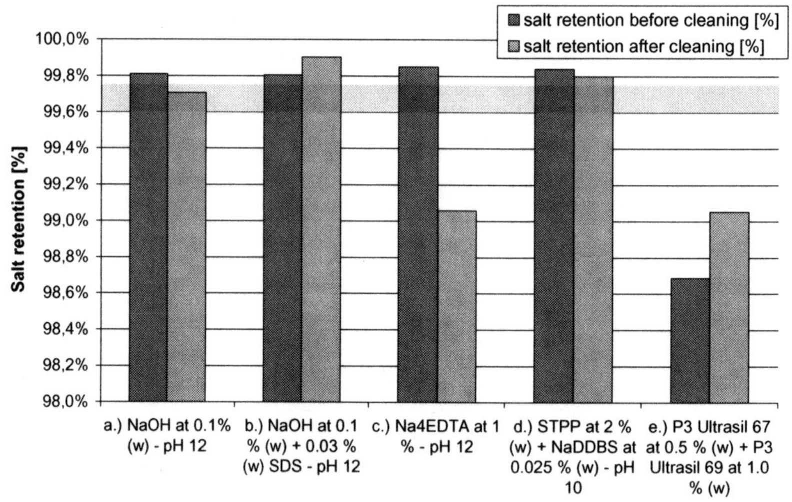

Figure 4.9 Salt retention of fouled membranes after alkaline/detergent cleaning. (The recommended salt retention as given by the manufacturer is 99.6%–99.75% and is marked as light grey bar in the figure)

Combined treatment viz. alkaline treatment followed by acidic cleaning was conducted by treating first either with NaOH or NaOH/SDS and subsequently by a variety of acidic cleaning steps listed in Tables 4.4 and 4.6. This combined treatment had only limited effect on the permeability of the membrane, either causing a decrease or an increase of about 10%. One exception was with NaOH/SDS treatment followed by sulfamic + citric acid, which gave a permeation reduction of 20%. Furthermore, the effects on salt retention were very limited and no decline of salt retention was observed with the treatments including oxalic acid. Chemical analysis of the fouling layers showed that treatment including oxalic acid gave the highest reduction of the concentration of iron. The overall results of combined cleaning presented in Table 4.6 show that best results were obtained with the cleaning including oxalic acid.

Table 4.6 Qualitative comparison of cleaning success of tested alkaline/detergent & acidic cleaner combinations

Cleaner combination	Improvement of permeability	Salt retention	Deposit removal (ICP)[*)]
NaOH & HCl			n.d.
NaOH/SDS & SBS			48,4%
NaOH/SDS & citric acid		OK	60,8%
NaOH/SDS & Nitric acid/phosphoric acid	X	OK	45,9%
NaOH/SDS & sodium hydrosulfite		OK	53,5%
NaOH/SDS & sulfamic/citric acid		(OK)	51,0%
NaOH/SDS & oxalic acid	X	OK	76,1%
Oxalic acid & NaOH/SDS	X	OK	75,6%

[*)]Deposit removal was calculated for the sum of all elements quantified by ICP-OES in comparison to the deposit before cleaning.

4.3.4.3.2 LCME contribution

Hydraulic performance recovery: Cleaning efficiencies of cleaners were measured in terms of permeability recovery (performed with synthetic seawater) and salt retention improvement. Permeability measurements performed with three coupons obtained from the same membrane sheet revealed that the permeability measurement is relatively well reproducible (±10%).

For Sequence-A, results (Table 4.7) showed a slight increase of the permeability of the membrane after each cleaning step however these improvements remain in the domain of error of the protocol. For Sequence-B, an opposite trend was observed with a slight decrease of the permeability of the membrane after alkaline and acidic cleanings. If the decrease in permeability after STPP remains in the domain of errors, the permeability recorded after the oxalic acid treatment has been significantly reduced. Changes in the cake layer density and micropore size distribution within the foulant layer (breakdown of larger complex molecules to lower molecules due to Fe decomplexation) may have occurred in the presence of oxalic acid. Notice that the impact of oxalic acid on the membrane polymer has not been evaluated based on our experimental conditions.

Table 4.7 Permeability and salt retention measurements during bench-scale experiments

		Permeability (LMH/bar)	Salt retention (%)
Sequence-A	Before cleaning	0.60 ± 0.06	95.2
	After NaOH	0.66 ± 0.07	97.7
	After HCl	0.72 ± 0.07	98.8
Sequence-B	Before cleaning	0.69 ± 0.07	95.0
	After STPP	0.66 ± 0.07	97.3
	After oxalic acid	0.35 ± 0.04	97.8

As shown in Table 4.7 salt rejection (as measured by conductivity) was improved after both sequences. Sequence-A appeared to be slightly more efficient than Sequence-B.

Foulant characterization: As expected from the results obtained in batch mode, the relative decreases in Ca and Fe concentrations were significantly larger when performing Sequence-B as compared to Sequence-A.

In fact Sequence-A had no effect on the relative concentration of all three elements (Figure 4.10). In accordance with the results obtained during batch experiments, both Sequences did not change the relative abundance of Si. For Sequence-A, one could have expected a stronger impact of the acidic cleaning on Ca, however the relative short contact time operated as compared to batch experiments (4 hours versus 24 hours) may explain our finding. The same remark can be addressed when looking at the reduction of Fe during Sequence A. The mode of operation, foulant suspension versus cake layer soaking can be another reason for lower removal efficiencies at RO bench scale. Bench scale experiments confirmed that Fe can be significantly removed (from 50 to 20 mg/g) from the foulant when applying STPP and oxalic acid in series.

The relative biopolymers distribution of the deposit determined by pyrolysis GC/MS (the relative abundance of each biopolymer is determined as the sum of the areas of their corresponding pyrolysis fragments divided by the sum of the areas of all the fragments considered in the chromatogram) was not significantly changed after conducting both chemical cleaning sequences (Figure 4.11). A small decrease of the relative proportion of polysaccharides balanced by an increase of proteins can be suggested after Sequence-A. This finding contradicts the results observed after batch treatments. Again contact times were severely reduced during the RO bench scale tests as compared to batch tests.

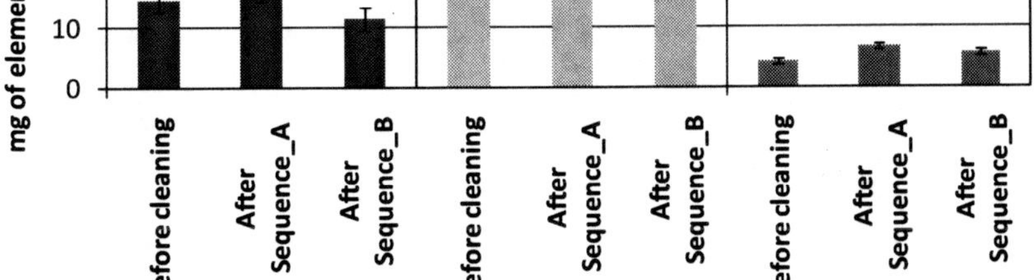

Figure 4.10 Metal concentrations (mg/g) in lyophilized deposit after cleaning sequences applied during bench-scale experiments

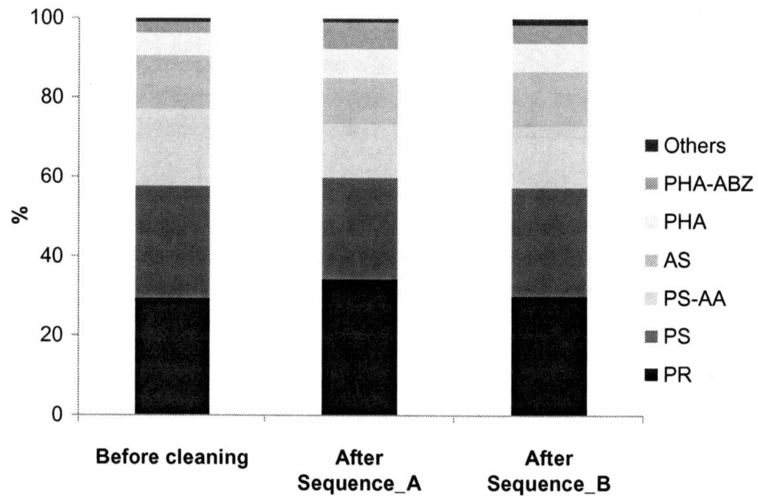

Figure 4.11 Biopolymers distribution in lyophilized deposit after cleaning sequences applied during bench-scale experiments; PR: Proteins, PS: Polysaccharides, PS-AA: Acetic Acid, AS: Amino sugars, PHA: Polyhydroxyaromatic, PHA-ABZ: Alkyl benzene. This figure is available in colour at http://www.iwapublishing.com/template.cfm?name=isbn9781843393214

Another approach for the interpretation of complex pyrochromatograms is to look at the relative ratio of some specific by-products. Pyrochromatograms of membrane foulants are always dominated by peaks identified as pyrolysis by-products of polysaccharides (i.e. acetic acid, hydroxypropanone and furfural) and aminosugars (i.e. acetamide produced from the pyrolysis of N-acetyl aminosugars present in bacterial cell walls). The pyrolysis of model structures (not showed) revealed that the relative abundance of these by-products depend on the polymeric nature of the studied structures. A high molecular weight polymer of alginic acid leads to preferential formation of furfural (F), while the alginic dimer is mainly producing hydroxypropanone (H). Similarly acetamide (A) is the dominant peak of the pyrochromatogram of coulomic acid (homopolymer of N-acetylneuraminic acid) while the pyrolysis of neuraminic acid produced acetic acid and acetamide to the same extent. Detailed analysis of the pyrochromatograms showed significant increases of the F/H ratios (surface areas) after all alkaline/acidic sequences operated under batch conditions. As compared to the large increase of the F/H ratio obtained after the successive STPP and Oxalic acid batch cleaning (before: F/H = 1.35; after: F/H = 17.33) only a small increase was found at RO bench scale for Sequence-B (before: F/H = 1.40; after: F/H = 1.60). Shorter contact times is again a possible explanation of the weak impact of the cleaning sequence. Nevertheless, whatever the experimental condition applied (batch or RO bench scale), results showed that the removal of Fe is associated to an increase of the F/H ratio attributed to a preferential removal of lower molecular weight (LMW) sugars. Fe-LMW sugar complexes could be preferentially removed during the STPP/Oxalic acid cleaning sequence.

4.3.5 Conclusions

Characterization of RO foulant by different analytical tools revealed that foulants are mainly organic in nature (proteins, polysaccharides) and the presence of multivalent ions (especially Fe and Ca) diagnose their major role in the formation of organic foulant layer by undergoing complexation and/or bridging with negatively charged humic molecules. Batch experiments (foulant suspension) showed that among the different alkaline cleaners applied to remove these cations, STPP was the most efficient towards Ca while EDTA led to the best results for Fe removal. Oxalic acid appeared to be highly efficient in removing Fe after STPP treatment.

The cleaning efficacy by means of deposit removal was much higher in case of mixed model fouling layers than on membrane samples from a full-scale RO plant. This observation suggests that the generated mixed model fouling layers differ from naturally originated fouling layers. Although chemical composition of the mixed model fouling layer is mimicking natural fouling deposits, the structure of the layer and/or the adhesion to the membrane surface may be different. In addition, the binding forces between the mixed model fouling layer components and the membrane surface can be weaker due to the accelerated deposition process during the generation of the mixed model fouling layer. These differences may result in an altered susceptibility to cleaning agents. Removal of the deposits of the mixed fouling layers was strongest with NaOH/SDS & oxalic acid, but treatment including oxalic acid (1% w/v) caused a decrease of salt retention, suggesting that membrane damage could occur with this treatment.

Experiments conducted with fouled membrane harvested from full scale plant showed that the application of alkaline chemicals gave a clear increase in the permeability; the strongest increase was observed with NaOH/SDS and with EDTA (pH 12). Some increase of the salt retention was observed with NaOH/SDS, EDTA caused a reduction of the salt retention. Application of acid cleaning after alkaline cleaning gave the best deposit removal, when oxalic acid was included in the treatment. However, the effect of the combination of alkaline and acidic cleaning only had a limited effect on permeability, and the combination of NaOH/SDS with sulfamic/citric acid even showed a decline in permeability. With the membranes from site D, no reduction of salt retention was observed after the cleaning with oxalic acid. The STPP/Oxalic acid sequence showed higher efficiency as compared to the NaOH/HCl sequence. Lower molecular weight sugars seem to be preferentially eliminated during the STPP/Oxalic acid sequence, observation that may correspond to the break down of Fe-sugar complexes. However, results showed some alteration of the hydraulic performances (decrease of permeability and salt rejection) of the membrane when applying the STPP/Oxalic acid cleaning sequence.

Obviously, foulant removal depends on the combination of foulant composition and type of chemical. The tests were conducted on membranes from a specific site. Despite the clear effect of oxalic acid on the removal of inorganic constituent (Fe), the effect on permeability was limited. These observations suggest that the removal of other foulants, e.g. organic compounds and biomass, was too limited for performance improvement. Furthermore, the use of oxalic acid for deposits removal may result in membrane damage. The effects of citric acid suggest that this chemical can be used for the removal of inorganic compounds. However, removal of biomass/organic foulants is needed to obtain the desired performance.

4.4 EFFECTS OF CLEANING ON MEMBRANE PERFORMANCE IN A PILOT PLANT

4.4.1 Introduction

Effects of cleaning procedures were tested under laboratory conditions to assess the impact of chemicals on foulants and membrane permeability (see Sections 4.2 and 4.3). In these tests, samples of plastic tubing with attached biomass and membrane samples of modules from full-scale plants were used. These test conditions differed from conditions applied during cleaning in full–scale plants. Furthermore, the impact on the pressure drop in the feed channel was not investigated. Therefore, Anjou-Recherche–Veolia Water designed and constructed a pilot plant for testing the effect of cleaning procedures as used in practice (cleaning in place, CIP) on 8 inch modules from full-scale plants. These investigations were conducted with membranes which were also tested in the laboratory (see Strathmann *et al.*, 2010, Section 4.3).

4.4.2 Pilot plant, membranes and test procedures

4.4.2.1 Pilot plant

A test rig was designed, built and used (i) for characterization of the performances of 8 inch modules (NF or RO including membranes for seawater desalination) under standard conditions and (ii) for conducting (and determine the effects of) cleaning tests on single 8 inch modules. All equipment and instruments were adapted to high pressure operation (75 bars max.), at constant temperature (regulation to 25°C) and for a large range of chemical products (pH: 1–13). The test rig can operate at constant pressure (55 bars to perform RO standard test), with a fixed recovery (range from 8 to 20%), at an adapted feed flow and permeate flux (between 3 to 18 m^3/h and 5 to 40 LMH, respectively). Whatever the trial (characterization or cleaning), the water used for rinsing or for preparing cleaning reagents presents a conductivity between 250 and 300 $\mu S/cm$. Trans-membrane pressure (TMP), flow rate, temperature, pH, and conductivity of the feed, permeate and concentrate were measured and recorded for each trial. Conductivity meters in the feed, concentrate and permeate streams enable to determine permeate quality and salt rejection rate. Flow meters and pressure sensors enable to determine hydraulic performances such as permeability and pressure drops.

4.4.2.2 Membrane elements

Four membrane modules (Dow Filmtec SW30HR; active area of 37 m^2) were collected from a full-scale seawater RO plant (site D). Pre-treatment of seawater at this plant consists of in-line coagulation followed by dual media filtration. All modules originate from the first pass, front position of the plant and underwent similar operational conditions. One module was used as a reference; the other 3 modules were subjected to different cleaning sequences as mentioned below.

4.4.2.3 Test protocols

4.4.2.3.1 Hydraulic characterization of modules

Standard test conditions for the RO element were selected from the membrane supplier's manual. These characterization tests allow: (i) measurement of the salt rejection of an element (NaCl), (ii) permeability determination, (iii) evaluation of the pressure drop in the feed channel (of a fouled module). Consequently, these tests enable to evaluate the efficiency of chemical cleaning regarding hydraulic performances when conducted before and after CIP. The applied cleaning sequences and analysis are listed in Table 4.8.

Table 4.8 Cleaning sequences and analyses performed at lab scale and in the pilot scale*

Chemicals	Temperature	Mode	Flow rate	Duration
Permeat water rinsing	35°C	Flushing	1.1 × Q**	*20 min*
Membrane characterization (flux and salt retention)				
Alkaline/detergent step	35°C	Flushing	½ Q	*5 min*
	35°C	Soaking	None	*60 min*
Cleaning solution analysis – Purge				
	35°C	Flushing	½ Q	*5 min*
	35°C	Recirculation	½ Q	*30 min*
	20°C	Soaking	None	*15–18 hrs*
	20°C	Recirculation	Q	*60 min*
Cleaning solution analysis – Purge				
Permeate water rinsing	20°C	Flushing	1.1 × Q	*20 min*
Membrane characterization (flux and salt retention)				
Acidic step	20°C	Flushing	½ Q	*5 min*
	20°C	Soaking	None	*60 min*
Cleaning solution analysis – Purge				
	20°C	Flushing	½ Q	*5 min*
	20°C	Recirculation	½ Q	*30 min*
	20°C	Soaking	None	*90 min*
	20°C	Recirculation	Q	*60 min*
Cleaning solution analysis – Purge				
Permeate water rinsing	20°C	Flushing	1.1 × Q	*20 min*
Membrane characterization (flux and salt retention)				
Membrane autopsy				

**no permeate was produced during cleaning; **Q corresponds to the regular flow rate of an 8 inch module (~8 m3/h); pH must be maintained. If pH varied by more than 0.5, it was re-adjusted.*

4.4.2.3.2 Principle

The element to be characterized is supplied with a salt water standard solution. The trials are carried out on a closed loop, with recirculation of the permeate and the concentrate in the feed tank. After stabilization of the operating parameters, the conductivity and flow rate of the permeate, concentrate and feed streams are measured to determine the salt rejection rate and permeability. Moreover, the measurement of differential pressure gives access to the pressure drop in the feed channel.

4.4.2.4 Cleaning procedures

The various CIP sequences tested are presented in Table 4.9. Sequences 1 and 2 are typical cleanings performed on full-scale plants. Sequence 3 was selected from the experiments at laboratory scale (see Strathmann *et al.*, 2010, Section 4.3).

Table 4.9 Chemicals used in the applied cleaning sequences

Sequence*	Alkaline/detergent step	Acid step	Biocide step
Sequence 1	NaOH – 0.1% (w) – pH 12	HCl – 0.2–0.5% (w) – pH 2	
Sequence 2	STPP (2% (w)) + NaDDBS (0.025% (w)) – pH 10	Citric acid – 2% (w); pH 2	DBNPA at 0.05%
Sequence 3	NaOH – 0.1% (w) + SDS – 0.03% (w) – pH 12	Oxalic acid – 1% (w) – pH 1.5	–

**Sequence 1 was applied to module 1, sequence 2 was applied to module 2 and cleaning sequence 3 was applied to module 3.*

4.4.2.5 Analytical procedures

Cleaning solutions were analyzed with ICP-OES (blank control) after soaking. Inorganic matter, organic matter, and microbiological analyses were performed on scrapped deposits from membranes surfaces.

4.4.3 Results

4.4.3.1 Permeability, pressure drop and salts retention

Permeability was determined after warm (35°C) water rinsing, before and after cleaning. Warm-water rinsing had no effect on the permeability of the reference module. The various modules presented different permeabilities after warm-water rinsing, indicating that these modules were not similar. However, the initial values of pressure drops and salt retention showed that the modules were not heavily fouled. The effect of the cleaning sequences on permeation generally was limited ($<10\%$) and the initial permeability, as specified by the membrane manufacturer, was not restored. Also, no significant decrease of the feed channel pressure drop was observed after the application of the cleaning sequences. For modules 1 and 2 the pressure drop remained clearly above the value specified by DOW for a new module. No significant change in salt retention was observed after cleaning. These observations thus show very limited effects of cleaning sequences on the membrane performance. The monitoring of hydraulic performances on the full-scale plant shows that cleaning efficiency could be more easily observed on a RO train than on a single module. A possible explanation is a better sensitivity of the measurements due to the presence of multiple pressure vessels and modules.

4.4.3.2 Cleaning solution analysis

Very low concentrations of inorganic compounds were observed after the alkaline cleaning (results not shown). Acid cleaning yielded clearly higher concentrations of metals (Figure 4.12) in cleaning solution after soaking. HCl gave the highest removal of Ca and oxalic acid treatment gave the highest removal of iron. The high concentration of P after citric acid cleaning might have been due to the previous treatment with STPP, suggesting insufficient rinsing after alkaline treatment.

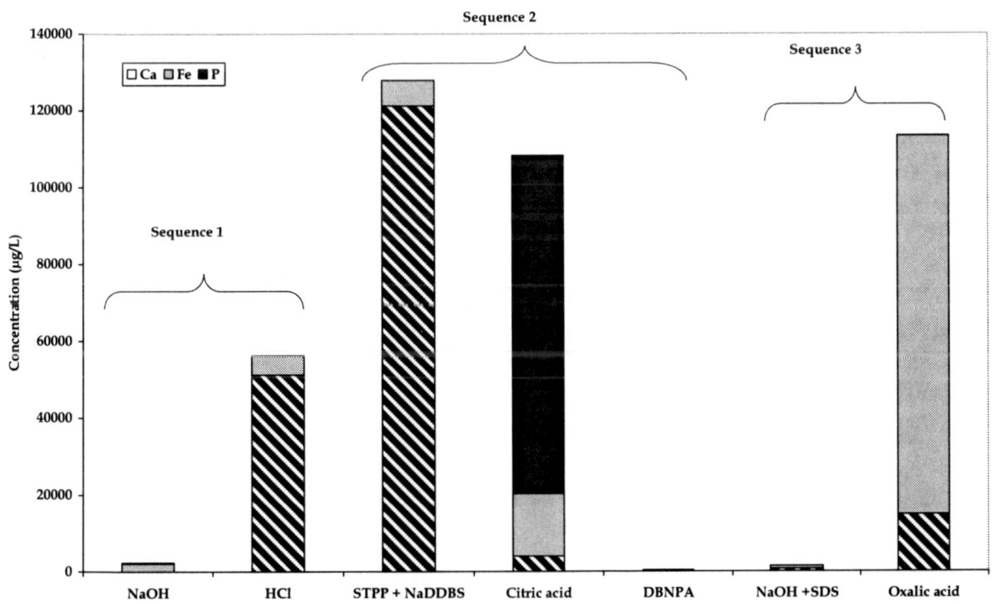

Figure 4.12 Concentrations of a selection of metals present in the acid cleaning solutions after soaking

After cleaning, reference and cleaned modules were autopsied. Visual inspection clearly revealed the presence of brownish deposits on the membrane surface, except for module 3, which presented a greenish colour attributed to complexation of Fe with oxalic acid. Permeability testing performed at lab scale (cf. Section 4.3), revealed no effects of the cleaning sequences 1, 2 and 3. No significant change of salt retention was observed whatever the tested sequence.

Microscopic analysis of the deposits revealed the presence of bacterial aggregates and brownish iron precipitates. Dry matter tests showed that about 50% of the dry matter consisted of compounds which were volatile at 550°C (Figure 4.13), i.e. organic compounds. The concentration of inorganic matter on the reference membrane was about 60 μg/cm² and the concentration of organic matter was about 65 μg/cm². Figure 4.13 suggests that cleaning sequence 1 was not effective in terms of organic and inorganic compounds removal, whereas sequence 3 seems more effective, especially for inorganics removal. However, this interpretation is based on the assumption that the reference represents the amount of initial foulants present on modules 1, 2 and 3 (before cleaning). The autopsy of 2 additional reference modules confirmed this assumption (results not shown).

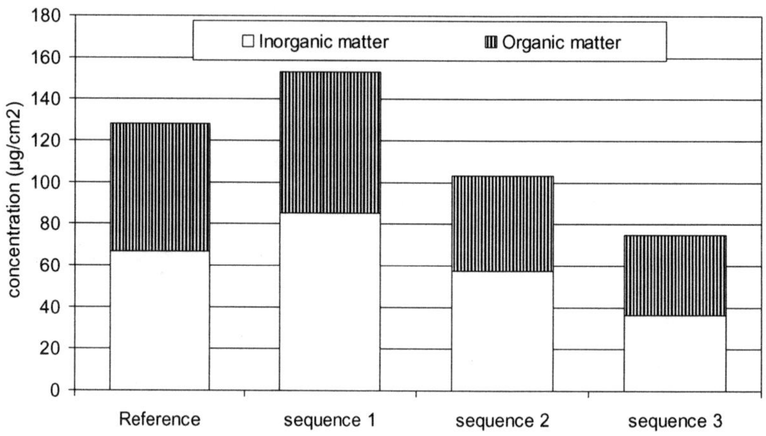

Figure 4.13 Concentrations of inorganic and organic matter on the membranes. Modules 1 to 3 had been cleaned as described above

ATR-FTIR spectra of the fouled and cleaned membranes demonstrated the presence of carbohydrates and proteins. Oxalic acid was also observed, suggesting that rinsing after soaking was insufficient to remove this compound. Analysis of the inorganic compounds on the membranes deposits revealed that iron was a significant proportion of the deposits on the membranes, with about 8.6 μg/cm² on the reference membrane. In module 3, the iron concentration (after cleaning) was about 2 μg/cm², indicating that cleaning sequence 3 was effective in terms of iron removal. This result confirms the previous observations.

Total direct cells counts (TDC) showed that the number of bacterial cells on the reference membrane was 3.9×10^7 cells/cm². The TDC values after cleaning were 6.1×10^7 cells/cm² (module 1); 3.6×10^7 cells/cm² (module 2) and 2.1×10^7 (module 3), suggesting that sequence 3 was the most effective treatment for the removal of micro-organisms. The concentration of ATP on the reference element was relatively low (1.3 ng/cm²), but the concentration of carbohydrates (CH) was relatively high (>100 μg/cm²). Cleaning sequence 2 (CIP2) had the largest effect on ATP reduction, but removal of CH was strongest with CIP3 (Figure 4.14). Comparison of CH concentrations in the reference membrane and those on module 3 show that CH removal was about 50%.

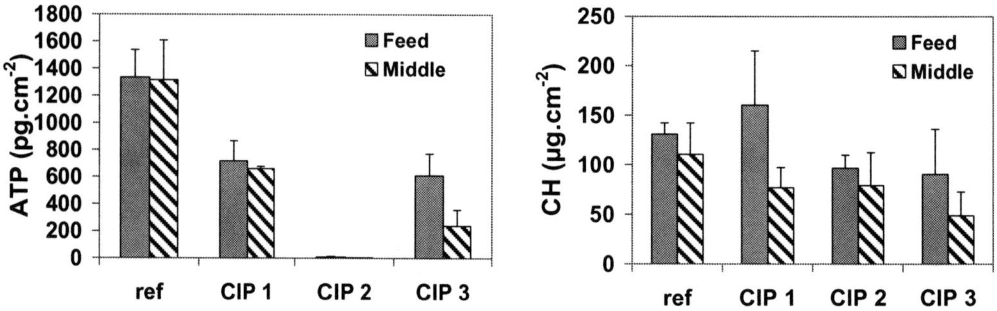

Figure 4.14 ATP and CH concentrations in the elements and the cleaning efficiencies for both parameters (error bars is range of two values). CIP 1 = cleaning sequence 1, etc. Feed = membrane sample from feed side; middle = sample from middle of element

4.4.4 Discussion and conclusions

4.4.4.1 Discussion

The observations presented above demonstrate that the developed test system can be used for determining the membrane performance and the effects of CIP on membrane performance. The effect of the applied cleaning sequences on permeability and feed channel pressure drop was limited despite the effective removal of iron with sequence 3. The variations of these parameters are probably too slight, ranging in the domain of error, which is explained by the low fouling rate of the initial modules.

A relatively high concentration of organic compounds was observed on the membranes before (reference) and also after cleaning. FTIR analyses and the high concentration of carbohydrates on the membranes show that the organic compounds were of biological origin, suggesting that biofouling had occurred in the involved full-scale plant. The ATP concentration however, was relatively low in relation to the CH concentration, but the activity of the microorganisms on the membranes probably had decreased during transportation and storage of the membrane elements before and after cleaning.

4.4.4.2 Conclusions

- The designed test system is well suited for determining the hydraulic performance of membranes from full-scale plants and for conducting cleaning procedures under conditions similar to those performed in practice;
- Three different tested cleaning procedures had limited effect on permeability and on feed channel pressure loss in the membrane elements from a full-scale plant (site D); this could be explained by the fact that the selected modules were not heavily fouled;
- Cleaning with NaOH (0.1%) + SDS (0.03%), followed by exposure to oxalic acid (1%) at pH 1.5 showed the most effective removal of iron from the membrane;
- Cleaning solution analysis may give relevant information about the removal of foulants such as metals, e.g. iron;

REFERENCES

Bos R., van der Mei H. C. and Busscher H. J. (1999). Physico-chemistry of initial adhesive interactions – its mechanisms and methods for study. *FEMS Microbiology Rev.*, **23**, 179–230.

Bruchet A., Rousseau C. and Mallevialle J. (1990). *Journal AWWA*, **82**, 66–74.

Cornelissen E. R., Vrouwenvelder J. S., Heijman S. G. J., Viallefont X. D., van der Kooij D. and Wessels L. P. (2007). Periodic air/water cleaning for control of biofouling in spiral wound membrane elements. *J. Membr. Sci.*, **287**, 94–101.

Corpe W. A. (1974). Detachment of marine periphytic bacteria from surfaces of glass slides. *Dev. Indust. Microbiol.*, **15**, 281–287.

Dubois M., Gilles K. A., Hamilton J. K., Rebers P. A. and Smith F. (1956). Colorimetric method for determination of sugars and related substances. *Analytical Chemistry*, **28**(3), 350–356.

Flemming H.-C., Neu T. R. and Wozniak D. J. (2007). The EPS Matrix: The "House of Biofilm Cells". *J. Bacteriol.*, **189**(22), 7945–7947.

Flemming H.-C., Schaule G. and McDonough R. (1993). How do performance parameters respond to initial biofilm formation on separation membranes. *Vom Wasser*, **80**, 177–186.

Grasshoff K. (1976). *Methods of Seawater Analysis*. (ed.), Verlag-Chemie, Weinheim, New York, p. 317.

Greenlee L. F., Lawler D. F., Freeman B. D., Marrot B. and Moulin P. (2009). Reverse osmosis desalination: water sources, technology, and today's challenges. *Water Research*, **43**, 2317–2348.

Hijnen W. A. M., Biraud D., Cornelissen E. R. and van der Kooij D. (2009). Threshold concentration of easily assimilable organic carbon in feedwater for biofouling of spiral-wound membranes. *Environ. Sci. Technol.*, **43**, 4890–4895.

Hijnen W. A. M., Cornelissen E.R. and van der Kooij D. (2010). Development of a laboratory method for testing membrane cleaning procedures. *The Medina Project*. IWA Book.

Hijnen, W.A.M., Cornelissen E.R. and van der kooij D. (2011). Threshold concentrations of biomass and iron for pressure drop increase in spiral-wound membrane elements. *Wat Res.* **45**, 1607–1616.

Hoenig M. (2001). Preparation steps in environmental trace element analysis – facts and traps. *Talanta*, **54**, 1021–1038.

Hong S. and Elimelech M. (1997). Chemical and physical aspects of natural organic matter (NOM) fouling of nanofiltration membranes. *Journal of Membrane Science*, **132**, 159–181.

Khedr M. (2000). Membrane fouling problems in reverse osmosis desalination applications. *Desalination & Water Reuse*, **10**(3), 8–17.

Kwon K. D., Vadillo-Rodriguez V., Logan B. E. and Kubicki J. D. (2006). Interactions of biopolymers with silica surfaces: force measurements and electronic structure calculation studies. *Geochimica et Cosmochimica Acta*, **70**, 3803–3819.

Lattemenn S. (2010). *Development of An Environmental Impact Assessment and Decision Support System for Seawater Desalination Plants*. Dissertation Unescco IHE, Delft, CRC Press/Balkema Leiden, the Netherlands.

Mondamert L., Labanowski J., Berjeaud J. M., Rapenne S. and Croué J.-P. (2009). Autopsy of RO desalination membrane: Part 2. Chemical characterisation of the foulant. *Desalination and Water Treatment*, **9**, 73–81.

Omoike A., Chorover J., Kwon K. D. and Kubicki J. D. (2004). Adhesion of bacterial exopolymers to alpha-FeOOH: inner-sphere complexation of phosphodiester groups. *Langmuir*, **20**, 11108–11114.

Ridgway H. F. and Flemming H.-C. (1996). Membrane biofouling. In: *Water Treatment Membrane Processes*, J. Mallevialle, P. E. Odendaal and M. R. Wiesner (ed.), McGraw-Hill, New York, 6.1–6.62.

Ridgway H. F., Kelly A., Justice C. and Olson B. H. (1983). Microbial fouling of Reverse-Osmosis Membranes used in advanced wastewater treatment technology: chemical, bacteriological and ultrastructural analyses. *Appl. Environ. Microbiol.*, **45**, 1066–1084.

Schaule G., Rapenne S., Strathmann M., Grobe S., Robert C. and Jacquemet V. (2009). Autopsy of RO desalination membranes: Part 1 microbial characterization of foulants. *Desalination and Water Treatment*, **9**, 66–72.

Schippers J. C., Kruithof J. C., Nederlof M. M., Hofman J. A. M. H. and Taylor J. S. (2004). *Integrated Membrane Systems*, Report American Water Works Research Foundation, Denver, CO.

Spettmann D., Eppmann S., Flemming H.-C. and Wingender J. (2008). Visualization of membrane cleaning using confocal laser scanning microscopy. *Desalination*, **224**, 195–200.

Strathmann M., Pütz D., Flemming H.-C., Mondamert L. and Croué J. P. (2010). Effects of chemicals on membrane permeability and foulants in laboratory tests. IWA Book Medina 2010.

Van der Hoek J. P., Hofman J. A. M. H., Bonne P. A. C., Nederlof M. M. and Vrouwenvelder J. S. (2000). RO treatment: selection of a pre-treatment scheme based on fouling characteristics and operating conditions based environmental impact. *Desalination*, **127**, 89–101.

Van der Kooij D. and Veenendaal H. R. (2001). Biomass production potential of materials in contact with drinking water: method and practical importance. *Water Sci. Technol. Water Supply*, **1**(3), 39–45.

Vrouwenvelder H. R., Van Paassen J. A. M., Folmer H. C., Hofman J. A. M. H., Nederlof M. M. and Van der Kooij D. (1998). Biofouling of membranes for drinking water production. *Desalination*, **118**, 157–166.

Whittaker C., Ridgway H. and Olson B. H. (1984). Evaluation of cleaning strategies for removal of biofilms from reverse-osmosis membranes. *Appl. Environ. Microbiol.*, **48**(3), 395–403.

Zhu P., Long G., Ni J. and Tong M. (2009). Deposition kinetics of extracellullar polymeric substances (EPS) on silica in monovalent and divalent salts. *Environmental Science and Technology*, **43**(15), 5699–5704.

Zobell C. E. (1943). The effect of solid surfaces upon bacterial activity. *J. Bacteriol.*, **46**, 39–56.

Chapter 5

Process strategies for mitigation of impact of concentrates on the environment

F. Macedonio[1,2], E. Drioli[1,2], K. J. Raat[3], P. J. Stuyfzand[3,4], S. van de Wetering[5], M. Groenendijk[5], S. Laborie[6], C. Cabassud[6], Jean-Pierre Méricq[6] and J. Gilron[7]

[1]University of Calabria (Italy)
[2]ITM-CNR (Italy)
[3]KWR Watercycle Research Institute (The Netherlands)
[4]VU University (The Netherlands)
[5]Water supply company Brabant Water's-Hertogenbosch (The Netherlands)
[6]INSA Toulouse (FRANCE)
[7]ZIWR, Ben-Gurion University (Israel)

In this chapter, the most interesting results carried out in Work Package 5 of MEDINA project are summarized. The WP5 leader has been Prof. Enrico Drioli. The contributors of this chapter have been as follows:

- UNICAL (Francesca Macedonio, Enrico Drioli);
- KWR (Klaasjan J. Raat, Pieter J. Stuyfzand, Stephan van de Wetering and Martijn Groenendijk);
- BGU (Jack Gilron);
- INSA (Stéphanie Laborie, Corinne Cabassud), Jean-Pierre Méricq);

WP5 was structured into three complementary Work Tasks:

- ***WT 5.1: Reduction of Brine Volume***, which includes the study of the performance of Fresh holder/Brakish water RO (mainly focused on low recovery RO in combination of injection of concentrate in deeper & saltier aquifers) and of Vacuum Membrane Distillation (VMD) for the treatment of RO retentate stream.
- ***WT 5.2: Recovery of dissolved salts as crystalline product***. In this work task, Membrane Crystallization (MCr) and WAIV(wind intensified enhanced evaporation) are studied as reasonable technologies for securing the major benefits to sustainable water management in terms of increased water recovery factor from marginal resources, reduction of brine volumes and of disposal costs.
- ***WT 5.3: Economic Evaluation***, in which the economical analysis of the investigated technologies are carried out.

In the following, the research activities carried out in the 39 months and the final results are reported.

5.1 INTRODUCTION

Work package 5 (WP5) of MEDINA project was focused on the development of effective separation methods able to reduce the volume of concentrated brines discharged from the desalination plants and to recover some of the salts that are present (calcium sulphate, sodium chloride, epsomite, etc.). In a few words, WP5 was aiming at the improving of the productivity of a desalination system, and to the reduction of their energy consumption and environmental impact. For reaching this goal, the following different technologies were evaluated: injection of RO concentrate in the subsurface, Vacuum Membrane Distillation (VMD), Membrane Crystallization (MCr) and/or Wind Intensified enhAnced eVaporation (WAIV). In particular, the carried out research activities considered: (i) a pilot study aiming to test the deep well injection of supersaturated membrane concentrate; (ii) the development of appropriate tools for the analysis and simulation of Membrane Distillation and Membrane Crystallization units and for anticipating the effects of the main operative parameters on the performance of each stage; (iii) the experimental evaluation of the potentiality and the reliability of these unit operations; (iv) the analysis of the distribution of precipitation on a WAIV setup.

The contractors involved in WP5 and their roles have been as follows:

- UNICAL, whose involvement was related to the study and development of the membrane crystallizer technology (WT 5.2 and WT 5.3);
- AR-VW, responsible to provide all partners with samples of seawater;
- KWR, aiming to test *Fresh holder/Brackish water RO* (WT 5.1);
- BGU, whose involvement was related to the application of Wind-Aided Intensified eVaporation (WT 5.2 and WT 5.3);
- INSA, involved in WT 5.1 in the application of vacuum membrane distillation for the treatment of seawater RO concentrates;
- GVS, who supplied several membranes for VMD experiments and who supported the partners with the analysis of results of tests that have been carried out (WT 5.1).

5.2 WT 5.1: REDUCTION OF BRINE VOLUME

Seawater and brackish water desalination processes are contributing in a significant way to solve the problem of water shortage. However, recently, the increasing number and capacity of installed desalination plants exacerbated the problem of the negative environmental impacts of concentrate wastes which are often discharged into surface waters or into the ocean.

The sensitivity of the marine habitat to the discharged brines vary from place to place, and its magnitude depends on environmental and hydrogeological factors characteristic of the sea: bathymetry, waves, currents etc. In general, the high concentrations and the corresponding high densities cause the brines to sink to the sea floor, creating a saline desert in the vicinity of the discharge pipe; in addition, chemical added in pre-treatment may harm the marine life. For what concerns brackish water desalination, the impact of the concentrate on the environment depends on its concentration since it is often injected into deeper groundwater.

Therefore, alternative designs and engineering approaches need to be developed not only to drive down the cost of producing water and to ensure the quality of water supply, but also to increase the sustainability of the desalination systems and to assure a more rational use of the limited water resource.

In WT 5.1, Fresh holder/Brakish water RO and Vacuum Membrane Distillation are studied as possible methods for reducing the effect of brine volume on the environment.

5.2.1 BWRO concentrate disposal by deep well injection: design criteria for BWRO plants and field test results

Brackish water reverse osmosis (BWRO) is an interesting source treatment option, even for countries in temperate climates. The major obstacle for large scale application of BWRO is the disposal of the saline concentrate. Deep well injection is the solution to this problem, though mineral precipitation is an operational risk, as it may clog the injection well and its surroundings.

Careful selection of source and disposal aquifer and proper design of the RO installation and injection well, aid to the success of BWRO plants. The source water should preferably be low in ions less soluble than Na, K and Cl, to prevent scaling of membranes or clogging of the injection wells. Furthermore, groundwater is preferred that, after concentration, fits perfectly into the natural chemical environment of the target aquifer, because its deep well injection is then more likely to be permitted by regulators.

Deep well injection was tested in practice in BWRO pilot Zevenbergen (Brabant Water), where concentrate was injected into a deep, anoxic aquifer, while water quality changes in the target (disposal) aquifer were closely monitored.

Field test results indicate that supersaturation of BWRO concentrate is allowed, without causing operational problems in the RO or injection well. Precipitation of carbonate minerals had place in the target aquifer, but up to date this did not lead to clogging of the injection well.

5.2.1.1 Introduction

Countries in temperate climates are today seriously considering reverse osmosis of brackish groundwater (BWRO) as an option to prepare drinking water (Stuyfzand and Raat, 2010). The drivers consist of environmental problems like the salinization of fresh water wells, drawdown of water tables in phreatic aquifers, effects of climate change that may render surface waters less fit for drinking water production (due to contaminant levels and temperatures exceeding limits for water intake), and increasing costs to produce drinking water from heavily polluted, fresh groundwater.

An obstacle for the application of BWRO at inland sites is the disposal of the membrane concentrate. Many of the disposal options mentioned by Mickley (2006), like evaporation ponds and discharge to surface water, are deemed to fail or to be rejected by environmental regulations. Deep well injection into a more saline, confined aquifer is considered to be the best solution to the waste water problem (Stuyfzand and Raat, 2009), but even then disposal may be problematic because: (i) the concentrate may contain antiscalants, (ii) several constituents of the feed water (notably Fe, NH_4, PO_4, As, Ba, Cr, F and Ni) can be concentrated above natural background or even above maximum contaminant levels, and (iii) the regulator has a bias towards lower salinity levels and as a consequence it is difficult to get approval for injection of water higher in salinity than the ambient water.

The advantages of BWRO have triggered two water utilities in the Netherlands (Brabant Water and Vitens) to start BWRO pilot plants, using deep well injection for concentrate disposal (Oosterhof *et al.*, 2009). The pilots aim to test whether deep well injection is technically feasible, especially considering the injection of supersaturated solutions and the risk of clogging of the injection well and its surroundings. Furthermore, the effects of concentrate injection on the hydrochemistry of the target (disposal) aquifer are studied. To a large extent, the success of the pilots is determined by the selection of source and target aquifers and the recovery level of the reverse osmosis (RO) installation. Together, these determine the mineral saturation levels in the concentrate and the possible risks for mineral precipitation in the RO or near the injection well.

This section describes the set up and first field test results of the BWRO pilot Zevenbergen, which was initiated and operated by water supply company Brabant Water, the Netherlands. In addition, criteria are presented for the selection of source and target aquifers and the design of the injection well.

5.2.1.2 Criteria for aquifer selection and BWRO design

A careful selection of source and target aquifer is important for the realization of a sustainable BWRO plant. Constraints are set by the local geohydrology and water chemistry, which determines both the quality of the permeate (drinking) water and the level of mineral supersaturation in the concentrate. An extensive review of criteria and constraints for the selection of aquifers is provided by Stuyfzand and Raat (2009). A summary is presented here, together with design criteria for the RO installation and injection well.

5.2.1.2.1 Selection criteria for source aquifers

Constraints to the source aquifer are set by the RO process, by the geohydrology, and by the water quality of the source aquifer (RO feed water). This water quality is important both in terms of the product (permeate) and the waste water (concentrate). Selection criteria for BWRO source aquifers are listed in Table 5.1.

Scaling of the membranes or clogging of the injection well due to the formation of mineral precipitates is an operational risk for BWRO plants. An indicator for this risk is the mineral saturation index (SI_M) that describes the tendency of a mineral to dissolve ($SI_M < 0$, undersaturated) or to precipitate ($SI_M > 0$, supersaturated). Preferably, the membrane concentrate should have $SI_M \leqslant 0$, which puts severe constraints to the groundwater types suited as BWRO feed water and to the RO efficiency. For instance, many groundwaters in the Netherlands are in equilibrium with calcite and their RO concentrate would, by definition, be supersaturated towards calcite ($SI_{calcite} > 0$). Also, to be economically feasible, BWRO systems should operate with an efficiency of 50% or higher, which implies that solutes will be concentrated with a factor 2 or higher.

In practice, however, many natural groundwater waters have $SI_M > 0$, without precipitates being present or being formed. Possible causes are complexation of cations by dissolved organic matter, which lowers the effective IAP, inhibition of crystallization by other solutes, kinetic hindrances and sluggish crystallization (Stuyfzand and Raat, 2009). In addition, for many minerals, the calculated SI is very sensitive to the pH value used and as a consequence inaccuracies in pH measurements may result in an overestimation (or underestimation) of the SI. As such, the mineral saturation index is a *risk* indicator for precipitation, and subsequent problems with chemical clogging of membranes and injection wells. When $SI < 0$, these problems are not to be expected; when $SI > 0$, the index indicates there is a chance that clogging may occur. Field pilots, like at Zevenbergen, are needed to further quantify this chance and to see at which saturation levels precipitates start to form and become an operational problem.

Table 5.1 Selection criteria for BWRO source aquifers (modified after Stuyfzand and Raat, 2009)

Salinity	The salinity must be favorable for optimum membrane purification at reduced costs: preferably chloride <10,000 mg/L or total dissolved solids (TDS) <20,000 mg/L. BWRO systems normally operate between 1,000 and 7,000 mg Cl/L.
Feed water quality	Pumping may result in temporal and spatial quality variations in the feed water that either adversely affect the RO process or require a sophisticated RO adaptation strategy. A solution to this problem is offered by pumping several wells that vary in salinity (both higher and lower than the desired level) at a rate autocontrolled by on-line electrical conductivity (EC) measurements in each well and their mixture
Geohydrology	The abstraction of brackish feed water should not result in salinization nor freshening of the source aquifer. Like any source aquifer, permeability and transmissivity should be sufficient for water recharge.
Clogging risk	The concentration of ions less soluble than Na, K and Cl should be low enough to prevent scaling of membranes or clogging of the injection wells with, for instance, mineral precipitates of silicates ($SiO_2 \cdot nH_2O$, $MgSi_2(OH)_6$), sulphates ($BaSO_4$, $CaSO_4 . 2H_2O$), carbonates ($CaCO_3$, $CaMg(CO_3)_2$, $FeCO_3$), or phosphates ($Ca_5(PO_4)_3OH$; $NH_4MgPO_4 \cdot 6H_2O$).
Permeate water quality	The concentration of ions in the permeate should not exceed the maximum permissible concentrations for drinking water. Simple post-treatment like aeration is adequate to get rid of gases like CH_4, H_2S and CO_2. Uncharged dissolved species with a small size, like boron (H_3BO_3) may pose serious problems. Boron is very hard to remove; it can attain rather high concentrations in brackish groundwater, may pass membranes and may then exceed the World Health Organization (WHO) and the European Union (EU) drinking water standards of 500 and 1,000 μg/L, respectively.
Permitting	Groundwater is preferred that, after membrane concentration, remains below the levels in the target aquifer for parameters like salinity, nutrients and heavy metals, because its deep well injection is then more likely to be permitted by regulators. Ideally, the membrane concentrate fits perfectly into the natural chemical environment of the target aquifer. Therefore, the use of antiscalants should be avoided.

5.2.1.2.2 Selection criteria for disposal aquifers

A sustainable, deep well injection of membrane concentrate can be realized when the hydrogeological and hydrogeochemical properties of the target aquifer meet the criteria listed in Table 5.2. Despite the fact that deep well injection is the most environmental-friendly disposal option (often even better than disposal to sea, where it can have an effect on local marine life), permitting is still problematic. To date, in the Netherlands, clear policy and legislation towards concentrate deep well disposal has yet to be defined.

Table 5.2 Selection criteria for BWRO disposal aquifer (modified after Stuyfzand and Raat, 2009)

Geohydrology	Hydraulic conductivity and storativity should be high enough to store a large quantity of waste water for a long time;
Geohydrology	The aquifer must be as confined as possible, so that the water injected remains contained in a limited space well below land surface and the source aquifer;
Water quality	To obtain the necessary permits, the native groundwater should preferably have a higher salinity and higher concentration of critical compounds (nutrients, heavy metals) than the injectate. Ideally, the source aquifer would be selected such that the membrane concentrate fits perfectly into the natural chemical environment of the target aquifer.
Redox conditions	Brackish groundwaters in the Netherlands are (deep) anoxic and often high in ferrous iron. To prevent precipitation of ironhydroxides, the disposal aquifer should be anoxic as well (which will generally be the case when a deeper, more saline aquifer is selected).
Sodium adsorption ratio (SAR)	The tendency of clay minerals in the aquifer to mobilize by deflocculation should be low, because it could provoke mechanical clogging of the aquifer when the particles strand in the pore necks. This tendency is low when the sodium adsorption ratio (SAR) of the injectate and of the native groundwater is low, their salinity is high, the clay mineral content is very low, and the dominant type of clay minerals is favorable (kaolinite > illite > smectite).

The Dutch Soil Protection Act allows the disposal of membrane concentrate in the soil when the disposal has no negative effects on the soil and groundwater quality in the receiving aquifer. In current practice, the concentrate water quality is evaluated against the maximum concentrations listed in the Soil Protection Act, for groups like organic pollutants, metals, nutrients and halogens. Being a paleo-water, the brackish water used as RO feed water is free of organic pollutants, but these waters can, by nature, be high in some metals (a.o. arsenic), nutrients and halogens (chloride, bromide and fluoride). Concentrations for (some of) these natural constituents may thus already be above maximum allowable concentrations in the feed water, and are even increased more in the RO concentrate.

From a hydro-technological viewpoint, it may be advisable to inject higher salinity water, as this water is more likely to migrate downwards by density driven flow and reside in the lower regions of the target aquifer. In fact, upward migration of injected, lower salinity water has been documented at several wastewater injection sites in Florida, USA (Maliva *et al.*, 2007). In the Netherlands, however, the regulator has a bias towards lower salinity levels, which stems from current problems with salinization of aquifers and surface waters in the Netherlands.

Only mild interactions between concentrate and the target aquifer are expected. The mixing of the injectate with native groundwater will probably not pose serious clogging phenomena, because the water quantity involved is too small, especially in single porosity aquifers. When refraining from antiscalants and acids in the RO process, mineral phases in the target aquifer will hardly or not dissolve, as the injected solution is supersaturated by itself. Redox reactions will be minimal as well, because both the injected water and native aquifer are (deep) anoxic. Water-rock interactions are therefore expected to remain limited to cation exchange and the deposition of some minerals like calcite and apatite, and possibly also dolomite, siderite, vivianite, barite and amorphous silica. Some heavy metals in the injectate may co-precipitate with these minerals, which prevents spreading of these metals into the target aquifer.

5.2.1.2.3 Design criteria for the RO installation and injection wells

Membrane scaling and clogging of the injection well are important operational risks for BWRO plants. As discussed above, a careful selection of the source aquifer is important to lower the chance of mineral precipitation and subsequent scaling. In addition, scaling/clogging may occur due to filtration of suspended solids, biological fouling and intrusion of oxygen. The design of the RO installation and the injection well can be adjusted to lower the risk for these types of clogging (Table 5.3).

Table 5.3 Some design criteria for RO installations and injection wells

Oxygen	To prevent precipitation of ironhydroxides (iron flocs) from the anoxic feed water, the intrusion of oxygen in the RO installation and injection well should be avoided at all costs. When present in the water, iron flocs will effectively clog the injection well. It is advised to monitor O_2 in the RO constantly and to automatically shut down the RO and well injection when O_2 is detected.
Suspended solids	Suspended solids (particles) in the injected water may cause mechanical (physical) clogging of the injection well. This type of clogging is very difficult to remove and may cause irreparable damage to the injection well. It is therefore advised to install filters in the RO to remove suspended solids from the feed water.
Microbiological fouling	Microbiological fouling is a known problem when injecting aerobic water in anaerobic aquifers. Some anoxic (feed) waters may be relatively high in nutrients like nitrogen and phosphorous, which could enhance microbial growth as well. However, at this point, it is unclear if microbial fouling is a serious threat when injecting anoxic water in an anoxic aquifer. Microbiological fouling can be prevented by chlorination, but that would reduce the chance of getting a disposal permit.
Bubble formation 1	Anoxic waters often contain high concentrations of dissolved gasses like CH_4, H_2S and CO_2. These gasses may form bubbles when the water pressure decreases. These bubbles are, again, very effective in clogging the injection well. To prevent the formation of gas bubbles, the membrane concentrate should be kept pressurized. This pressure should be higher than the sum of the partial pressures of the gasses in solution.
Bubble formation 2	In the injection well, water may accelerate as in a free fall, which could cause a vacuum and the subsequent formation of gas bubbles. This can be tackled by decreasing the width of the injection pipe, such that the friction between water and injection well surfaces prevents a free fall.
RO efficiency	Mineral precipitation determines the maximum obtainable RO efficiency in a BWRO installation. In the pilots starting in the Netherlands, an efficiency of 50% is possible, while keeping oversaturation at acceptable levels (SI $\approx$ 1.0) for critical minerals like calcite and apatite.
Antiscalants and acids	A higher RO efficiency could be achieved by using anti-scalants, like polyphosphates, or acids. The use of anti-scalants is not an option, as regulators do not accept deep well injection when the concentrate contains anti-scalants. There is still debate on the use of HCl to lower saturation levels for carbonate minerals like calcite and siderite.

Scaling of membranes is a nuisance, but fouled membranes often can be rehabilitated or, in the worse case, be replaced. In that respect, well clogging may form a larger problem, especially when rehabilitations are not successful. If well clogging occurs, (periodical) mechanical well rehabilitation methods are preferred over chemical methods. Houben and Treskatis (2007) provide an extensive overview of available methods. In addition, periodic backpumping (extracting) of the injection well could be an effective way to remove filtered particles (mechanical clogging) from the well's gravel pack and the borehole wall.

5.2.1.3 BWRO pilot plant Zevenbergen: setup

The BWRO pilot plant Zevenbergen is located in the southwestern coastal zone of the Netherlands. A schematic of the aquifer system and the pilot setup is presented in Figure 5.1. The brackish groundwater in the sedimentary aquifers originates from Pleistocene and Holocene intrusions of the North Sea, and the repetitive trans- and regressions have resulted in a complex pattern of chloride concentrations, with an inversion in chloride concentration present between aquifer 2 and 3.

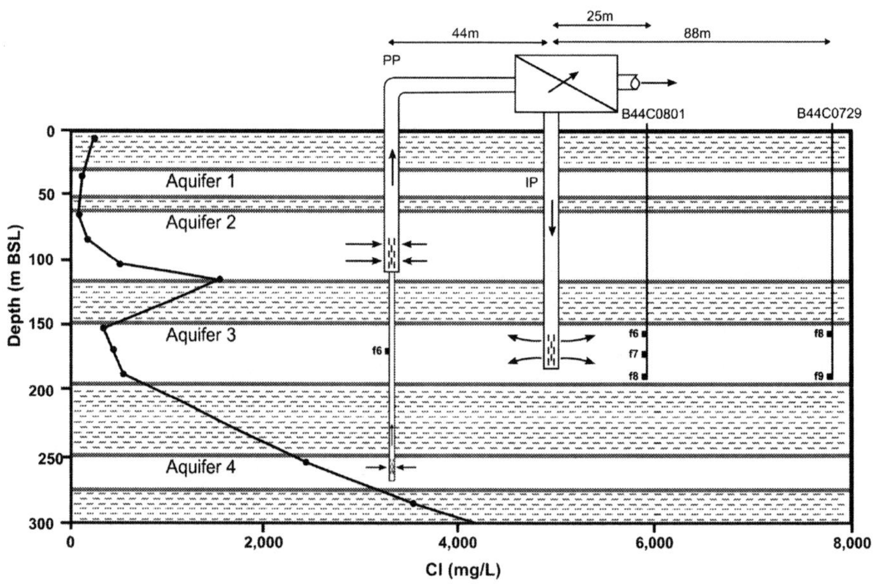

Figure 5.1 Schematic of the aquifer system at the Zevenbergen BWRO pilot, the observed chloride depth profile, and the position of the pumping, injection and observation wells. mBSL = meter below sea level. Aquifer 2 = fluvial, coarse grained sands of early Pleistocene age; Aquifer 3 = marine, fine grained sands of late Tertiary and early Pleistocene age, with clay intercalations

The pilot plant has been in operation since December 2009. BWRO source water is pumped from Aquifer 2 at a rate of 50 m^3/h. The transmissivity of this aquifer is large enough to supply this water, though an important question at the start of the pilot was whether the water quality of the pumped water would stay constant, i.e. whether the salinity of the abstracted water would stay high enough (>300 mg-Cl/L). To compensate for a possible decrease in salinity during the pilot, an extra pumping well was installed in the more saline Aquifer 4. The RO is operated at an efficiency of 50%, resulting in an injection of membrane concentrate of approximately 25 m^3/h. The concentrate is injected in Aquifer 3, at a depth of 168–182 mBSL. This aquifer is separated from the source aquifer by a thick aquitard, preventing short-circuiting of injected water with the source aquifer. Water quality in the target aquifer is monitored closely, through observation wells at 25, 44 and 88 m distance from the injection well. The RO installation uses a one stage staging with 30 spiral wound elements, to reach a recovery of 50%.

5.2.1.3.1 Initial water quality

Water quality of the BWRO concentrate and of the native groundwater in the disposal aquifers at the start of the pilots is presented in Table 5.4. Relative to the target aquifer, the membrane concentrate is higher in Ca, Fe, Si, HCO_3 and salinity (Cl), though levels of especially Fe are low compared to many other Dutch anoxic, brackish groundwaters. Even though salinity of the concentrate is higher than in the target aquifer, both water types fall within the range for slightly brackish water (300–1,000 mg-Cl/L).

The 50% recovery results in a doubling of concentrations in the BWRO concentrate as compared to the BWRO feed water and a logical increase in mineral saturation levels. The concentrate is supersaturated towards carbonates (calcite, dolomite and siderite) and phosphates (hydroxyapatitie, vivianite). Carbonate supersaturation is largely accounted for by the high CO_3 concentration (a result of the relatively high pH and bicarbonate levels) and not so much because of Ca, Mg and Fe concentrations. Moderate supersaturation ($SI \leq 0.5$) of carbonate minerals has been observed in many natural groundwaters, which indicates that moderate supersaturation will not necessarily lead to mineral precipitation. As PO_4 levels are low, it is not expected that precipitation of phosphates will result in significant scaling or clogging. Furthermore, precipitation of phosphate minerals is uncertain because of a high sensitivity of the calculated SI to analytical inaccuracies (pH, difficulties in distinguishing inorganic from organic PO4), and variations in mineral composition and degree of crystallization (Stuyfzand and Raat, 2010).

Table 5.4 Water quality of BWRO concentrate and target aquifer at the start of the BWRO pilot Zevenbergen

		BWRO pilot Zevenbergen			
		Concentrate	**B44C0801f6**	**B44C0801f7**	**B44C0801f8**
Distance from injection well (m)		**n.a.**	**24**	**24**	**24**
filter depth (m)		**n.a.**	**167–169**	**175–177**	**181–182**
pH		7.54	7.59	7.67	7.75
EC	mS/m	270#	164	177	191
Cl	mg/l	650	385	435	465
Br	mg/l	2.1#	1.2	1.3	1.5
F	mg/l	0.3	0.9	1.0	1.0
HCO_3	mg/l	736	444	452	450
SO_4	mg SO_4/l	9	3	2	2
PO_4-P	mg PO_4/l	0.6#	0.1	0.1	0.1
TOC	mg C/l	7#	3.1	3.9	2.9
CH_4	mg/l	1.2#	0.1	0.1	0.0
Na	mg/l	330	270	290	300
K	mg/l	12	14	20	26
NH_4	mg NH_4/l	2.3	1.0	1.3	1.8
Ca	mg/l	240	64	52	42
Mg	mg/l	23	28	43	52
Sr	mg/l	1.9	4.6	4.1	3.0
Fe	mg/l	3.3#	0.69	0.56	0.5
Mn	mg/l	0.08#	<0.01	0.01	0.01
Si	mg/l	20	10	10	9
Mineral saturation indices (SIs)					
Calcite	$CaCO_3$	1.1	0.5	0.5	0.4
Dolomite	$CaMg(CO_3)_2$	1.6	0.9	1.2	1.3
Hydroxyapatite	$Ca_5(PO_4)_3OH$	3.2#	−1.5	−1.5	−1.1
Rhodochrosite	$MnCO_3$	−0.1#	n.a.	−1.0	−1.0
Siderite	$FeCO_3$	1.3#	0.6	0.6	0.6
Vivianite	$Fe_3(PO_4)_2{:}8H_2O$	0.3#	−2.8	−2.9	−2.6

n.a. = not applicable; # = estimated from data collected two weeks after start of injection.

5.2.1.4 Water quality changes following concentrate injection

Injection of the BWRO concentrate resulted in a complex pattern of hydrochemical interactions. Figure 5.2 shows the breakthrough curves for a selection of parameters at the first (25 m distance; WP B44C0801) and second (43 m distance; PP001) observation wells. EC and chloride were good indicators of injectate breakthrough, though EC levels in the observation wells did not increase to the levels of the concentrate fully. This also applies to HCO_3. pH values, which were higher for the native groundwater than for the BWRO concentrate, dropped and stabilized at levels lower than the BWRO concentrate after breakthrough. Ca and Fe, which were lower in the initial groundwater than in the concentrate, increased after breakthrough, but then stabilized at levels below the concentrate. Contrary, magnesium concentrations increased to levels well above the concentrate levels after breakthrough, indicating a release of Mg from the aquifer exchange complex, but decreased again later on.

The observed patterns were explained by a combination of cation exchange and precipitation reactions. The magnesium pattern is a clear example of cation exchange. At breakthrough, Mg is released from the exchange complex and replaced by Ca and Fe, which are present in the injectate at concentrations higher than the native groundwater. After depletion of adsorbed Mg, levels drop again. Loss through adsorption only partly explains why levels of Ca and Fe do not mimic the concentrate after breakthrough; precipitation as calcite and siderite is a more likely cause. Other indicators of precipitation are EC and HCO_3, which, after breakthrough, do not rise to the same levels as in the concentrate. Finally, the drop in pH is explained by precipitation of carbonates, which induces a shift from HCO_3 to CO_3 thereby releasing protons (H^+). Hydrochemical modelling (Phreeqc; Parkhurst and Appelo, 1999) confirmed the above reasoning. Observed patterns, including the pH drop, can be modelled only when including cation exchange and precipitation of calcite and siderite.

It was concluded that at Zevenbergen injection of supersaturated BWRO concentrate led to precipitation of calcite and siderite. To date, however, this has not caused any clogging of the injection well, as indicated by the injection pressure, which has remained stable since the start of injection 7 months ago. It is therefore thought that due to slow kinetics, precipitation is taking place at some distance from the injection well. Note that precipitation was not observed in the RO installation, which also indicates that there is a time lag between attaining mineral supersaturation and precipitates actually forming.

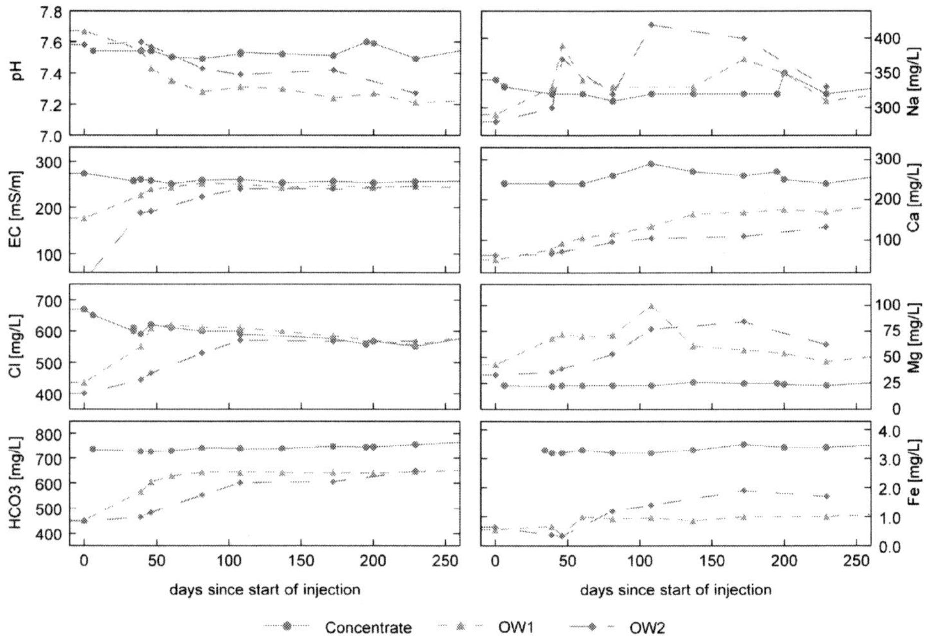

Figure 5.2 Breakthrough patterns for a selection of parameters at the first (OW1, 25 m distance; WP B44C0801-f7) and sec ond (OW2, 43 m distance; PP-f6) observation wells at BWRO pilot Zevenbergen

5.2.1.5 Conclusions

Deep well injection provides a solution for disposal of BWRO concentrate, being environmentally safe and technically feasible. Crucial for the success is a careful selection of source and disposal aquifer and proper design of the RO installation and injection well. Field test results from BWRO pilot plant Zevenbergen (Brabant Water) indicate that supersaturation of BWRO concentrate is allowed, without causing operational problems in the RO or injection well. Precipitation of carbonate minerals had place in the target aquifer, but up to date this did not lead to clogging of the injection well.

5.2.1.6 Acknowledgements

The BWRO pilot Zevenbergen is operated by water company Brabant Water and supported financially by the Province of Brabant. This study was carried out within the framework of the Joint Research Programme (BTO) of the Dutch Waterworks, which is carried out by KWR Watercycle Research Institute. Additional funding was received from the EU Medina-project.

5.2.2 Vacum membrane distillation

The feasibility and limits of vacuum membrane distillation (VMD) for the treatment of seawater reverse osmosis concentrates was studied by INSA in WT 5.1.

The objectives are to concentrate the RO retentates from about 50 g.L^{-1} to about 300 g.L^{-1} (concentration factor of 6). The work performed by INSA can be divided in three parts:

(i) Simulations of the influence of operating conditions (permeate pressure, feed temperature, and hydrodynamics) and membrane characteristics (Knudsen permeability) on the permeate vapour flux. These simulations will allow the selection of favourable operating conditions.

(ii) Experiments with synthetic mineral seawater at different concentrations in order to study and to check if fouling or polarisation effects are significant and can be a limitation of the process. Fouling will be also characterised by an analysis of fouled membrane (membrane autopsies).

(iii) Experiments with real RO retentates. These experiments will allow the verification of previous conclusions with a real effluent. The importance of the organic and bio-fouling will be also discussed.

5.2.2.1 Progress towards objectives: Simulations of the influence of operating conditions

The Deliverable 5.1.2, Vacuum Membrane Distillation (VMD) Modelling Software version 2, was achieved and delivered on April 2008. It is composed of (1) a software, (2) a small report describing model hypothesis, main equations, validation of results and indications to help using the software. It runs for flatsheet membranes but also for hollow fibre membranes with inside-outside configuration. Using inlet parameters given by users (membrane geometry and operating conditions), it calculates the different profiles alongside membrane (temperature, salt concentration and water vapour flow). It can be used with high salt concentrations and it computes local and global values as well. Comparison between modelling and experimental results shows a good accuracy.

Thanks to this Modelling Software, simulations were performed to study the variation of the permeate vapour flux within the following range of operating conditions:

- total salt concentration $\boldsymbol{C_f}$ (0 to 291 g.L^{-1}),
- permeate pressure $\boldsymbol{P_p}$ (0 to 10,000 Pa),
- feed temperature $\boldsymbol{T_f}$ (20 to 70°C)
- Reynolds number ***Re*** (0 to 7,000)

The simulations take only into consideration the mineral part of the RO retentates and were performed for different levels of the concentration of a RO retentate: 50 g.L^{-1} (Real RO retentates), 148.6 g.L^{-1} (Synthetic mineral seawater, SW 150) and 291.1 g.L^{-1} (Synthetic mineral seawater, SW 300). Two membranes were considered in these simulations: the experimental Fluoropore membrane ($\boldsymbol{K_M}$ = 3.26 × 10^{-6} s.mol$^{1/2}$.m^{-1}.kg$^{-1/2}$) and a theoretical membrane with a higher Knudsen permeability ($\boldsymbol{K_M}$ = 1.85 × 10^{-5} s.mol$^{1/2}$.m^{-1}.kg$^{-1/2}$).

5.2.2.1.1 Effect of the feed concentration

A simulation was performed to study the variation of the permeate vapour flux during the concentration of a RO retentate (Figure 5.3). As expected, the permeate flux decreases when the bulk salt concentration increases.

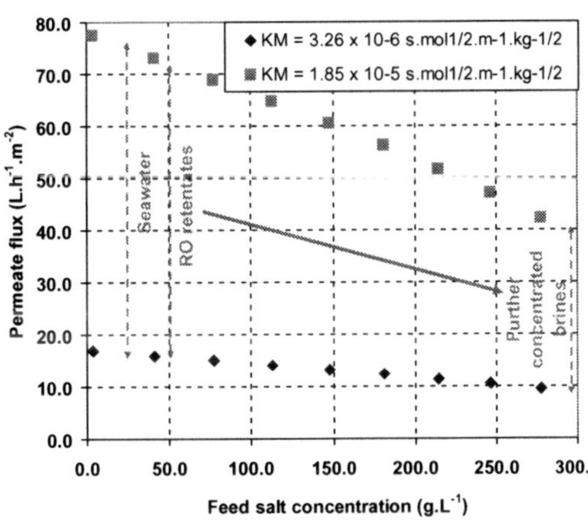

Figure 5.3 Variation of permeate flux during the concentration of RO Retentates (P_p = 500 Pa, T_f = 50°C, Re = 4,500)

The simulated permeate flux obtained for RO retentates (C_f = 50 g.L^{-1}) is very close to the permeate flux obtained for a natural seawater (C_f = 39 g.L^{-1}) with less than 2 % difference. The decrease in the permeate flux is only of about 40–42 % for a concentration factor of 6 corresponding to a retentate concentration Cf = 300 g.L^{-1}. VMD can thus be operated at high salt concentration without loosing too much water production.

5.2.2.1.2 Effect of permeate pressure, feed temperature and Reynolds number

The next step aimed to determine, on the basis of the simulation, the best operating conditions (permeate pressure, feed temperature and hydrodynamics). Figure 5.4 to Figure 5.9 present the results of simulation obtained with the two membranes respectively for a variation of permeate pressure (Figure 5.4 and Figure 5.5), feed temperature (Figure 5.6 and Figure 5.7) and Reynolds number (Figure 5.8 and Figure 5.9).

A low permeate pressure and a high feed temperature results in a high permeate flux.

For a given $\boldsymbol{T_f}$ and ***Re***, when $\boldsymbol{P_p}$ varies from 6,100 to 600 Pa, the permeate flux is more than doubled whereas the specific energy requirement is nearly the same. Indeed, vacuum pressure energy requirement is only a small part of the total energy requirement (less than 2%). The effect of the pressure is more pronounced for the high concentration: the permeate flux is multiplied by nearly 4 for the SW 300 and only by 2 for the RO retentates. At high concentration, the permeate pressure represents a more significant part in the transmembrane pressure difference. Indeed, the feed water vapour partial pressure is strongly reduced at high concentration and as consequence the permeate pressure becomes a more sensitive parameter.

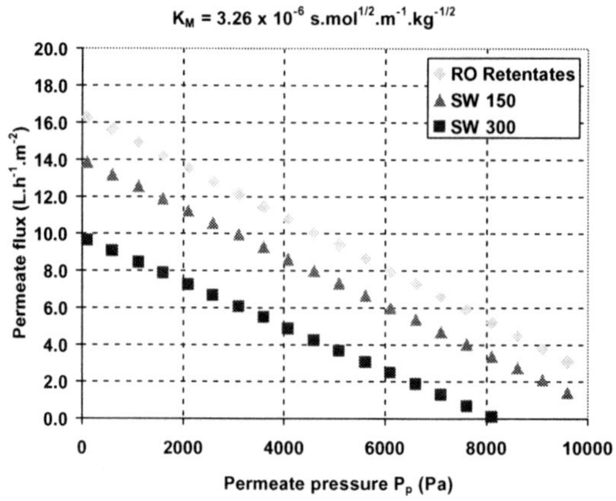

Figure 5.4 Variation of permeate flux vs. Permeate pressure P_p (T_f = 50°C, Re = 4,500) for a $K_M = 3.26 \times 10^{-6}$ s.mol$^{1/2}$.m^{-1}.kg$^{-1/2}$ membrane

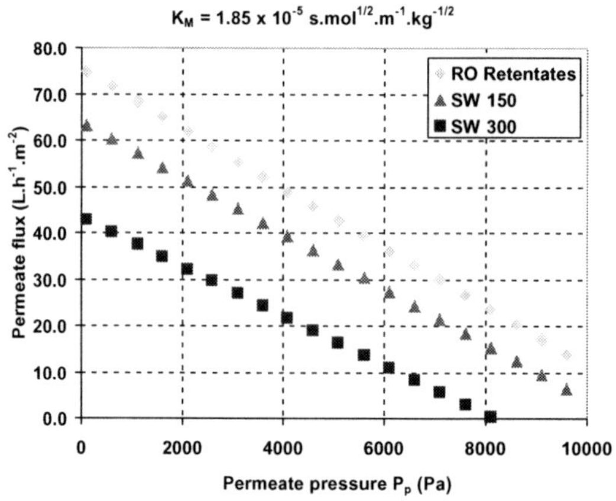

Figure 5.5 Variation of permeate flux vs. Permeate pressure P_p (T_f = 50°C, Re = 4,500) for a $K_M = 1.85 \times 10^{-5}$ s.mol$^{1/2}$.m^{-1}.kg$^{-1/2}$ membrane

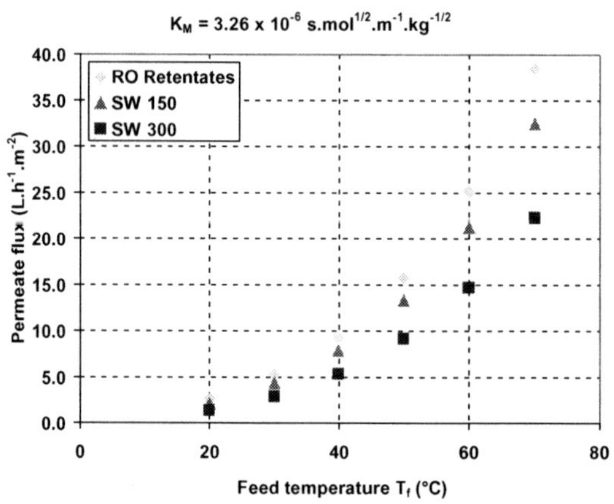

Figure 5.6 Variation of permeate flux vs. Feed temperature T_f (P_p = 500 Pa, Re = 4,000) for a $K_M = 3.26 \times 10^{-6}$ s.mol$^{1/2}$.m^{-1}.kg$^{-1/2}$ membrane

Feed temperature is a very sensitive operating parameter which significantly influences both permeate flux and total energy requirement. It has a major influence on the water vapour partial pressure and thus on the transmembrane pressure difference.

The effect of hydrodynamics on the permeate flux is strongly dependent on the Knudsen permeability of the membrane. From a laminar flow (Re = 350) to a turbulent flow (Re = 6,100), the permeate flux increases of 13–15% for the low permeable membrane and of 44–50% for the more permeable membrane.

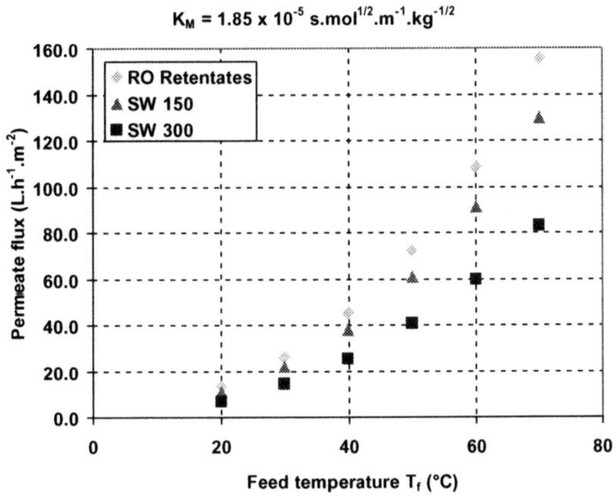

Figure 5.7 Variation of permeate flux vs. Feed temperature T_f (P_p = 500 Pa, Re = 4,000) for a K_M = 1.85 × 10^{-5} s.mol$^{1/2}$.m^{-1}.kg$^{-1/2}$ membrane

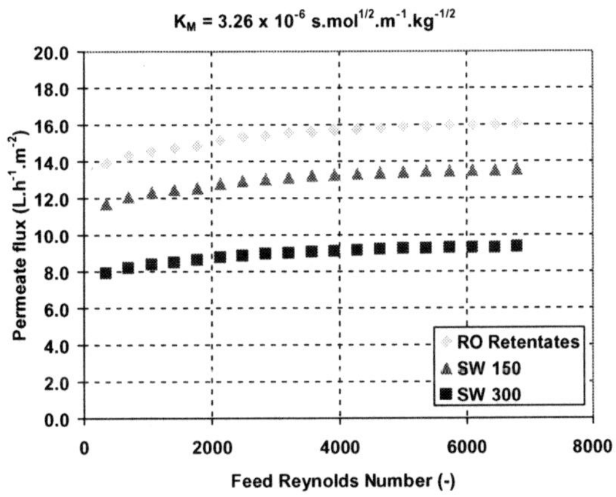

Figure 5.8 Variation of permeate flux vs. Re (P_p = 500 Pa, T_f = 50°C) for a K_M = 3.26 × 10^{-6}s.mol$^{1/2}$.m^{-1}.kg$^{-1/2}$ membrane

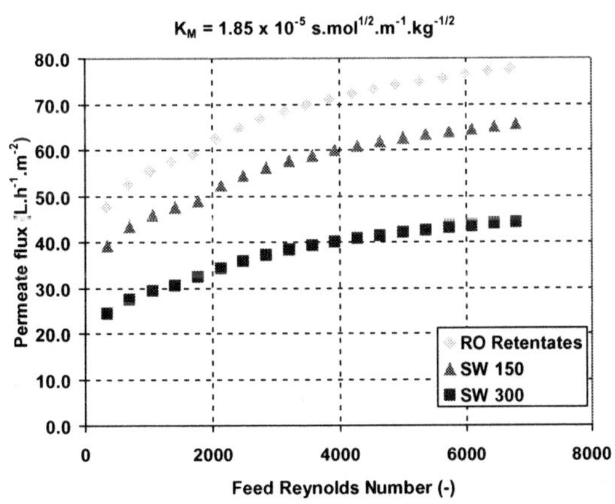

Figure 5.9 Variation of permeate flux vs. Re (P_p = 500 Pa, T_f = 50°C) for a K_M = 1.85 × 10^{-5}s.mol$^{1/2}$.m^{-1}.kg$^{-1/2}$ membrane

For the low permeable membrane, the permeate flux increases slightly when Reynolds number increases (in laminar flow regime) and then reaches a stable value in the turbulent flow regime (Re > 3,000). For the more permeable membrane, the permeate flux increases strongly with Reynolds number (in both laminar and turbulent flow regime). However, the effect of hydrodynamics on the permeate flux are more pronounced in laminar flow regime than in turbulent flow regime and the permeate flux seems to reach slowly an asymptotic value (Re > 6,000).

For the low permeable membrane, the temperature and concentration polarisation have been shown to have nearly no effect which is confirmed by the independence between the permeate flux and hydrodynamics. However, when the membrane permeability is higher, the permeate flux is higher and so concentration and temperature polarisation might be higher, depending on Knudsen permeability value and Re value. Due to these polarisation effects, temperature at the membrane wall might be lower than in the bulk and concentration higher than in the bulk, and these combined effects will decrease the permeate flux. Increasing turbulence of the flow will increase the heat and mass transfer coefficients in the boundary layer near the membrane. Difference between concentration and temperature in the feed and close to the membrane will decrease and so the permeate flux will increase.

5.2.2.1.3 Effect of the membrane permeability

The membrane permeability is a key parameter for the VMD performances. Indeed, without apparently increasing the energy requirement, the permeate flux will drastically increase when the membrane permeability is increased. Figure 5.10 shows the effect of the Knudsen permeability on the permeate flux for a set of operating conditions ($\boldsymbol{P_p}$ = 500 Pa, $\boldsymbol{T_f}$ = 50°C, $\boldsymbol{Re}$ = 4,500).

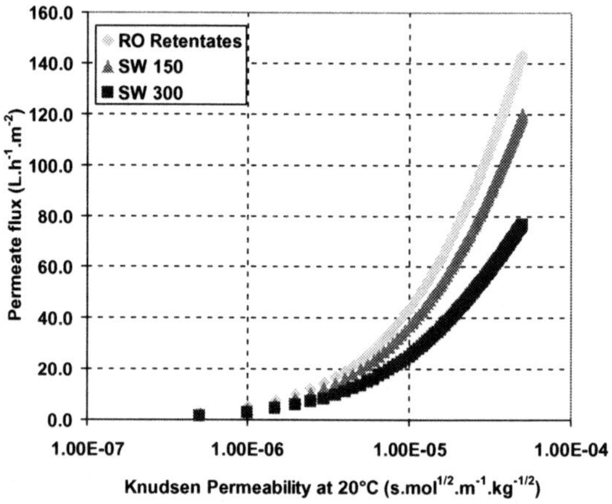

Figure 5.10 Computed variation of permeate flux vs. Knudsen permeability for synthetic mineral seawaters and RO retentates (P_p = 500 Pa, T_f = 50°C, Re = 4,500)

For highly salty solutions, higher fluxes can be reached with high Knudsen permeability membranes. Knudsen permeability depends on membrane structure. High Knudsen permeability can be obtained by using a very porous and thin membrane. However, even a slight increase in Knudsen permeability value drastically improves water permeate vapour flux, it may also decrease membrane hydrophobicity and no longer prevent water in the liquid phase from passing through membrane pores anymore. Wetting is indeed a sensitive point in membrane distillation. A balance must be found between the Knudsen permeability value and membrane hydrophobicity.

5.2.2.2 Experiments with synthetic mineral seawater

Previous experiments performed on GVS membranes confirmed that these membranes were less permeable than some commercial ones. For that reason, INSA decided to carry out all the experiments with concentrated solutions only with Fluoropore (Millipore) membranes ($K_M = 3.26 \times 10^{-6}$ s.mol$^{1/2}$.m^{-1}.kg$^{-1/2}$).

Table 5.5 introduces the operating conditions for the different sets of experiments (A to F): feed temperature inlet $\boldsymbol{T_{fi}}$, permeate pressure $\boldsymbol{P_p}$, feed Reynolds number $\boldsymbol{Re}$. Initial permeate vapour flux from experiments $\boldsymbol{J_{initial}}$ and computed $\boldsymbol{J_{th}}$ are also presented.

Table 5.5 Initial experimental operating conditions and permeate fluxes

Experiment	Feed water	T_{fi} (°C)	P_p (Pa)	Re	$J_{initial}$ (L.h^{-1}.m^{-2})	$J_{th.}$ (L.h^{-1}.m^{-2})
A	SW 95	52	5,000	3,800	9.3	9.8
B	SW 150	53	7,000	3,500	8.0	6.8
C	SW 150	48	4,500	3,500	7.9	6.7
D	SW 300	31	900	2,700	2.6	2.6
E	SW 300	54	5,700	3,900	7.5	4.5
F	RO Retentates	53	7,000	3,900	10.1	9.0

Figure 5.11 presents values of normalized permeate flux versus the produced water volume by membrane surface (recovered permeate volume) for the three previous experiments: the SW 95 (Experiment A), SW 150 (Experiment C) and SW 300 solutions (Experiment E).

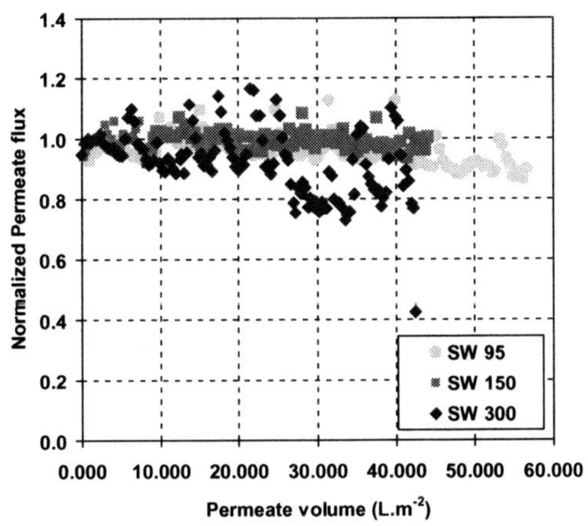

Figure 5.11 Normalized permeate flux vs. Permeate volume: SW 95 (Experiment A from Table 5.1); SW 150 (Experiment C from Table 5.1); SW 300 (Experiment E from Table 5.1)

The permeate flux decline was measured for a permeate volume of 43 $L.m^{-2}$. The decline is respectively of 11%, 8% and 24% for the SW 95, SW 150 and SW 300.

The decrease of the permeate flux can have two explanations. Firstly, it can be due to a modification of the feed water vapour partial pressure caused by the concentration of the feed solution linked to the permeate volume filtrated. Indeed, increasing seawater concentration decreases the water molar fraction X_{water} and the feed water activity coefficient α_{water}. The permeate flux J_{water} is so reduced when the transmembrane pressure difference decreases. Secondly, it can be due to a fouling on the membrane surface which can modify the membrane properties (pore size, porosity, hydrophobicity. . .). In order to isolate these two phenomena, a new Knudsen permeability, called the apparent Knudsen permeability, was calculated versus time:

$$K_M = \frac{J_{water}\sqrt{M_{water}}}{(\alpha_{water}X_{water}p_m^*(T_m) - P_p)} \tag{5.1}$$

Calculation of this apparent Knudsen permeability is based on some experimental data with some hypothesises validated previously in this paper:

- T_m is equal to the feed bulk temperature T_f (no temperature polarisation effects)
- X_{water} is calculated by a mass balance using the permeate vapour flux.
- X_{water} close to the membrane surface is equal to X_{water} in the bulk (no concentration polarisation effects).
- α_{water} is calculated with the PHREEQC software and the VMD modelling.

The apparent Knudsen permeability represents the permeability of the membrane during the experiment. This permeability takes into consideration a possible fouling which modifies membrane properties (reduction of the surface porosity, reduction of pore size. . .).

Figure 5.12 shows the apparent Knudsen permeability versus the permeate volume. After a filtration of 43 $L.m^{-2}$, the decline of the apparent Knudsen permeability was calculated. The decline is respectively of 8%, 9% and 18% for the SW 95, SW 150 and SW 300. The Knudsen permeability variations are very close to those of the permeate flux which seems to indicate that the decrease of the permeate flux is only linked to the modification of the apparent Knudsen permeability. The effect of the concentration can be neglected for these experiments. Nevertheless, it must be noted that the decline of the permeate flux seems very limited in all cases.

The observed modification of the Knudsen permeability can be linked to a modification of membrane properties by scaling. Salt precipitation can partially or totally block membrane pores and thus decrease the area used for supporting the liquid-vapour interface. It can also form a cake on the membrane surface which increases the transfer resistance in the feed water or damage the membrane physically or by wetting. Observations of the fouled membrane (after drying) with SEM and analysis with EDS probe allow visualising the membrane surface and identifying some deposited components. Precipitation may also have occurred during the drying of membrane but this point was not considered here. Figure 5.13 presents a view of the virgin membrane whereas Figures 5.14, 5.15 and 5.16 show the used membranes after permeation of respectively the SW 95, SW 150 and SW 300 waters.

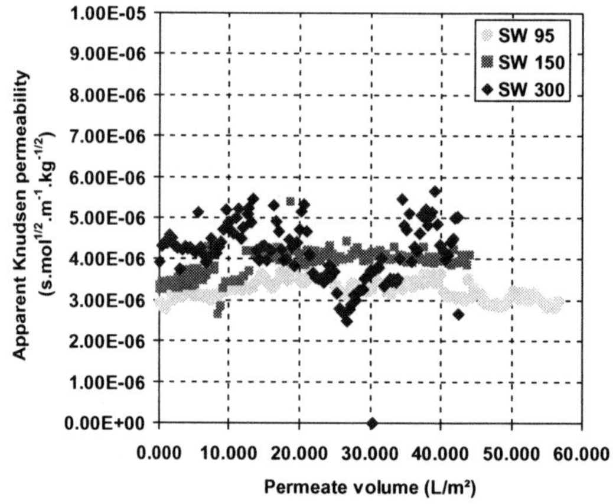

Figure 5.12 Apparent Knudsen permeability vs. Permeate volume: SW 95 (Experiment A from Table 5.1); SW 150 (Experiment C from Table 5.1); SW 300 (Experiment E from Table 5.1)

Figure 5.13 New membrane (×1500)

Figure 5.14 Membrane after SW 95 experiment (×1500)

5.2.2.3 Experiments with real RO retentates

A five day experiment was performed with real RO retentates.

Figures 5.17 and 5.18 describe respectively the variation of the normalized permeate flux and of the apparent Knudsen permeability versus the permeate volume/m^2. For all the experiments, the obtained permeate was fresh water with conductivity lower than 40 μS.cm^{-1}.

A decrease of the permeate flux was observed the first two days linked to a decrease of the apparent permeability. However, after this two first days, the permeate flux and the apparent permeability remain constants. This seems to indicate that there is no effect of scaling, organic fouling or biofouling. It must be noted that salt concentration (about 50 g.L^{-1}) is

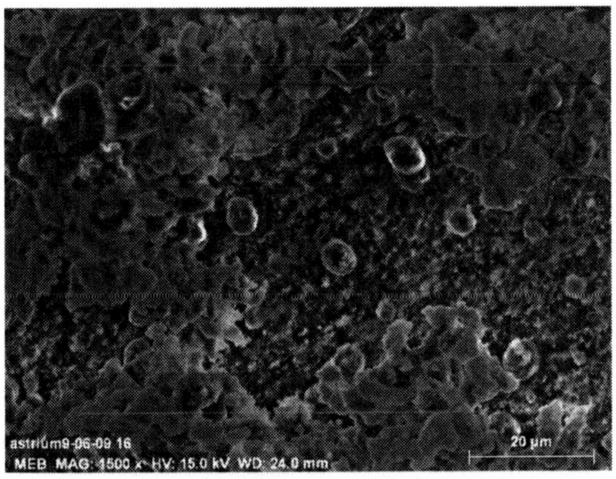

Figure 5.15 Membrane after SW 150 experiment (×1500)

Figure 5.16 Membrane after SW 300 experiment (×1500)

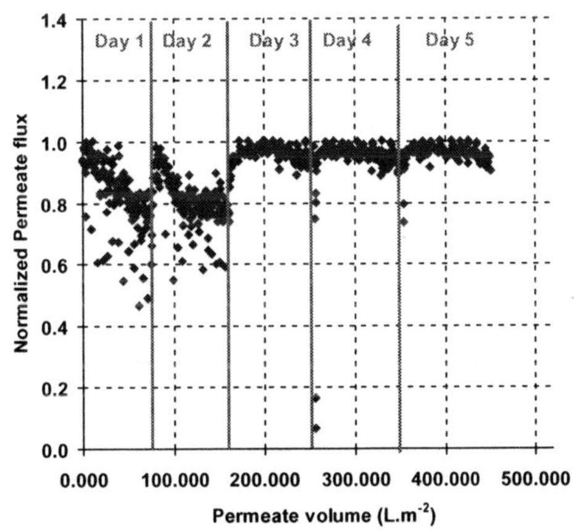

Figure 5.17 Normalized permeate flux versus permeate volume for a long term experiment with a real RO retentate

much lower than in the synthetic solutions and concentration in organic matter (1.3 and 1.6 mg.L^{-1} of TOC) is very low compared to the total salt concentration. TOC concentration in the RO retentates is very close to TOC concentration in seawater. The pre-treatment performed on site before RO (acidification and microfiltration membrane) seems to remove a large part of the TOC. During the RO process, salt and remaining TOC are concentrated. Very few salt precipitations (calcium carbonate) might occur. Another problem may nevertheless be the organic matter or biofouling. Organic matter is not enough concentrated to block the membrane pores but can modify the hydrophobicity of the membrane by adsorption.

5.2.2.4 Conclusion

A simulation of an industrial use of the VMD to further concentrate RO brines was performed. A 40,000 m^3/day RO plant was simulated with a 40% recovery factor and a 38.9 g L^{-1} feed seawater. It was considered that the VMD

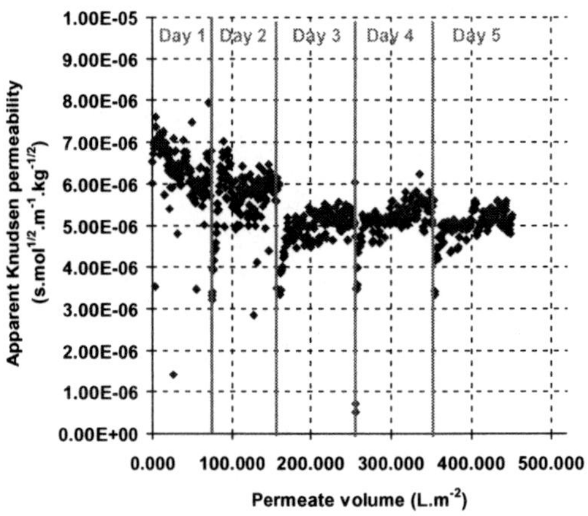

Figure 5.18 Apparent Knudsen permeability versus permeate volume for a long term experiment with a real RO retentate

Table 5.6 Simulation results from SWRO alone and SWRO coupled with VMD to concentrate the RO brines

	SWRO	SWRO + VMD
Water production (m^3/h)	1667	3714
Water recovery factor (%)	40	89
Feed seawater flow (m^3/h)	4167	4167
Seawater concentration (g/L)	38.9	38.9
Brine flow (m^3/h)	2500	453
Brine concentration (g/L)	64.8	297.4

further concentrates RO brines in a batch circulation during 1 h up to about 300 g L^{-1} concentration in given operating conditions (Pp = 6,000 Pa, Tf = 50ºC and Re = 4,000). Next Table 5.6 summarizes simulation results from the SWRO alone and the SWRO coupled with VMD to concentrate the RO brines.

With SWRO coupled with VMD, the water recovery increases from 40 to 89 % and the brine volume can be reduced by a 5.5 factor. The seawater was concentrated 7.6 times and the part of the water production by VMD is more important than the one by RO.

5.3 WT 5.2: RECOVERY OF DISSOLVED SALTS AS CRYSTALLINE PRODUCT

In WT 5.2 two others processes have been tested as reliable and efficient separation methods able to reduce the volume of concentrated streams usually discharged by the desalination plants: membrane crystallization (MCr) and wind intensified enhanced evaporation (WAIV).

Membrane crystallization is an extension of the membrane distillation process. In fact, MCr uses the evaporative mass transfer of volatile solvents through microporous hydrophobic membranes in order to concentrate feed solution above their saturation limit, thus attaining a supersatured environment where crystals may nucleate and grow. Since seawater is the most abundant aqueous solution on the earth (3.3% of its composition is represented by dissolved salts, and seven elements (Na, Mg, Ca, K, Cl, S and Br) account for 93.5% of the ionic species), concentrated streams of the desalination plants contain about the same components of seawater but at higher concentration. These ions can be recovered as high quality crystals through MCr. For inland desalination plants, WAIV can be used as way to bring concentrate to solid or near solid.

The use of Vacuum Membrane Distillation, Membrane Crystallization and/or WAIV operations is assumed to be a reasonable approach for securing the major benefits to sustainable water management in terms of increased water recovery factor from marginal resources, reduction of brine volumes and of disposal costs. In order to realize this goal, the proposed research project considered the following different design alternatives whose performance has been evaluated:

- Membrane Crystallizer operating on: i) RO retentate, ii) post-treated NF retentate;
- WAIV on RO concentrates;
- Membrane Crystallizer operating on WAIV super-concentrates.

5.3.1 Membrane crystallization

The analysis and the improvement of membrane crystallization technology for the exploitation of the components present in the retentate streams of the desalination plants has been studied by UNICAL in WT 5.2 and WT 5.3.

UNICAL's tasks:

a. building of a membrane crystallizer lab plant;
b. analytical description, fluid-dynamic analysis and kinetic study of the crystallizing equipment;
c. MCr tests on synthetic NF and RO retentate streams and estimation of the effect of organics on crystals size, coefficient of variation (CV) and crystals growth rate;
d. MCr lab experiments on real RO brine and real concentrated brackish water;
e. analysis of the methods utilized for removing cleanable deposits in the carried out experimental tests;
f. economical analysis of the desalination system in which membrane crystallizer operates on RO retentate or on post-treated NF retentate.

Topics enclosed in points a., b., c. and d. have been meticulously described in the deliverable *D5.2.1 – Protocols for salts crystallization*. The methods for removing cleanable deposits in the carried out experimental tests, together with the achieved results in term of permeation flux recovery have been reported in the *D 5.2.3 – Defined methods for removing cleanable deposits*, whereas the economical analysis of a membrane-based desalination system with membrane crystallizer units operating on the reverse osmosis (RO) or on the nanofiltration (NF) retentate streams (WT 5.3) has been illustrated in the deliverable *D 5.3 – Report on the economic analysis*. Moreover, the use of MCr to concentrate brine streams of the desalination plants was reported in the following articles:

- F. Macedonio, E. Drioli, E. Curcio, G. Di Profio, *Experimental and economical evaluation of a membrane crystallizer plant*, Desalination and Water Treatment, 9 (2009) 49–53.
- F. Macedonio, E. Drioli, *Hydrophobic Membranes For Salts Recovery from Desalination Plants*, Desalination and Water Treatment, 18 (2010) 224–234.
- F. Macedonio, L. Katzir, N. Geisma, S. Simone, E. Drioli, J. Gilron, *Wind-Aided Intensified eVaporation (WAIV) and Membrane Crystallizer (MCr) Integrated Brackish Water Desalination Process: Advantages and Drawbacks*. Desalination (2011). doi: 10.1016/j.desal.2010.12.002. *In press.*

In succession, a detailed description of the topics is presented.

a. *Building of a membrane crystallizer lab plant.* In the first year, an existing lab plant has been rehabilitated and improved in order to carry out the membrane crystallizer experimental activity with the control of the process.

In Figure 5.19 the schematic flow sheet of the put right MCr lab plant is shown.

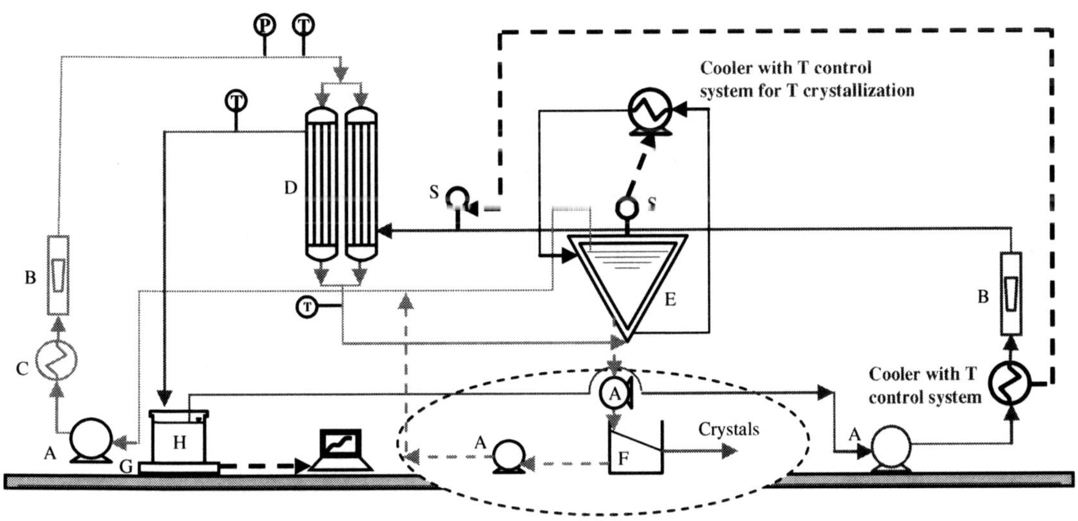

Figure 5.19 Schematic flow sheet of the MCr lab plant: (A) pump; (B) flow-meter; (C) heater; (D) membrane module; (E) crystallizer tank; (F) crystals separation system; (G) balance; (H) distillate tank; (P) manometer; (T) thermocouples; (S) external temperature sensor (Macedonio *et al.*, 2009; Macedonio *et al.*, 2010 and Macedonio *et al.*, 2011, in press)

The retentate and the permeate streams are the foremost lines: they converge in counter-current mode towards the membrane module. Once the crystals are formed, they are removed from the plant through a "*crystals recovery system*", highlighted in Figure 5.19 in the dotted area. The *crystals recovery system* has been designed and built in the laboratory. In the plant the control of the temperature inside the crystallization tank occurs through the use of a refrigerated bath supplied with external temperature sensor. The estimation of the trans-membrane flux occurs by evaluating weight variations in the distillate tank with a balance. A second refrigerated bath provides the maintenance of the inlet distillate temperature by removing the heat flux transferred from the retentate to the distillate side of the membrane modules.

b. *Analytical description, fluid-dynamic analysis and kinetic study of the crystallizing equipment.* The analytical description of the system included energy and mass balances.

The energy balance allowed to determine the total heat transferred across the membrane throught the following equation:

$$Q = U \cdot \Delta T \tag{5.2}$$

where U is the overall heat transfer coefficient that for the MD/MCr process is given by:

$$\frac{1}{U} = \frac{1}{h_f} + \frac{1}{h_m + h_v} + \frac{1}{h_p} = \frac{1}{h_f} + \frac{1}{\left(\frac{K_g \cdot \varepsilon + K_m(1-\varepsilon)}{\delta}\right) + \left(\frac{N \cdot \Delta H_V}{T_{fm} - T_{pm}}\right)} + \frac{1}{h_p} \tag{5.3}$$

where N is the rate of mass transfer, ΔH_V is the heat of vaporization, each *h* and each *T* represent the corresponding heat transfer coefficients and temperatures shown in Figure 5.20.

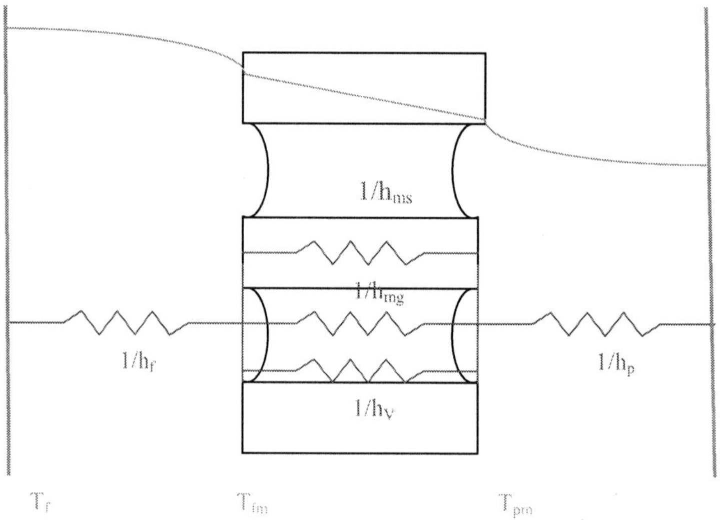

Figure 5.20 Heat transfer resistances in MD

The mass balance across the feed side boundary layer yields the relationship between molar flux N, the mass transfer coefficient k_x and the solute concentrations c_m and c_b at the interface and in the bulk, respectively:

$$\frac{N}{\rho} = k_x \ln\frac{c_m}{c_b} \tag{5.4}$$

where ρ is the solution density.

The mass transfer process through the membrane pores in MD is driven by the imposed vapour pressure gradient between both sides of the membrane. The mass transport mechanism is governed by three basic mechanisms known as Knudsen-diffusion, Poiseuille-flow, Molecular-diffusion or the combinations between them known as transition mechanism (excluding surface diffusion, negligible in MD because, by definition of the MD phenomenon, molecule-membrane interaction is low and the surface diffusion area in MD membranes is small compared to the pore area).

The Dusty–gas model is usually used as a general model taking into account the latter basic mechanisms:

$$\frac{N_i^D}{D_{ie}^k} + \sum_{j=1\neq i}^{n} \frac{p_j N_i^D - p_i N_j^D}{D_{ije}^0} = \frac{-1}{RT}\nabla p_i \tag{5.5}$$

$$N_i^v = -\frac{\varepsilon r^2 p_i}{8RT\tau\mu}\nabla P \tag{5.6}$$

$$D_{ie}^k = \frac{2\varepsilon r}{3\tau}\sqrt{\frac{8RT}{\pi M_i}} \tag{5.7}$$

$$D_{ije}^0 = \frac{\varepsilon}{\tau}D_{ij}^0 \tag{5.8}$$

where N^D is the diffusive flux, N^V is the viscous flux, D^k is Knudsen diffusion coefficient, D^0 is the ordinary diffusion coefficient, p_i is the partial pressure of the component i, P is the total pressure, M_i is the molecular weight of component i, r is the membrane pore radius, ε is the membrane porosity (assuming the membrane consists of uniform cylindrical pores), μ is the fluid viscosity, τ is the membrane tortuosity. The subscript "e" is indicative of the effective diffusion coefficient function of the membrane structure.

However, regardless of which mechanism involved in the mass transportation process, the molar flux, N, must be proportional to the vapour pressure difference across the membrane:

$$N = C \cdot \Delta P \tag{5.9}$$

where ΔP is the vapour pressure difference across the membrane (function of temperatures and compositions at the membrane surface), C is the membrane distillation coefficient that can be obtained experimentally. C is a function of temperature, pressure and composition within the membrane as well as membrane structure and depends on the MD configuration employed as well as on the Knudsen number (Kn, ratio of the mean free path of the transported gas molecules (λ) through the membrane pores to the mean pore diameter of the membrane (*d*)). Finally, mass transfer in permeate boundary layer is not taken into account since the mole fraction of the transporting species in the permeate stream is approximately equal to one.

As it can be seen from Eq. (5.2)–(5.9), simultaneous solution of the heat and mass transfer equations must be carried out via iteration because the heat transfer rate is dependent on mass flux and vice versa.

On the other side, the estimation of the total heat transferred across the membrane (Eq. (5.2)) has been of extreme importance for determining the required heat to be furnished to the MCr feed. In fact, a suitable heating or cooling of the MCr feed (T_f) can guarantee that the temperature of the solution flowing along the membrane is high or low enough to be always under saturation condition, thus guaranteeing crystals formation in the crystallization tank and not in the membrane module or on the membrane surface. Then, an iterative procedure has been put right for the calculation of T_f: At the end, $T_f \geq 34°C$ was obtained, both for NF and RO retentate.

c. *MCr tests on synthetic NF and RO retentate streams and estimation of the effect of organics on crystals size, coefficient of variation (CV) and crystals growth rate.* The experimental study has been conducted on (i) aqueous solution of sodium chloride, (ii) aqueous solution of epsomite ($MgSO_4 \cdot 7H_2O$) and (iii) synthetic aqueous solution representing the RO and the NF brine composition. Moreover, in order to evaluate the effect of organics on the crystallization process, (iv) MCr tests on synthetic NF/RO brine solutions with different humic acid concentrations have been also carried out. Finally, the results achieved from the crystallization of RO synthetic brine solutions have been compared with those obtained from the crystallization of (v) real sea/brackish water RO brine.

By analysing NF and RO brine composition, calcium sulphate, magnesium sulphate and sodium chloride are the salts that can be recovered. The preliminary objective has been to limit calcium sulphate precipitation that causes the reduction of SO_4^{2-} content in the solution and drastically limits the recovery of magnesium sulphate. Ca^{2+} ions have been precipitated as $CaCO_3$ through reactive precipitation with anhydrous sodium carbonate Na_2CO_3. The composition of the achieved crystallites have been analyzed by EDX method.

After the precipitation step, the so-treated retentate streams have been sent to a MCr in order to recover sodium chloride and magnesium sulphate. The crucial requirement of the MCr has been to prevent crystals deposition on membrane surface and inside the membrane module.

The objectives of the experimental tests have been:

- to check the crystallization conditions in which there are no crystal deposition inside the membrane module;
- to verify that the "crystals recovery system" is able to remove the produced salts and, in particular, to avoid their deposition inside the membrane module and in the retentate line;
- to test temperature and fluid-dynamic effect on membrane crystallization operation.

A tool which helps in detecting crystals deposition inside the membrane module is the trans-membrane flux. Figure 5.21 shows the trend of trans-membrane flux with time as obtained during lab tests at two different retentate flow rates and constant temperature, and using as MCr feed the NF retentate.

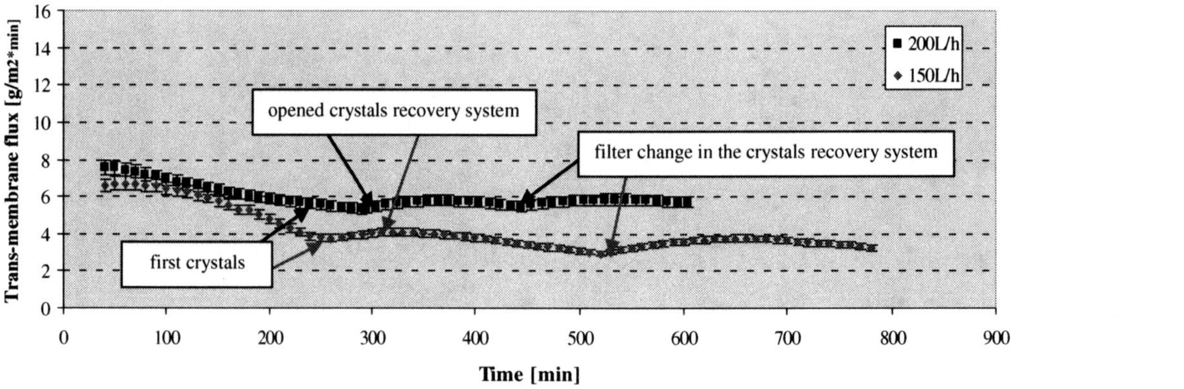

Figure 5.21 Trans-membrane flux vs time at different retentate flow rates and constant temperature ($T_{feed,\ retentate\ side}$ = 34 ± 1°C; $T_{feed,\ permeate\ side}$ = 16 ± 2°C) (Macedonio *et al.*, 2009 and Macedonio *et al.*, 2010)

In both cases, apart the initial transitory stage trans-membrane flux has shown almost a constant trend, therefore during the operative time no crystals deposition inside the membrane module has been observed. This happens because the chosen temperature of feed and the selected thermal difference between the two streams have been high enough to contain the decrease in driving force due to concentration rise. Moreover the slight decrease in trans-membrane flux observed when the crystals were formed has been solved putting into operation the home-made "*crystals recovery system*". This tool allows to separate the produced salts from the crystallizing solution: the crystals are kept on the paper filter while the remaining solution is sent to the membrane modules in order to be further concentrated.

Moreover, the figure shows that solvent trans-membrane flux enhances when feed flow rate increases. Feed flow rate is an important parameter in MCr operations. It influences both mass and energy transport phenomena: higher feed flow rate means higher Reynolds number and transport coefficients. As a consequence, polarization effects decrease and higher fluxes can be expected.

Figure 5.22 shows that the trans-membrane flux increases with the temperature of the retentate. This trend is characteristic of a temperature driven membrane operation (as membrane crystallization) in which the driven force grows when the temperature of the feed and/or the trans-membrane temperature difference increase.

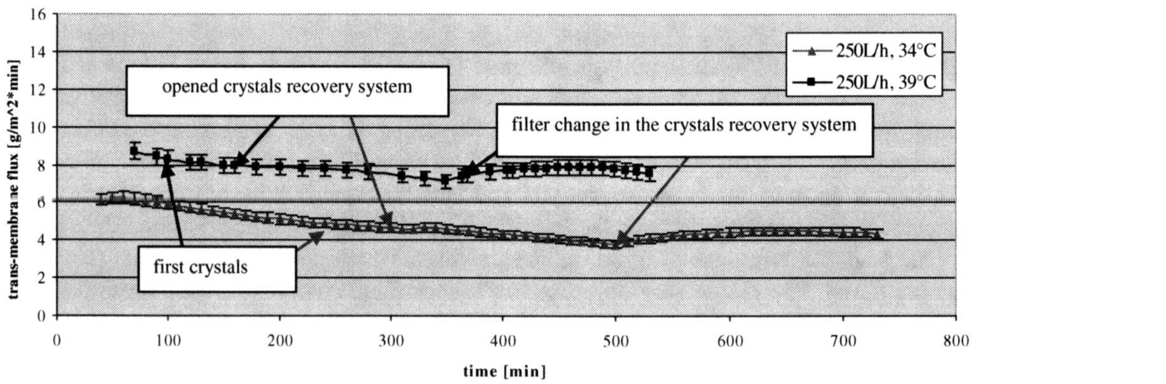

Figure 5.22 Trans-membrane flux vs time at different retentate temperature and constant flow rate (retentate flow rate = 250 L/h) (Macedonio *et al.*, 2010)

For what concerns the type of produced crystals, they have been analyzed by SEM (Scanning Electronic Microscopy), EDX (Energy Dispersive X-ray), low temperature DSC (Differential Scanning Calorimeters) and FT IR (Fourier Transform Infrared Spectroscopy) methods:

1) the low temperature DSC measures on the achieved crystals (maximum temperature 250ºC) clearly showed that no $MgCl_2 \cdot 6H_2O$ was formed during the MCr tests of NF/RO retentate;
2) the EDX (recorded using a Philips EDAX Analysis System) and FT IR analysis proved the formation of NaCl from RO retentate. Both NaCl and epsomite have been found in the salts produced from the crystallization of NF retentate.

Since the crystals appeared, every 30 minutes, a suspension sample of them has been extracted, particles have been filtered, weighted and examined visually with an optic microscope (ZEISS, model Axiovert 25) in order to determine the crystal size distribution (CSD) at various time intervals. Knowledge of the evolution of particle size distribution as function of time has allowed to evaluate quality, coefficient of variation (CV), middle diameter (d_m), growth rate of the produced crystals and to test the fluid-dynamic effect on crystals shape, dimension and growth rate.

Some experimental evaluations of the coefficient of variation (CV) and of the middle diameter (d_m) as obtained in the carried out NF retentate crystallization tests are reported in Table 5.7.

As can be seen in the Table 5.7, when crystals grown under milder conditions, CSDs are characterized by low CVs (characteristic of narrow distributions around the mean crystal size). In particular, for MCr feed flow rate equal to 120 L/h, the **target design of CV less than 35% has been reached**.

Table 5.7 Evolution in function of time of Coefficient of Variation (CV) and middle diameter (d_m) determined from the experimentally obtained CSDs at three different MCr feed flow rate (MCr feed = NF retentate; $T_{feed,\ retentate\ side} = 39 \pm 1°C$; $T_{feed,\ permeate\ side} = 16 \pm 2°C$)

120 L/h			150 L/h			250 L/h		
Time [min]	d_m [μm]	CV [%]	Time [min]	d_m [μm]	CV [%]	Time [min]	d_m [μm]	CV [%]
150	25.31	34.38	130	29.55	43.52	100	30.35	25.00
180	22.36	35.00	160	33.64	38.33	130	38.68	45.59
210	29.61	26.72	190	43.28	47.62	160	44.46	39.09

The obtained NaCl crystals, examined visually with an optic microscope, showed the characteristic cubic block-like form in accordance with the expected geometry of the NaCl crystals. Their cubic shape has been also confirmed by the length/width ratio distributions (reported in the histograms of Figure 5.23) that showed the most part of particles has a length/width ratio in the range 1.0–1.2, that is to have almost a cubic shape. Moreover, the number of crystals exhibiting a cubic block-like form increases when retentate flow rate decreases.

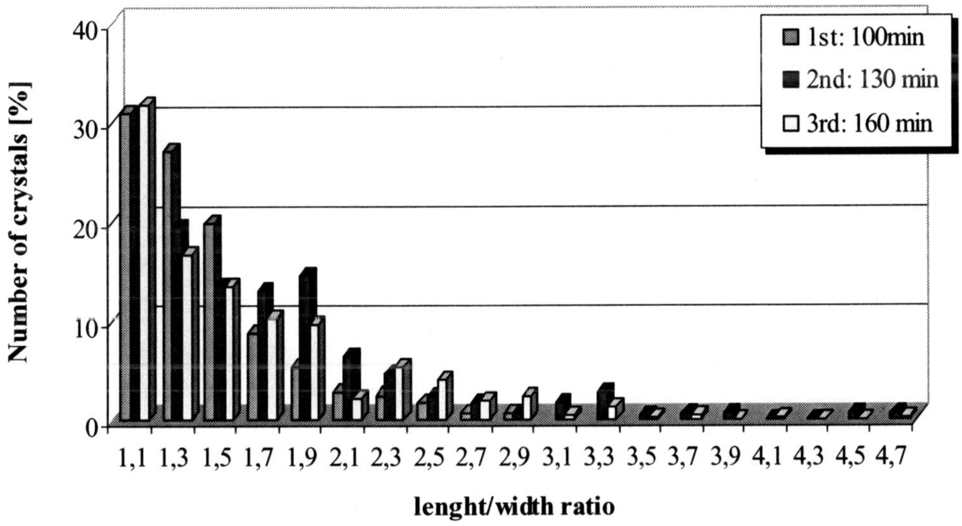

Figure 5.23 Number of crystals [%] vs length/width ratio. (MCr feed = NF retentate; $T_{feed,\ retentate\ side} = 39 \pm 1°C$; $T_{feed,\ permeate\ side} = 16 \pm 2°C$; Retentate flow rate = 250 L/h)

Conductivity measurements carried out on samples of solution taken out from the distillate tank demonstrated that the infiltration of salts through the membrane pores was negligible; therefore, polypropylene membranes preserved the crucial requisite of hydrophobicity.

5.3.1.1 Modelling of membrane crystallization

The Randolph-Larson general-population balance has been applied in order to predict nucleation and growth rates based on the experimental population density data. Plotting growth rate G versus retentate flow rate it has been possible to study the fluid-dynamic effect on crystals growth rate.

The obtained trend of the crystal growth rate versus feed flow rate (Figure 5.24) has proved that crystal growth is limited by the diffusional resistance to the movement of molecules to the growing crystal face.

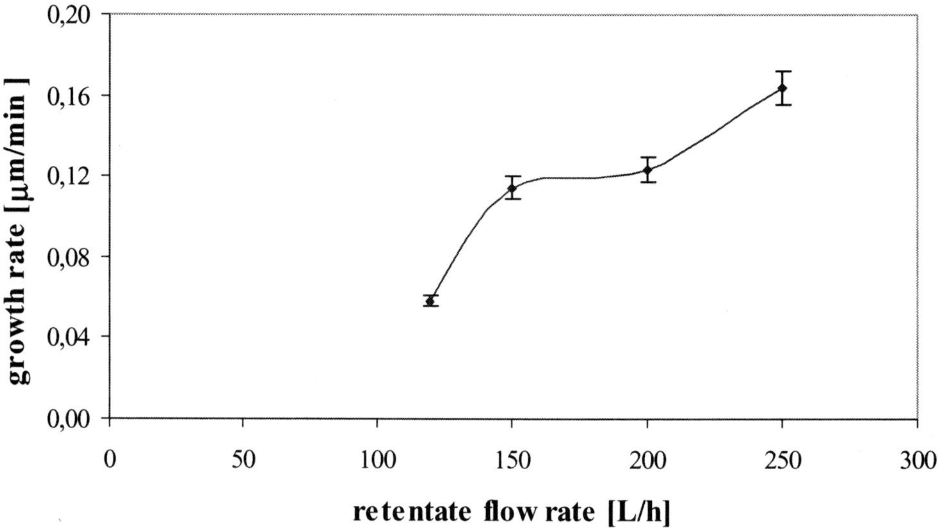

Figure 5.24 Growth rate vs retentate flow rate for NaCl crystals produced from the crystallization of NF brine (Macedonio *et al.*, 2010)

d. *MCr lab experiments on real RO brine and real concentrated brackish water.* Results achieved from the crystallization of *synthetic* NF/RO brine solutions have been used as starting point for the crystallization of *real* sea/brackish water RO brine.

In particular, the influence of organic components on membrane crystallizer process as well as crystals dimension, growth rate and composition have been analyzed and compared with the outcomes of the previous tests.

The MCr experimental runs on super-concentrated brackish water RO brine have been carried out on samples coming from a desalination plant in which feed water was first treated with four 4-inch BWRO elements (75% recovery) and the BWRO brine was then further concentrated through a Wind-Aided Intensified eVaporation (WAIV) step. ***This work has been carried out in collaboration with colleagues from Ben Gurion University who provided the super-concentrated brackish water RO brine to be submitted to MCr tests.***

Two different types of WAIV super-concentrated were further concentrated through the MCr process:

- a first sample named BGU 1 without antiscalant or organics,
- a second sample named BGU 2 with antiscalant and organics.

The aim of the analysis was, therefore, to study the effect of antiscalant and organics on the membrane crystallizer process.

Ca^{2+} ions contained in the two super-concentrated WAIV samples have been recovered as $CaCO_3$ through reactive precipitation with anhydrous sodium carbonate.

After the precipitation step, the so-treated streams have been sent to the MCr in order to recover the water and the salts contained.

In the case of BGU 1, the chosen operative conditions have guaranteed good stability to the MCr process (Figure 5.25). In fact, trans-membrane flux has shown almost a constant trend during the operative time.

In the concentration of the sample named BGU 2, despite the higher driven force with respect that used in the test called BGU 1, a decrease of flux during time was observed (Figure 5.25). This was due (i) to the formation of small amount of $CaSO_4$ and (ii) to the higher organic concentration which probably caused a fouling layer.

Some experimental evaluations of the coefficient of variation (CV) and of the middle diameter (d_m) as obtained in the carried out crystallization tests are reported in Table 5.8. The low achieved CVs are characteristic of narrow crystal size distributions and, therefore, of a qualitatively good product. It can also be seen that crystals pulled from the MCr run with BGU2 (containing antiscalant) have somewhat lower CVs than those from the MCr run with BGU1 (without antiscalant) because the temperature of the system (in the run with BGU2) was held higher than that in the run with BGU1.

The SEM/EDX analysis of the precipitated salts clearly showed the presence of NaCl and small amount of $CaSO_4$.

For what concerns the shape of the produced NaCl crystals, the most part of them showed the characteristic cubic block-like form.

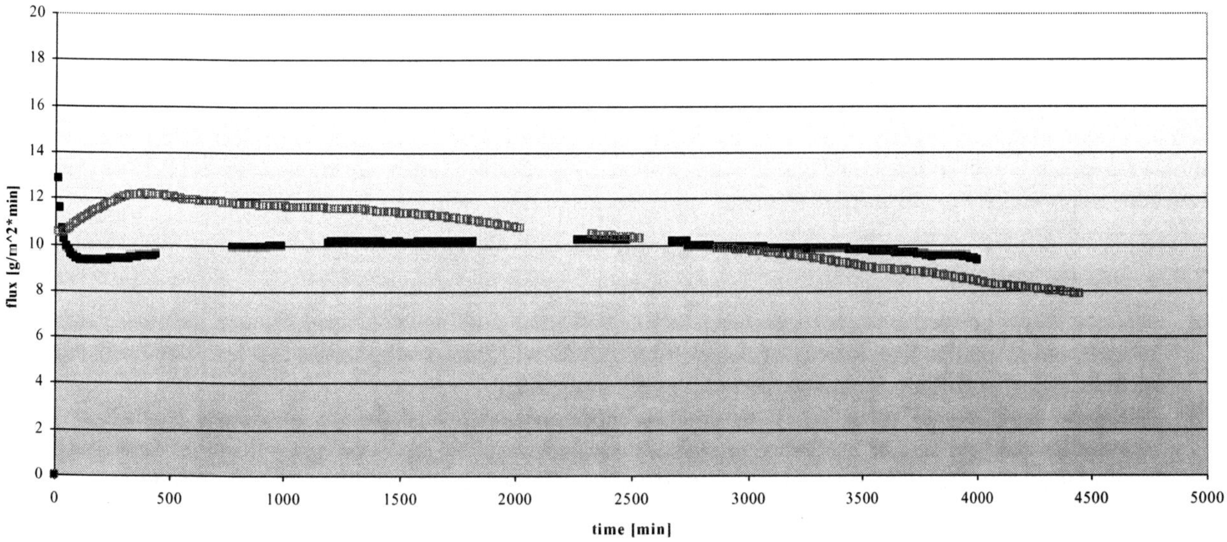

Figure 5.25 Trend of the trans-membrane flux vs time in the MCr test carried out using as feed water the WAIV super-concentrates

- Grey, BGU 2 (concentrated brackish water with organics. First crystals after 3720 min)
- Black, BGU 1 (concentrated brackish water without organics. First crystals after 3390 min).

Table 5.8 Evolution in function of time of Coefficient of Variation (CV) and middle diameter (d_m) determined from the experimentally obtained CSDs

BGU 1			BGU 2		
time [min]	d_m [μm]	CV [%]	time [min]	d_m [μm]	CV [%]
3450	35.1	40.2	4370	32.0	39.3
3660	37.4	45.0	4430	40.0	34.3

e. *Analysis of the methods utilized for removing cleanable deposits in the carried out experimental tests.* Some of the main problems in membrane separation systems, which can considerably affect the economy of the processes and whose control is a important problem in the design of membrane processes, are in fact membrane fouling and concentration polarization phenomena.

Although in MD/MCr membrane fouling and concentration polarization are not relevant problems than in other membrane separations, however, fouling particles/molecules attached to the membrane surface might cause plugging of the membrane pore entrances causing first some flux decay and, then, may lead to membrane pore wetting.

Since in WP5 the potentialities of a membrane-based desalination system with membrane crystallizer unit operating on the RO or on the NF retentate streams have been analyzed, *calcium sulphate, magnesium sulphate* and *sodium chloride* are the first and main salts that can be recovered from the crystallization step.

For what concerns *calcium sulphate*, as earlier stated, its precipitation during the concentration of the NF/RO brine has been avoided removing Ca^{2+} ions as $CaCO_3$.

In the membrane crystallizer, NaCl and $MgSO_4$ crystals formation and deposition on the membrane surface and in the membrane module were controlled through the following tools:

- by re-circulating the solution in order to remove particles eventually deposited on the membrane surface;
- by controlling the temperature of the solution flowing along the membrane module due to the fact that the solubility of solids in solution depends on temperature;
- by recovering continuously the produced crystals through a "crystals recovery system". This is a home made filtration system, suitably planned and built in the laboratory for removing the formed crystals, thus avoiding their deposition and/or accumulation inside the plant.

The cleaning of the membrane module and of the overall plant revealed itself necessary in order to try to restore the initial performance of the membrane fouled by humic acid and various inorganic components.

Cleaning was performed by re-circulating first of all pure water for 1 h, followed by 30 min of re-circulation of a citric acid solution (pH 3 ÷ 4), successively 30 min of pure water, then 30 min of a 0.1 M NaOH solution. Finally, a third rinse with distillate water was performed.

5.3.2 Wind-Aided Intensified eVaporation

The use of wind to evaporate desalination brines and reduce their volume in an economic way with minimal energy inputs was realized in WT 5.2 by BGU using Wind-Aided Intensified eVaporation (WAIV) and has previously been tested for evaporating desalination brines from the Ktziot desalination plant. In the present project, efforts were made to combine this process with an integrated desalination scheme by generating superconcentrated brines (10% or greater) to the membrane crystallizer to allow reduced energy inputs for recovering valuable minerals (WT 5.2).

5.3.2.1 BGU's task

a. evaluate WAIV performance on superconcentrates (10%) that could be developed from an integrated membrane process and determine how increasing compostion to 30% will impact evaporative driving forces under ambient meteorological conditions including thermodynamic modelling.
b. determine feasibility of using WAIV to produce superconcentrate feed for the membrane crystallizer so that evaporative and heat load of membrane crystallizer can be reduced when operating on brackish water desalination concentrates, as well as allowing reduction of calcium ion content to reduce sodium carbonate chemical consumption
c. Provide economic estimates for overall process economics

5.3.2.1.1 Tasks description

a. The use of WAIV to evaporate superconcentrates was reported in an article (Katzir, L. *et al.*, 2010). It was also reported at the EDS conference in Baden-Baden in May 2009 and at the membrane EU-MEDINA-China workshop in Qingdao in September 2009. As part of this work, experiments were carried out with a superconcentrate generated from a hybrid RO-ED process in which the brine began at 10% TDS and in another case using RO brine from a 75% recovery and 88% recovery brackish water source in the Negev highlands (Mashabe Sade). The compositions of desalination brines that were subjected to WAIV are shown in Table 5.9:

Table 5.9 Mashabe Sadeh well water and RO and ED concentrates used for WAIV concentration runs

Sample	Unit	Well water (RO feed)	RO conc. 75% = ED Feed	RO conc. 88%	ED Brine
pH		8.22	7.13	7.30	1.96
EC	dS m^{-1}	3.90	13.73	21.5	122.8
Cl	meq L^{-1}	33.0	119.0	160.95	1637.0
Ca	meq L^{-1}	9.44	56.50	89	91.57
Mg	meq L^{-1}	7.91	33.95	42.94	390.34
SO_4	meq L^{-1}	9.03	51.04	74.27	250.79
Sr	mg L^{-1}	6.07	21.27	28.84	77.3
K	meq L^{-1}	0.66	2.07	3.31	18.33
Na	meq L^{-1}	28.6	110	201.165	1227.0
SiO_2	mg L^{-1}	18.96	81.94	131.25	44.36

A thermodynamic software package (OLI^{TR} software, Morristown, NJ.) was employed for two different purposes. The first purpose was to use this software to create a model that will help us to calculate the driving force for this process, by evaluating the concentration factor that is attainable by evaporation from the RO and ED concentrate. The second purpose was to use this software to calculate the saturation index (SI) of different salts in the brine solution and predict which salts were supposed to precipitate during the time of the run.

The results of using this software were provided in terms osmotic pressure from which the relative humidity of the solution was calculated by the following equation:

$$R.H \equiv \frac{P}{P^0} = \exp(-\frac{V_W \Pi}{RT}) \tag{5.10}$$

The evaporative driving force was then provided by the following equation:

$$\Delta e \equiv e_s^* - e_a = [P_{W.B}^0 \left(\frac{P}{P^0}\right)_{T_{W.B}} - (P^0)_{air}(RH)_{air}] \tag{5.11}$$

The results of using this simulation are shown in Figure 5.26, in which the relative humidity of the brines are plotted along with the annual average relative humidity of the air at Sde Boker at different times of the day.

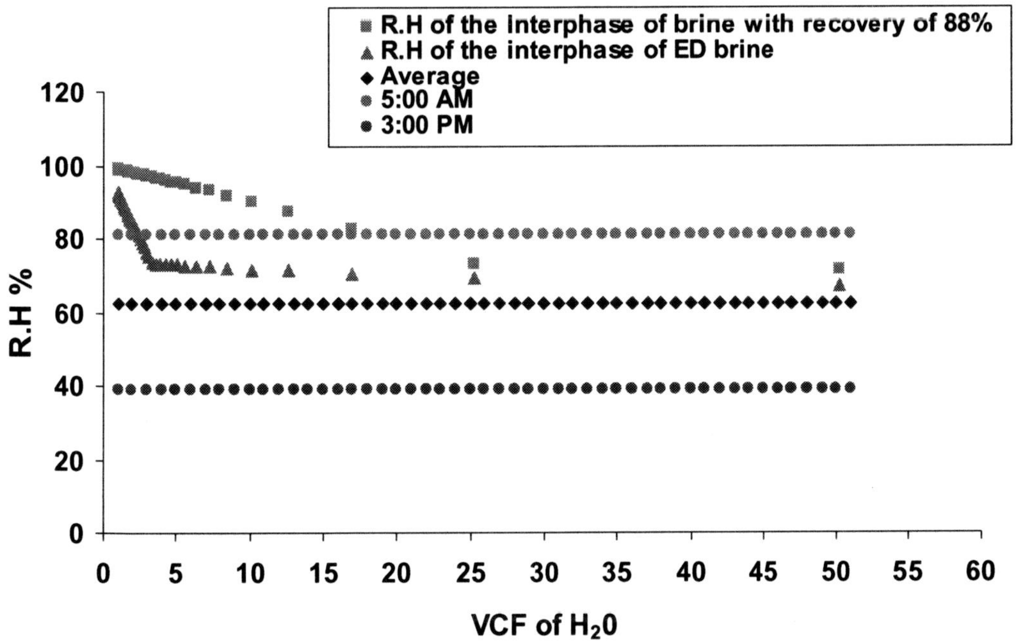

Figure 5.26 Projected relative humidity (R.H.) of RO brine (88% recovery) and RO-ED brine as it is concentrated by evaporation. Yearly relative humidities of Sde-Boker are shown for comparison, day average (diamond dotted line in dark grey), 5:00 AM (circular dotted line in light grey), and 3:00 PM (circular dotted line in mid grey)

The driving force of the evaporation expresses as the vapor pressure gradient (brine solution- air), and can also be expressed as the difference in the R.H. air verses solution (Eq. 5.10). As long as the R.H. of the super concentrated brine is higher then the R.H. of the air, evaporation will occur. If the gradient is equal to zero or negative there won't be evaporation, and moreover condensation will occur.

Figure 5.26 shows that while it is feasible to attain a VCF of 50, it is noticeable that the evaporation rate at VCF as high as 50 won't be efficient due to the significant reduction in the relative humidity gradient (Eq. 5.11). Therefore for RO brine with a recovery of 88% it is recommended to reach up to VCF of 20, and for RO-ED brine the limitation is even lower and it is recommended to reach up to a VCF of 7.

In addition, we have to take into account the change in the R.H. during the day, as early in the morning (5 AM.) the R.H. is about 80% and at the early afternoon (3 PM.) the R.H. is the lowest and reaches only 39%. That means that at early morning we would expect to get a negative evaporation (addition of water to the brine due to condensation) when the brine is very concentrated.

This phenomenon can be seen in Figure 5.27 which presents the evaporation rate of RO-ED brine from the end of batch 6 compared to the evaporation rate of tapwater during the same 48 hours (16/8/09–18/8/09). Every four hours a water sample was taken and the level of the solution was measured.

Figure 5.27 shows that what was assumed from the simulation is what actually happened in real time. During the nighttime when the R.H of the air is higher then the R.H. of the solution, the evaporation rate is negative, as condensation occurs instead of evaporation. This happens for the ED brine but not for the tapwater because the relative humidity of the ED brine is significantly lower.

This observation has significant implications for optimizing the hours of WAIV operation in order to increase the efficiency of the process. When the WAIV unit is operated with super concentrated brine such as RO-ED brine, it needs to be shut down during the nighttime to prevent condensation on the WAIV surfaces from night air that is often close to 100% relative humidity.

A meteorological correlation has been established for predicting evaporation rates as a function of wind speed parallel to the WAIV surfaces. This correlation takes into account local temperatures and relative humidities through the

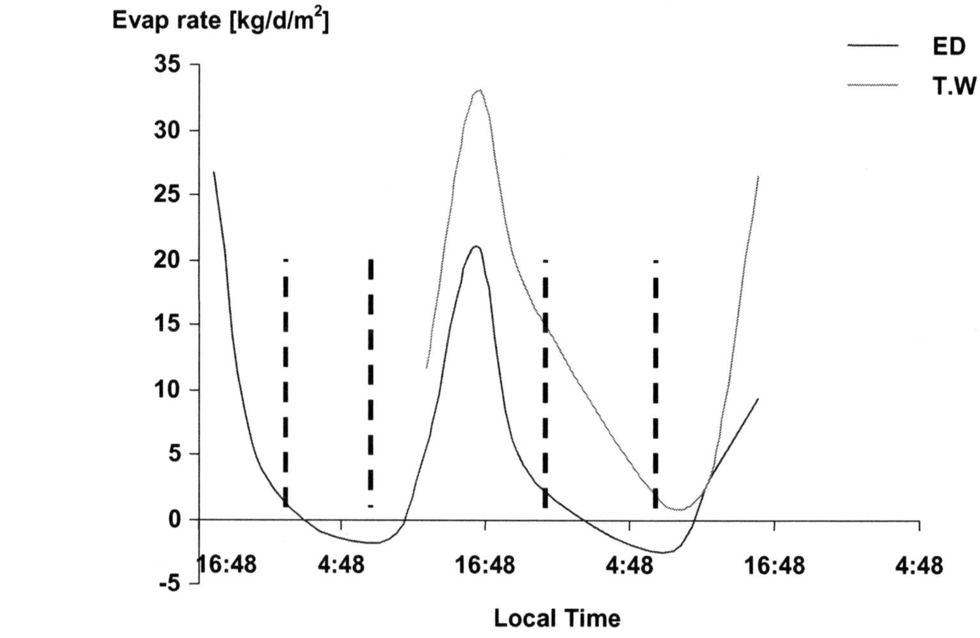

Figure 5.27 Evaporation rate over 48 hours (16/8/09-18/8/09) comparing super concentrated RO-ED brine with initial TDS of 26% vs. tap water

vapor pressure gradient with which the evaporation rate is normalized. This is shown in Figure 5.28 and 5.29 for 88% recovery RO brine and for RO-ED brine respectively. These show reasonably linear correlations between wind speed parallel to the evaporation surface and the specific evaporation rate defined as the evaporation rate divided by driving force of vapor pressure difference between brine surface and air.

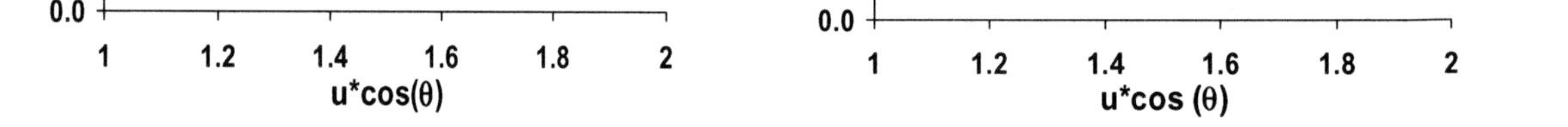

Figure 5.28 Mass transfer correlation for RO brine with 88% recovery. A- shows nominal evaporation rate while B- shows the evaporation rate based on Mg^{2+} ion mass balance

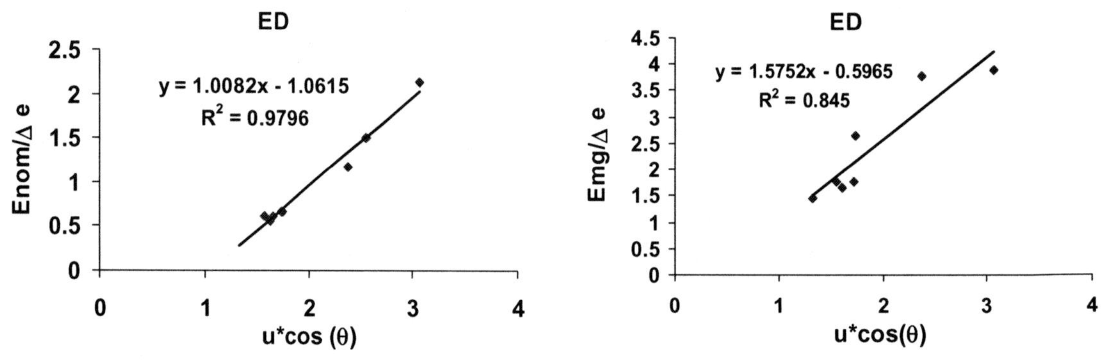

Figure 5.29 Mass transfer correlation for ED brine (A- shows nominal evaporation rate while B- shows the evaporation rate based on Mg^{2+} ion mass balance)

From the wind velocity correlation [f'(u)] used in the Penman equation and from the wind velocity mass transfer correlation found for evaporating RO brines, we see that even when the wind velocity is equal to zero [m/sec], evaporation is expected to occur. This is a result of the positive driving force existing even at low wind velocities and thanks to the natural air convection.

However this is not the case for ED brine. ED brine as discussed before is 10 times more concentrated then the RO brine (starting TDS of 10% and 1% respectively). As the brine is getting more concentrated the relative humidity of the solution reduces significantly. This reduction has a strong effect on relative humidity gradient, as shown in Figure 5.26.

During night time with an annual average of 80%, the relative humidity of the air will usually be higher than the relative humidity of the solution. Moreover at this time the wind velocity often drops to zero [m/sec], so that the combination of these two factors leads to negative evaporation, i.e. condensation as shown in Figure 5.27. This explains the negative intercept for the ED mass transfer correlation shown in Figure 5.29.

b. Over the summer of 2009, four batches of WAIV superconcentrate were provided to UNICAL for using in the membrane crystallizer. The compositions are summarized in Table 5.10. It should be noted that the calcium levels are drastically reduced by using the WAIV to preconcentrate the feed to the membrane crystallizer. This results in a drastically reduced requirement for sodium carbonate to remove the remaining calcium ion (see the economic section in item c following).

The predictions of compositions of the final WAIV concentrate using the OLI program as well as prediction from GWB version 6 carried out at Karlsruhe Research Institute by Ms. J. Scheiber are provided in Figure 5.30. In these simulations the volume concentration factor used for the simulation was the one based on mass balance of magnesium to prevent overestimating the extent of concentration based on the nominal volume changes that could be influenced by leaks or droplet loss. In general the OLI program tended to slightly over-predict while the GWB program tended to somewhat under predict experimentally found compositions.

5.4 WT 5.3: ECONOMIC EVALUATION

The investigated strategies have been also studied in terms of costs in order to establish how their introduction could impact on the overall process cost. In particular, the economical analysis has been carried out for the following different design alternatives:

(a) seawater desalination system with MCr operating on RO retentate;
(b) seawater desalination system with MCr operating on post-treated NF retentate;
(c) brackish water desalination system with WAIV and MCr operating on RO brine.

Cases (a) and (b) have been analyzed by UNICAL whereas system (iii) by BGU.

5.4.1 Description and results of the economical evaluation

(a) and (b). The economical evaluation of a membrane-based seawater desalination system with membrane crystallizer units operating on the reverse osmosis (RO) or on the nanofiltration (NF) retentate streams has been carried out on the basis of the results achieved in the experimental tests.

Figure 5.31 and 5.32 show, respectively, the total amount of salts and the total recovery factor that can be achieved from the desalination system MF/NF/RO with MCr unit on the NF or RO brine and at different MCr recovery factors. The values reported in these figures are of extreme importance for the understanding of the results of the economical evaluation reported subsequently. As it can be seen, whereas the amount of produced salts approximately doubles when MCr recovery factor increases from 90.2 to 97.0%, plant recovery factor improves of only 1.6% in the same range.

The economic evaluation of the proposed membrane systems for seawater desalination has been based on recent economic data and on the following equations extracted from actual field data and from design studies in literature:

- *Direct capital costs (DC)* include the cost for plant construction and for the acquisition of land and process equipments.
- *Indirect capital costs (A_{in})* include freight and insurance, construction overhead, owner's and contingency costs. These costs are usually estimated equal to 10% of total direct capital cost.
- *Annual operating costs*
  1. *Annual electric power cost*: $A_{electric} = c \cdot w. \cdot f \cdot m \cdot 365$ where c = electric cost (0,09 \$/KWh), w = specific consumption of electric power [KWh/m^3], f = plant availability (0.9), m = plant capacity [m^3/d].
  2. *Annual labour cost:* $A_{labor} = \gamma \cdot f \cdot m \cdot 365$ where γ = specific cost of operating labour (0.03 \$/m^3).
  3. *Annual membrane replacement cost, $A_{membrane}$.* Membrane replacement rate varies between 20% per year for membranes treating high-salinity seawater to 10 % per year for membranes treating low-salinity water supported by pre-treatment systems.

Table 5.10 Summary of composition of WAIV endbrines including those shipped to Unical

Batch Desc.	Completion Date	Condition	EC [mS/cm]	TDS [g/kg]	Mg [ICP] [ppm]	Mg [Titrat ppm	Ca [ICP] ppm	Ca [Titrati [ppm]	Na [ppm]	K [ppm]	SO4 [ICP] [ppm]	SO4 turb. [ppm]
75% RO no antiscalant no organic	22.10.08		109	79	3,020	3,358	1,370	1,783	24,080	756	9,407	8,133
75% RO SHMP antiscalant	6.4.09		78.7	90 (meas) 77.5 (calc)	2,897		1,026		24,168	679		12,511
75% RO PC-191	10.6.09		150.6	144 (meas) 124 (calc)	4,434	5,822	780	1,071	33,689	946	17,325	16,989
antiscalant	28.6.09		142	143 (meas) 130 (calc)	5,144	5,521	947	1,027	40,862	959	16,376	
88% RO PC-191 antiscalant	6.8.09		158	142 (meas) 126 (calc)	5,131	5,867	1,786	1,789	38,345	1,107	12,103	
	15.9.09			73	3,080	2,195	1,200	1,131	22,000	620	14,890	

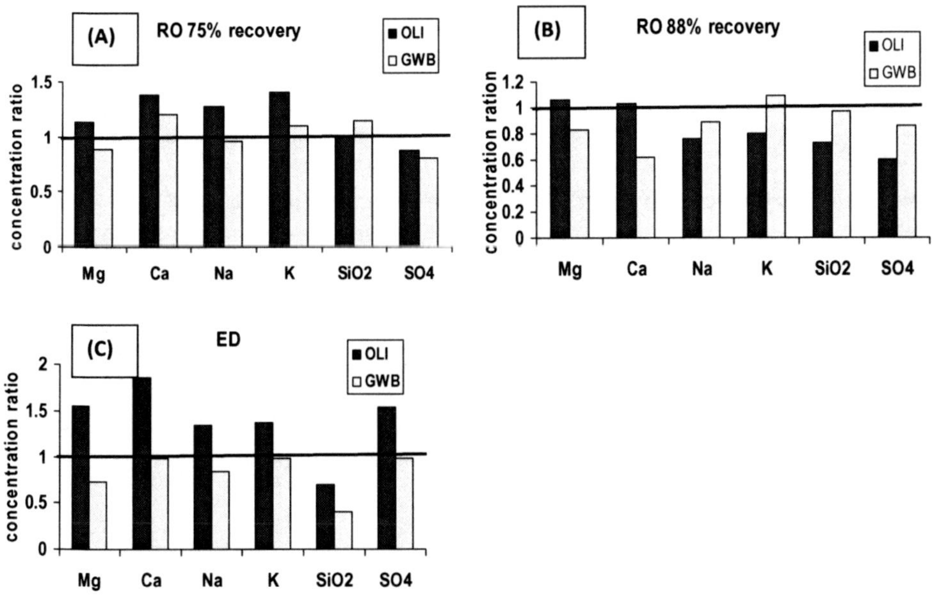

Figure 5.30 (A–C): Present the average ratio between both software predictions – OLI and GWB to the experimental data

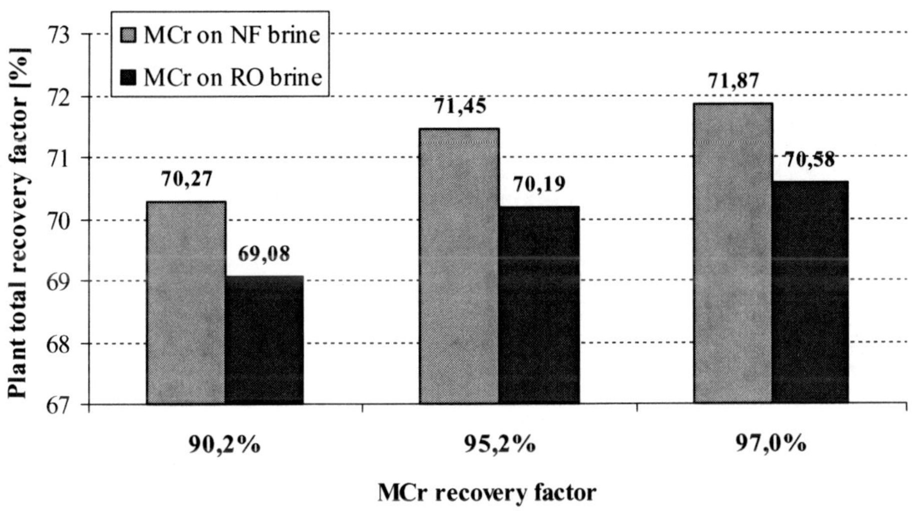

Figure 5.31 Amount of salts produced per m^3 of NF or RO brine at different MCr recovery factor

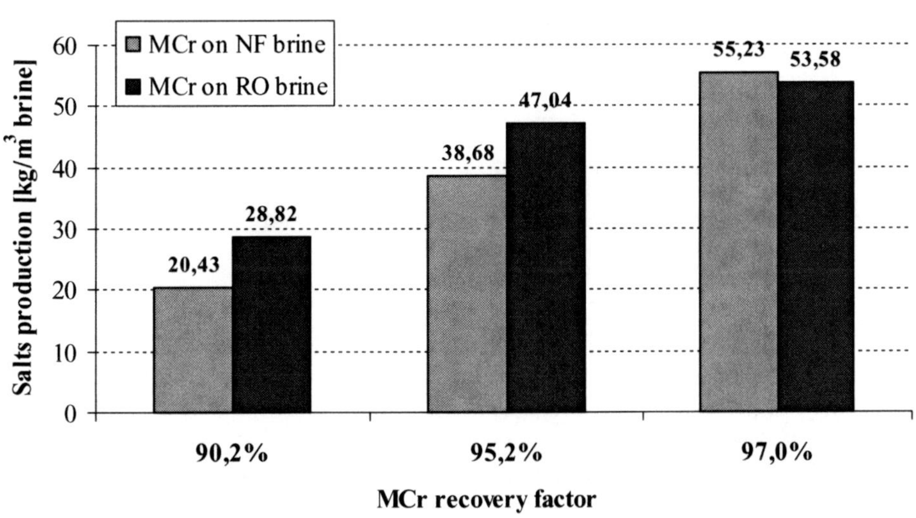

Figure 5.32 Recovery factor of the integrated MF/NF/RO system at different MCr recovery factor

4. *Annual maintenance and spare parts cost:* $A_{maintenance} = p \cdot m \cdot f \cdot 365$ where p = specific maintenance and spare parts cost (0,033 \$/m^3).
5. *Amortization or annual fixed charges*, calculated as $A_{fixed} = a \cdot DC$ where "a" is the amortization factor given by $a = \frac{i(1+i)^n}{(1+i)^n - 1} = [yr^{-1}]$; i = interest rate [%]; n = plant life [yr]. Design studies and experience in desalination industry indicate an amortization life of 30 years and an interest of 5%.
6. *Annual cost for chemicals* used in pre-treatment, cleaning operation and post-treatment: $A_{chemicals} = k \cdot f \cdot m \cdot 365$ where k= specific chemicals cost (0.018 \$/m^3).
7. *Annual brine disposal cost:* $A_{brine} = b \cdot B \cdot f \cdot 365$ where b= specific cost of brine disposal (0,0015 \$/m^3) and B= brine flow rate;
8. *Annual steam cost*: $A_{steam} = s \cdot G \cdot f \cdot 365$ where s = heating steam cost (0,0032 \$/lb).
9. *Annual Na_2CO_3 cost:* $A_{Na_2CO_3}$ = salt price · salt flow rate · f · 365
10. *Annual profit for the salts sale* equal to Annual profit = salt price · salt flow rate · plant availability · 365 [\$/yr].

Taking into account all the previous cost items, the *Total Annual Cost* and the *Unit Product Cost* are, respectively, given by:

$$A_{total} = A_{fixed} + A_{in} + A_{electric} + A_{chemicals} + A_{labor} + A_{steam} + A_{brine} + A_{maintenance} + \\ +A_{membrane} + A_{Na_2CO_3} - Annualprofitforthesaltsale$$

and $A_{unit,\,p} = A_{total}/(f \cdot m \cdot 365)$.

Cost of the desalted water produced with the analysed desalination systems with MCr operation on RO brine is reported in Figure 5.33. Desalted water cost has been estimated at increasing MCr recovery factor and at increasing temperature of the MCr feed. In addition, for all the investigated operative conditions, water cost has been estimated for the following different situations:

1. for the case in which the produced salts are not sold (blue bar of the histograms reported in Figure 5.33);
2. when the produced salts are not sold but thermal energy is available in the plant (viola bars);
3. when the salts are sold (pink bars);
4. when the salts are sold and thermal energy is available in the plant (blue sky bars).

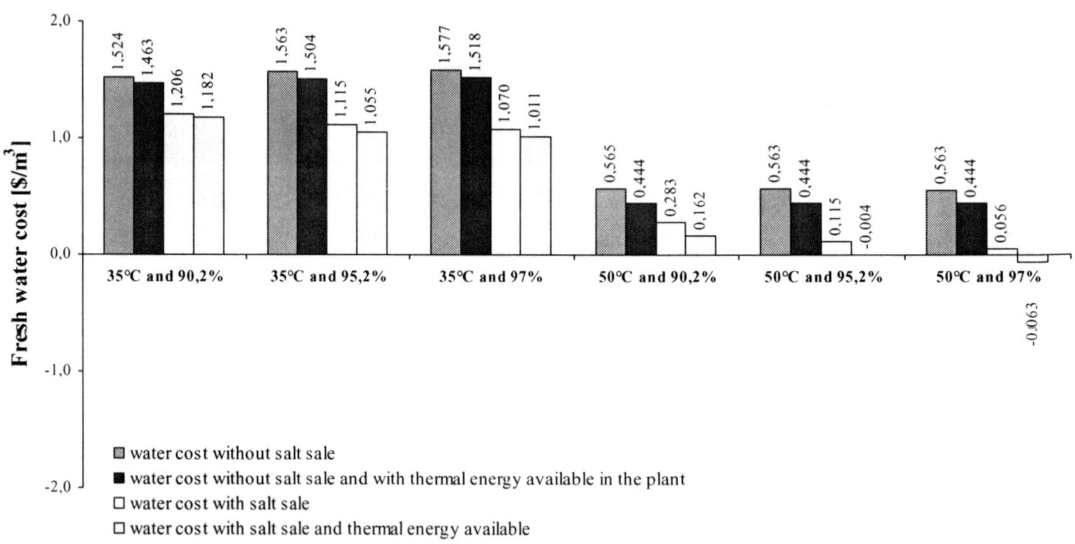

Figure 5.33 Water cost of the considered membrane desalination system with MCr operation on RO brine (at different MCr recovery factor and at different temperature of the MCr feed). This figure is available in colour at http://www.iwapublishing.com/template.cfm?name=isbn9781843393214

The lowest water cost was obtained for the desalination plant in which MCr works at a recovery factor equal to 95.2 ÷ 97.0%, and for the case in which the temperature of the MCr feed is equal to 50°C. Moreover, if the gain for the salts sale is considered, desalted water cost reduces so much to reach competitive values with those of the other desalination plant. In particular, in the analysed system with MCr operating on RO brine with a recovery factor equal to 95.2%, a water cost reduction of 28.7% (and 29.8% if thermal energy is available in the plant) can be reached at 35°C. For a MCr feed temperature equal to 50°C, water cost decreases 79.6% (and 100.9%, respectively).

(c) An economic evaluation can be made also with the WAIV unit that allows to calculate the benefit of using WAIV to preconcentrate the feed to the membrane crystallizer operated by UNICAL.

The capital cost for the WAIV unit in Table 5.12 was made based on an evaporation rate of 3 [mm/day]. As the evaporation rate increases the capital cost will decrease respectively. Three different evaporation rates were taken in order to compare the difference in the capital cost per cubic meter for each on of the disposal methods; the data is shown in Table 5.13.

Table 5.11 Capital cost for evaporation pond

Capital recovery factor (CRF)	0.078	
Annual evaporation rate- E_{pan}	5.19	[mm/day]
Correction factor E_{pan} to E_{pond}	0.7	
α based on CRF	0.83	
Pond capital cost	22–25	[USD/m^2]
Evaporation pond capital cost	1.56–1.77	[USD/m^3]

Table 5.12 Capital cost for WAIV unit

Capital recovery factor (CRF)	0.078	
Evaporation rate	3	[mm/day]
Evaporation surface area	5440	[m^2/module]
Evaporation per module	16.32	[m^3/day/module]
Total operating- (pumping, labor, and replacement)	0.265	[USD/m^3]
WAIV capital cost	0.99	[USD/m^3]

Table 5.13 Capital cost WAIV verses evaporation ponds

Capital cost for evaporation rate of	WAIV	Evaporation ponds	
1.5 [mm/day]	1.98	5.4–6.13	[USD/m^3]
3 [mm/day]	0.99	2.7–3.06	[USD/m^3]
5 [mm/day]	0.56	1.56–1.77	[USD/m^3]

WAIV shows a tremendous reduction of 64% from the capital cost in compare to conventional ponds with PCC = 22 [USD/m^2]. This reduction is even higher while PCC = 25 [USD/m^2] and it is equal to 68%. This savings is an addition to the saving by the reduction of the footprint for brine disposal by an order of magnitude or more.

In addition if WAIV reduces the amount of calcium and the amount of thermal energy needed to evaporate the feed in a membrane crystallizer, then it will provide additional costs savings. To this end we used 160 Euro/ton of sodium carbonate and 5 Euro/mt of waste steam. We assume a gained output ratio (m^3 water produced/mt steam supplied) of 2.55 under the conditions reported by UNICAL, and a water value of 1 Euro/m^3 for the produced water. If the value of the sodium chloride is 120 Euro/mt and value of precipitated calcium carbonate is 75 Euro/mt (ICIS pricelist for 4–8 micron calcium carbonate), we then obtained the following savings on operating the membrane crystallizer after the WAIV unit (see Table 5.14).

From Table 5.14 we can clearly see that there is an advantage to using WAIV before MCr but there is no advantage in going from VCF = 10 to VCF = 20 with WAIV. The advantage of using MCr alone is marginal compared to WAIV provided one can recover the value of chemicals. It is clearly advantageous to use MCr after WAIV if one CAN get credit for the recovered chemicals. Analogous conclusions were achieved also in the case of MCr working on the RO brine of a seawater (and not brackish water) desalination plant (Figure 5.33). In that case, significantly reduced desalination costs were obtained when the gain for the salts sale was considered as here significantly reduced brine disposal costs were obtained. It should be noted that the lower mineral content in brackish water (5 kg/mt BWRO brine vs. 22–50 kg/mt SWRO brine) means that the reduction in costs here will not be as great as in the case of MCr on SWRO brine. On the other hand the extent of evaporation required is much less because of the higher TDS (10–20%) feed to the MCr if it is preceded by WAIV.

Table 5.14 Costs for treating brackish water RO brine from an 88% recovery BWRO plant

Case	MCr on RO brine	WAIV at VCF = 10 + MCr	WAIV at VCF = 20 + MCr
WAIV water evaporation costs	€ 0.00	€ 0.84	€ 0.94
MCr Water evaporation costs	€ 1.17	€ 0.10	€ 0.04
Chemical costs for Ca precipitation	€ 0.75	€ 0.08	€ 0.04
Total costs	**€ 1.93**	**€ 1.02**	**€ 1.02**
Chemicals recovered			
Calcium carbonate, kg/m^3	4.45	0.45	0.22
Sodium chloride, kg/mt	n 5.10	5.10	5.10
Value of chemicals, Euro/mt			
calcium carbonate, @ 75 Euro/mt	€ 0.33	€ 0.03	€ 0.02
sodium chloride, @ 120 Euro/mt	€ 0.61	€ 0.61	€ 0.61
Net cost of treating RO brine	**€ 0.98**	**€ 0.38**	**€ 0.39**

5.5 CONCLUSIONS

Seawater and brackish groundwater desalination processes contribute positively to solve the problem of water shortage but, at the same time, cause locally some negative impacts on the environment that need to be minimized due to the discharge of their concentrate waste streams into surface waters or into the ocean. The objective of WP5 was to develop efficient separation methods able to improve the productivity of desalination systems, reducing their environmental impact and recovering some of the valuable ions that are present in the highly concentrated streams of the desalination plants. For reaching this aim, three technologies were evaluated: disposal of membrane concentrate into the subsurface, Vacuum Membrane Distillation (VMD), Membrane Crystallization (MCr) and/or Wind Intensified enhAnced eVaporation (WAIV).

Deep well injection was shown to be an environmentally safe and technically feasible technique to dispose BWRO membrane concentrate. Crucial for the success is a careful selection of source and disposal aquifer and proper design of the RO installation and injection well. Field test results indicate that supersaturation of BWRO concentrate is allowed, without causing operational problems in the RO or injection well. Moreover, interesting results were obtained on recourse to Vacuum Membrane Distillation (VMD), Membrane Crystallization (MCr) and/or Wind Intensified enhAnced eVaporation (WAIV) operations as means for securing the major benefits to sustainable water management in terms of increased water recovery factor, reduction of brine volumes and of disposal costs. In particular, (1) a Vacuum Membrane Distillation (VMD) Modelling Software was achieved (which runs both for flatsheet membranes and for hollow fibre membranes with inside-outside configuration); (2) the analytical description, fluid-dynamic analysis and kinetic study of the crystallizing equipment has been carried out; (3) the effect of organics on crystals size, coefficient of variation (CV) and crystals growth rate has been analyzed; (4) the economical evaluation of desalination systems with membrane crystallizers working on RO or on NF retentate was carried out; (5) the use of wind to evaporate desalination brines and reduce their volume in an economic way with minimal energy inputs was realized using Wind-Aided Intensified eVaporation (WAIV). In the present project, efforts were also made to combine WAIV with an integrated desalination scheme by generating superconcentrated brines (10% or greater) to the membrane crystallizer to allow reduced energy inputs for recovering valuable minerals.

REFERENCES

Curcio E., Criscuoli A. and Drioli E. (2001). Membrane crystallizers. *Ind. Eng. Chem. Res.*, **40**, 2679–2684.

Houben G. and Treskatis Chr. (2007). *Water Well Rehabilitation and Reconstruction*. McGraw-Hill, New York.

Macedonio F., Drioli E., Curcio E. and Di Profio G. (2009). Experimental and economical evaluation of a membrane crystallizer plant, *Desalination and Water Treatment*, **9**, 49–53.

Macedonio F. and Drioli E. (2010). Hydrophobic membranes for salts recovery from desalination plants, *Desalination and Water Treatment*, **18**, 224–234.

Macedonio F., Katzir L., Geisma N., Simone S., Drioli E. and Gilron J. (2011). *Wind-Aided Intensified eVaporation (WAIV) and Membrane Crystallizer (MCr) Integrated Brackish Water Desalination Process: Advantages and Drawbacks*. Desalination. doi: 10.1016/j.desal.2010.12.002. *In press*.

Maliva R.G., Guo W. and Missimer T. (2007). Vertical movement of municipal wastewater in deep injection well systems, South Florida, USA. *Hydrogeol J* **15**, 1387–1396.

Mericq J. P., Laborie S. and Cabassud C. (2009). Vacuum membrane distillation for an integrated seawater desalination process, *Desalination and Water Treatment*, **9**, 287–296.

Mericq J. P., Laborie S. and Cabassud C. (2010). Vacuum membrane distillation of seawater reverse osmosis brines, *Water Research*, **44**, 5260–5273.

Mickley M. C. (2006). Membrane concentration disposal: practices and regulation. 2nd ed., U.S. Department of the Interior, Bureau of Reclamation, Denver, Report No. 123, National Technical Information Service (NTIS), Springfield, 303p.

Oosterhof A. T., Wolthek N. B. A., Van der Meer W., Groenendijk M., Van de Wetering S., Boukes H., Raat K. J. and en Eerhart J. (2009). Doorbraak voor gebruik van brak grondwater als alternatieve bron voor drinkwatervoorziening. H2O **42**(14/15): 14–17. (in Dutch).

Parkhurst D. L. and Appelo C. A. J. (1999). User's guide to PHREEQC (version 2) – A computer program for speciation, batch-reaction, one-dimensional transport, and inverse geochemical calculations. Water-Resources Investigations Report 99-4259. USGS, Denver, Co.

Stuyfzand P. J. and Raat K. J. (2010). Benefits and hurdles of using brackish groundwater as a drinking water source in the Netherlands. Geohydrology Journal. DOI: 10.1007/s10040-009-0527-y.

Chapter 6

Innovative technologies to reduce energy consumption in seawater desalination facilities

C. Cabassud[1], S. Laborie[1], J. P. Méricq[1], S. Gabsi[2], S. Gaeta[3], E. Drioli[4,5] and A. Criscuoli[5]

[1] INSA Toulouse (France)
[2] ENIG (Tunisia)
[3] GVS (Italy)
[4] University of Calabria (Italy)
[5] ITM-CNR (Italy)

6.1 INTRODUCTION

As energy resources will be more and more expensive in the future, and a lack in fossil energy resources will be observed in the next decades, in the frame of a sustainable development it is necessary to reinforce the efforts

1. to improve the energy efficiency of well established existing processes like reverse osmosis and
2. to develop new systems and concepts based on renewable energies.

Concerning reverse osmosis systems, to date, different energy recovery devices have been developed to recover a portion of the energy from the brine stream of a reverse osmosis unit. These units operate in combination with a conventional high pressure pump system. Conventional energy recovery devices include Pelton wheels, work exchangers and pressure exchangers.

Concerning the use of renewable energies for water desalination, some possible processes are to couple reverse osmosis with wind or solar energy. This kind of processes are yet tested at semi-industrial scale for example in Australia for small communities. The limitation of these systems is that the solar or wind energy has to be converted to mechanical pumping energy, which generates energy losses. One of the main interests of membrane distillation systems is that their productivity is clearly linked to the feed temperature and that the specific energy costs could be decreased if operation is performed at high temperatures directly obtained by solar heating of the feed water. On another side, the interest of submerged membranes for reducing feed circulation costs for water clarification or in membrane bioreactors is now well known. However no tentative was done today on using submerged distillation systems. So new systems based on membrane distillation using submerged distillation membranes and solar energy have to be imagined, designed, tested and evaluated. For that purpose a complementary approach is necessary that will require competences in: membrane development, manufacturing and characterisation, module design and manufacturing, process testing and optimisation, process modelling, solar systems modelling and design, technical-economical evaluation.

Objectives of the work presented in this Chapter was to develop, characterise new membranes and modules and to provide information on the process operation for a distillation system, based also on submerged distillation membranes, coupled to solar energy for different configurations. The first step was to compare these configurations on the basis of modelling, simulation and lab-scale studies and then to fully experiment one of these configurations.

6.2 PREPARATION AND CHARACTERIZATION OF MEMBRANES

Different microporous hydrophobic flat membranes were developed and characterized. Their main properties are reported in Table 6.1. The PVDF membranes were prepared by phase inversion from a polymeric solution of PVDF coagulated in a non-solvent bath. The PTFE membranes and the acrylic membranes were acquired on the market and post-treated to tailor their surface tension by using proprietary technologies partially developed in the framework of this project. Membrane M09G0020 was produced by GVS according to patented know-how. Membrane M09G0020E was the membrane M09G0020 superficially treated to increase the hydrorepellency grade of the membrane. Membrane type M07R0020 and membrane type M05E0020 were acquired by GVS and superficially treated by GVS to tailor their surface properties according to proprietary know-how.

Table 6.1 Main characteristics of the microporous hydrophobic flat membranes

Membrane code	M09G0020	M07R0020	M05E0020	M09G0020E	M09G008
Membrane material	PVDF	Acrylic Based	PTFE	PVDF	PVDF
Pore size	0,2 μm	0,2 μm	0,2 μm	0,2 μm	0,8 μm
Thickness	199 μm	202 μm	218 μm	199 μm	199
Air flow [@441 mmHg & 3,7 cm^2]	5 l/min	5 l/min	11 l/min	5 l/min	>30 l/min
Water break trough	24 mH_2O	23 mH_2O	38 mH_2O	24 mH_2O	5,5 mH_2O
Oleorepellency grade [AATCC Test Method 118-2002]	<3	5,5	<3	3,5	<3
Idrorepellency grade [Dupont Teflon Test]	3	11	4,5	9,5	3
Membrane support	PET	Nylon	PET	PET	PET
Size of samples	21,2 × 29,8 cm	10,8 × 29,8 cm	18,8 × 29,8 cm	21,2 × 29,8 cm	21,2 × 29,8 cm
Number of samples	5	10	6	5	5

Membranes were characterised in terms of contact angle and SEM. SEM pictures show a symmetric structure, with the exception of M050020, that presents two completely different surfaces because of its composite type. As an example, the SEM picture of M050020 is shown in Figure 6.1.

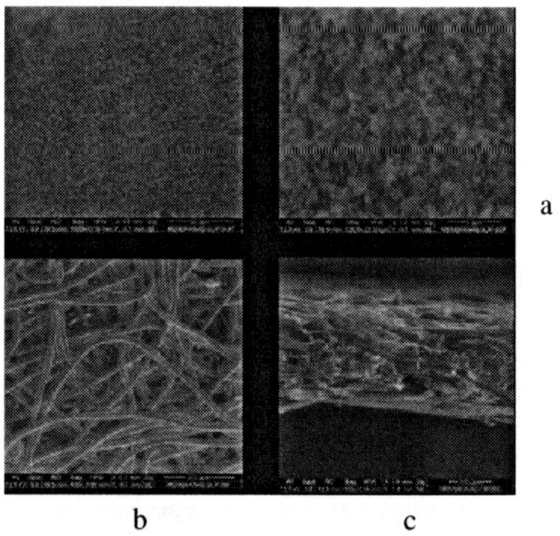

Figure 6.1 SEM pictures of the M050020 membrane (a. surface 1; b. surface 2; c. cross-section)

Membranes have also been characterized in terms of water vapor flux through VMD tests using distilled water as feed. In particular, for each membrane tests have been carried out by exposing each side (side a-the more hydrophobic; side b-the less hydrophobic) to the liquid phase and the fluxes achieved in the two cases have been compared. Different operating conditions have been varied and the most significant results are summarized in Figures 6.2–6.4.

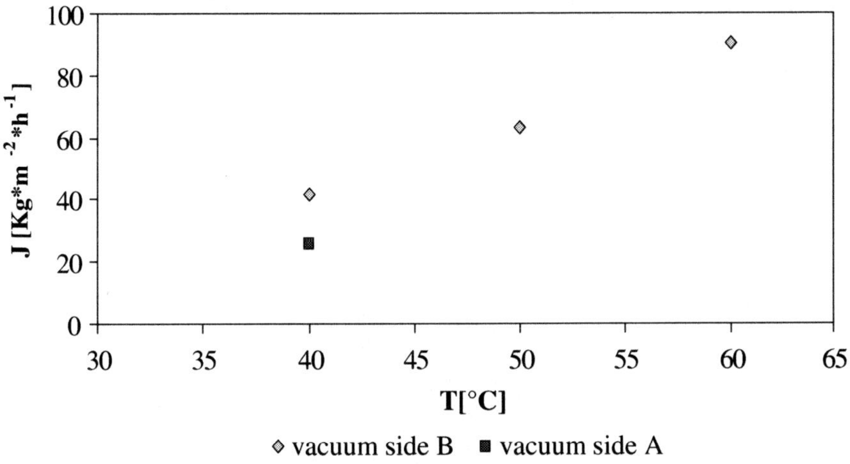

Figure 6.2 Distillate flux vs temperature. M05. Qfeed = 200 L/h; Pv = 20 mbar; Am = 40 cm^2

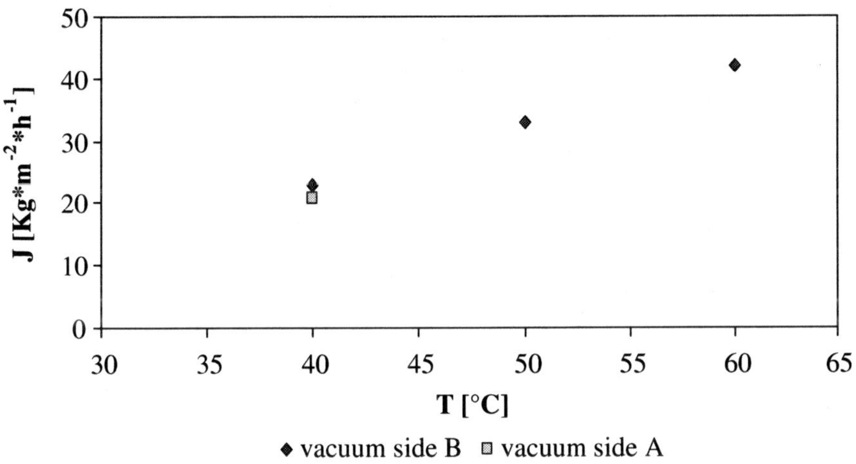

Figure 6.3 Distillate flux vs temperature. M07. Qfeed = 200 L/h; Pv = 20 mbar; Am = 40 cm^2

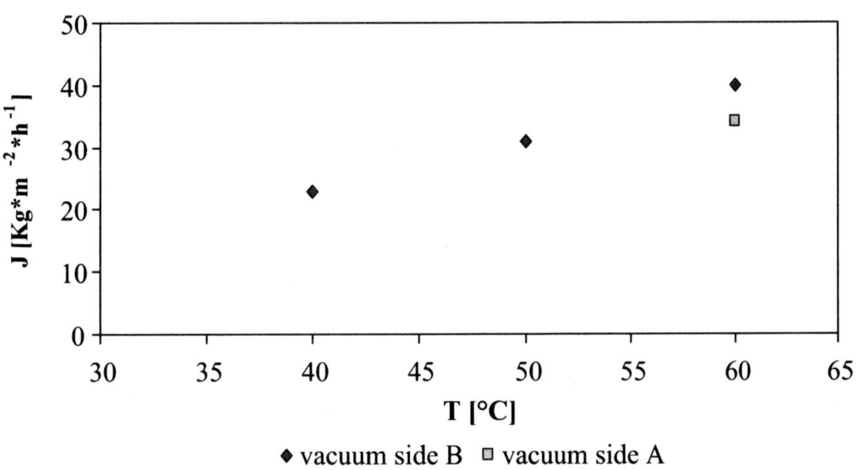

Figure 6.4 Distillate flux vs temperature. M09. Qfeed = 200 L/h; Pv = 20 mbar; Am = 40 cm^2

For all membranes the fluxes increase with the feed temperature, because of the increase of the vapor pressure. Moreover, the highest fluxes are obtained when the more hydrophobic side (side a) is in contact with the liquid while the vacuum is applied at the side b.

Figure 6.5 compares the performance of the different membranes tested. The highest flux is obtained with the M05, followed by the M09 treated (M09t), the M07 and the M09.

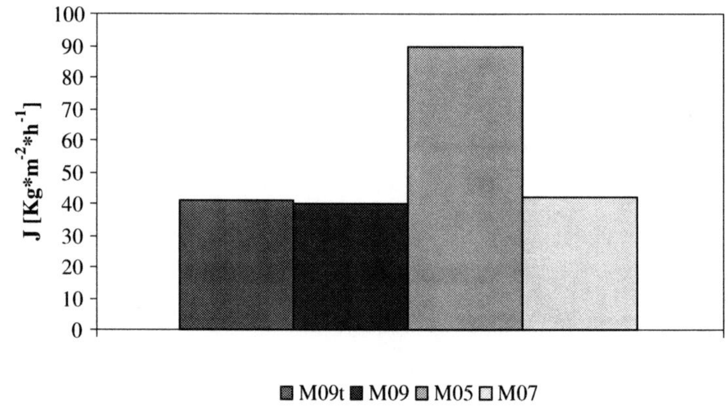

Figure 6.5 Comparison of the fluxes achieved with the different membranes

VMD tests in lab-scale modules on the M09G008 membrane was carried out by using distilled water as feed. During experiments, different operating conditions were varied, such as temperature, feed flow rate, vacuum pressure. However, not reproducible data were achieved because, due to the big pore size of the membrane, some liquid water permeated through pores by convection together with vapor.

Besides the flat membranes, PVDF hollow fibres have also been prepared with the dry/wet spinning technique.

The materials used were:

- *Polymer:* Poly(vinylidene fluoride) (PVDF Solef® 6010, 6012 and/or 6020);
- *Solvent:* N,N-Dimethylformamide (DMF);
- *Pore formers:* Water; (N-vinylpyrrolidone) K17 (Mw 7–11 KDa).

As post treatment, the produced hollow fibres were treated with NaClO 4000 ppm pH 7 in order to remove PVP and ensure hydrophobicity. They were then washed throughout with fresh water and afterwards treated with ethanol 50%, followed by isopropanol 50%, or with ethanol 50%, pure ethanol and a mixture ethanol/n-Hexane 1:1. All these steps were undertaken overnight. Fibres were, then, immersed in an aqueous solution containing glycerol at 30% w/w for 3–4 hours and dried at room temperature for one day. The fibres were characterized by SEM, mechanical properties and bubble point. An average pore size of 0.2 micron and a porosity of about 80% have been found with Young's modulus values falling in a wide interval comprised roughly between 70 and 130 N/mm^2.

Once prepared, the modules (more than 100) containing different type of fibres have been tested for VMD, using distilled water as feed. These tests had the objective to firstly check if the membranes were suitable for VMD and, secondly, to improve membrane fluxes by adjusting experimental parameters during hollow fibres spinning i.e. increase the percentage of pore-former additives. Figure 6.6 shows how the fluxes have been improved during the optimization step.

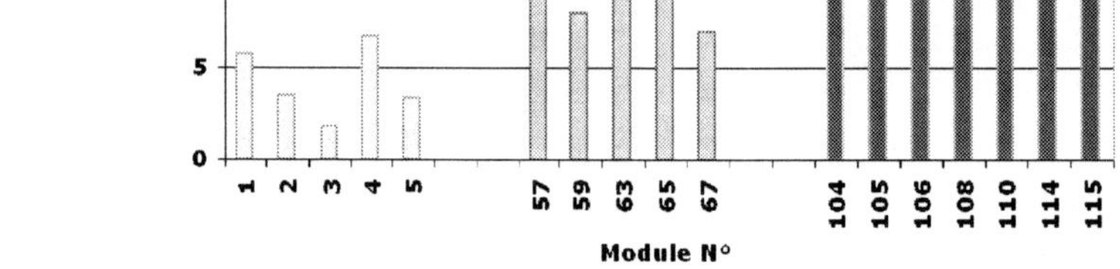

Figure 6.6 How the fluxes have increased during the optimization step

The VMD tests on these membranes were performed on lab-scale modules. Among all the prepared membranes, good mechanical properties together with acceptable trans-membrane fluxes (of the same order of those present in literature) were achieved in VMD tests carried out on distilled water with fibres obtained by using PVP at 14% in the dope solution and DMF at 25% as bore fluid. Figure 6.7 shows the trend of the trans-membrane flux with time for this type of fibre.

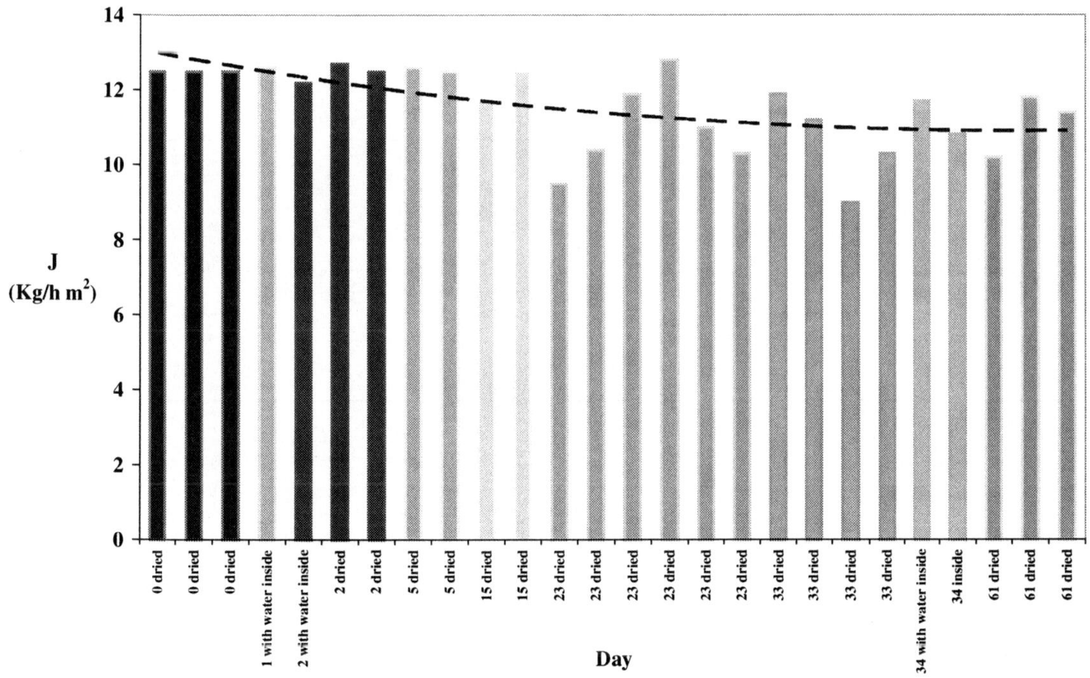

Figure 6.7 Performance of the module with time (DMF, 25 wt.%; PVP, 14 wt.%)

Preliminary VMD tests on synthetic seawater solutions were also encouraging, as reported in Figure 6.8.

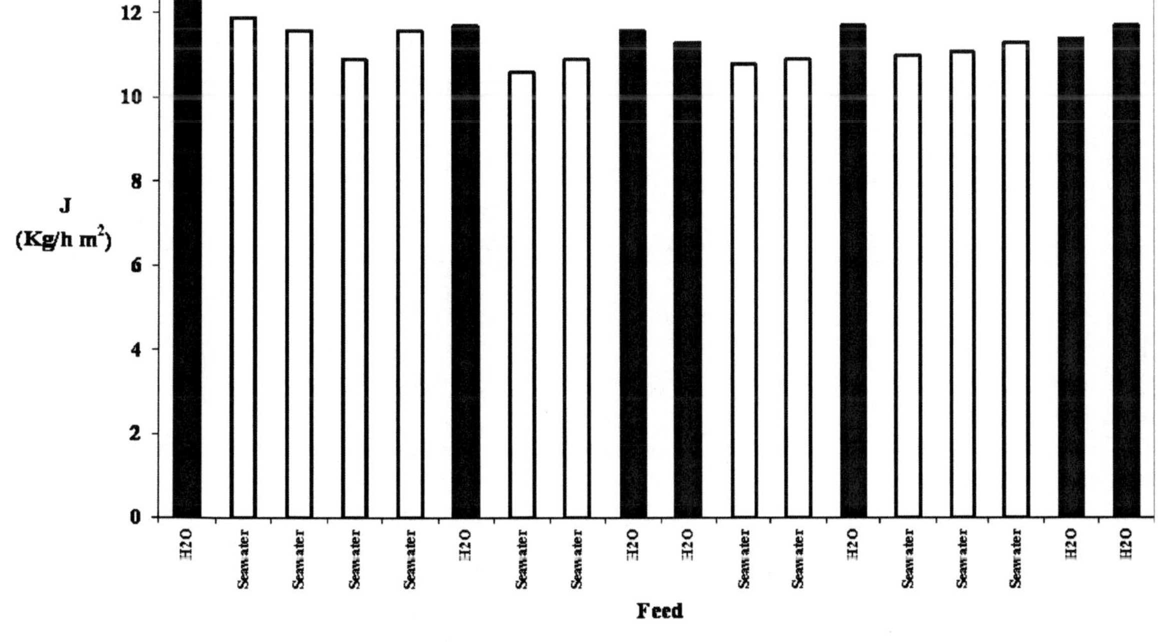

Figure 6.8 VMD tests with synthetic water (DMF, 25 wt.%; PVP, 14 wt.%)

6.3 MEMBRANE DISTILLATION AND SOLAR ENERGY

Four configurations to couple VMD and solar technologies were studied (Figure 6.9):

i) Configuration A: a membrane module submerged in the SGSP (salinity gradient solar pond)(Figure 6.9a)
ii) Configuration B: a membrane module fed by water taken from a SGSP (Figure 6.9b)
iii) Configuration C: a membrane module fed by seawater heated by solar collectors (Figure 6.9c)
iv) Configuration D: a membrane module fed by seawater directly heated with solar captors on the membrane module (Figure 6.9d)

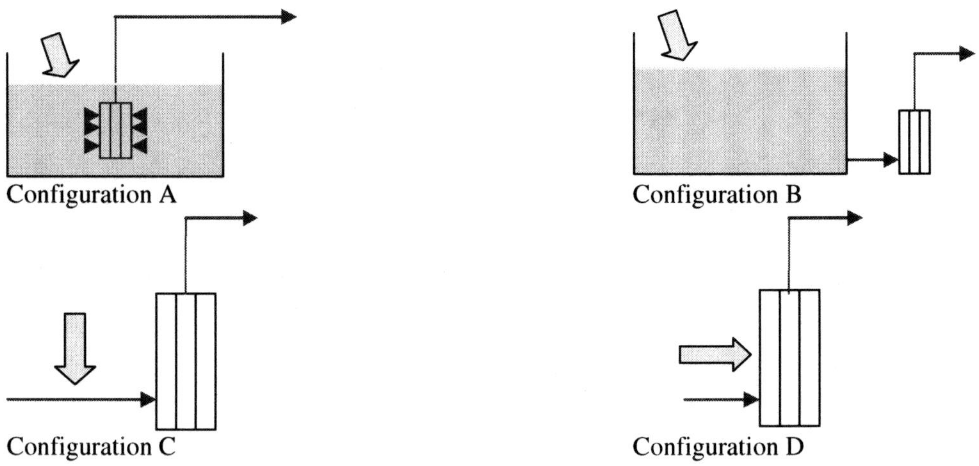

Figure 6.9 The four investigated configurations for coupling VMD to solar collectors

The approach of the work is based on simulations performed with the VMD modelling software that was adapted to consider the different configurations. Some flat sheet membranes with a high permeability were considered for the computations (Knudsen permeability of 1.85 10^{-5} s.mol$^{1/2}$.m^{-1}.kg$^{-1/2}$).

An estimation of the system performances – in terms of permeate flux, temperature and concentration polarisation was performed for the four configurations. The influence of membrane module properties such as membrane Knudsen permeability, membrane length, position of the module in the SGSP (in the Low convective Zone or in the High Convective Zone), position of the water sampling was studied and systems were compared for their optimal operating conditions. Table 6.2 shows a synthesis of results obtained for the four configurations.

Table 6.2 Comparison of the four configurations

Configuration	A (in LCZ)	B (in LCZ)	C	D
Feed temperature (°C)	62.2	62.2	70	70
Permeate pressure (Pa)	500	500	500	500
Feed velocity (m/s)	0	1.14	1.14	1.14
Permeate flux (L/h/m^2)	3.4	25	37.4	38.7
TPC	0.86	0.99	0.99	0.99
CPC	0.22	0.93	0.90	0.90

With

TPC: temperature polarisation coefficient is the ratio between the temperature at the membrane surface and the feed seawater mean temperature.

CPC: concentration polarisation coefficient is the ratio between the feed seawater mean concentration and the concentration at the membrane.

Concerning the solar pond (configuration A and B), results show that some important polarisation effects reduce strongly the permeate flux if the membrane is submerged (configuration A) in the low convective zone (LCZ), which is in the deeper layer of a solar pond. Nearly no energy requirement is needed but a permeate flux of less than 4 L/h/m^2 can be obtained. One solution to limit this polarisation would be to generate an artificial stirring close to the membrane but the risk to disturb the stability of the SGSP is important.

To reduce the polarisation, the LCZ can be used as feed water for the membrane module (thus in configuration B). In that case, due to the circulation of the feed in the membrane module, concentration polarisation effects are reduced and higher fluxes can be obtained compared to the previous configuration. A permeate flux of 25 L/h/m^2 can be obtained for a permeate pressure of 500 Pa.

Configurations C and D allow reaching the highest permeate flux with very similar values for C and D. Use of solar collectors seems to be a much more simple solution to operate and less expensive than a SGSP. Configuration D gives also interesting results in terms of permeate flux, concentration polarisation and energy requirement. However, simulations of configuration D do not take into consideration for the moment the radial profile for several membrane layers and so the flux is overestimated for this configuration. The Configuration C seems to avoid this radial profile problem since all membrane layers are in contact with feed seawater at the same temperature. Moreover the technology to build a captor module with good heat collection and transfer and a correct adequation between membrane area installed and heat collector area seems difficult to design and to manufacture. Technological feasibility of configuration D appears as limited.

For all these reasons, the coupling of VMD and solar collectors as pre-heating of the feed seawater was chosen (Configuration C).

6.3.1 Development and validation of a meteorogical model

A meteorological model was selected to describe solar heating in function of captor location, orientation and properties – this model was validated by experimental results. It was used to predict temperature levels for different kind of collectors (flat, vacuum, cylindro-paraboloical) and mode of operation (one pass or closed loop). It allows to predict that a temperature level of 65°C could be obtained for a serpentine in closed loop or of 90°C for a vacuum heat collector in closed loop, which allow to be fed with 300 l/h/m^2 of solar area.

In particular, the work was:

- to model solar radiation and to validate the model chosen with weather data of the of Gabès (from weather station installed at the National School Engineers of GABES and data of Borj Cédria in the north of the country). Three models were used (Model of Perrin de Brichambaut , Model of Liu Jordan, Model of Eufrat)

The three models take into account several climatic parameters (humidity of the site, slope of the collector, latitude of the city, longitude. . . .) and make it possible to determine the various solar radiations: total, diffuse and direct receipts by a horizontal flat plate and a tilted plan of slope Į and orientation $\tilde{\gamma}$ The comparison of the results of simulations obtained by the three models and the experimental values (taken by means of a weather station installed in ENIG) shows a satisfactory agreement. The selected model was that of Brichambaut because it takes into account the orientation of the solar collector (see Figure 6.10).

- to model the temperature provided by flat plate collectors, vacuum collectors and a cylindro parabolic collector and to calculate the duration for a temperature average of 60°C and lower than 75°C in order to determine that which ensures this objective with a flow of corresponding liquid.

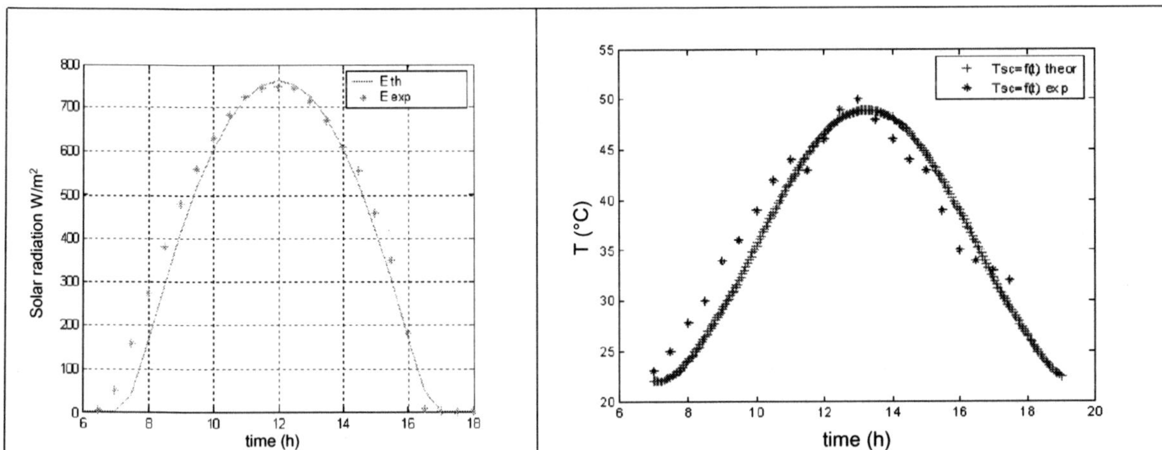

Example of comparison with the experimental values December 21 slope = 45°

Example of comparison with the experimental values Day 02/05/2008 slope = 30°

Figure 6.10 Comparison between simulation and experimental results. This figure is available in colour at http://www.iwapublishing.com/template.cfm?name=isbn9781843393214

The resolution of the equations of heat transfer established on the various components of the collector lead to the instantaneous evolution of the temperature of the absorber, the coolant and the glass (see Figure 6.11).

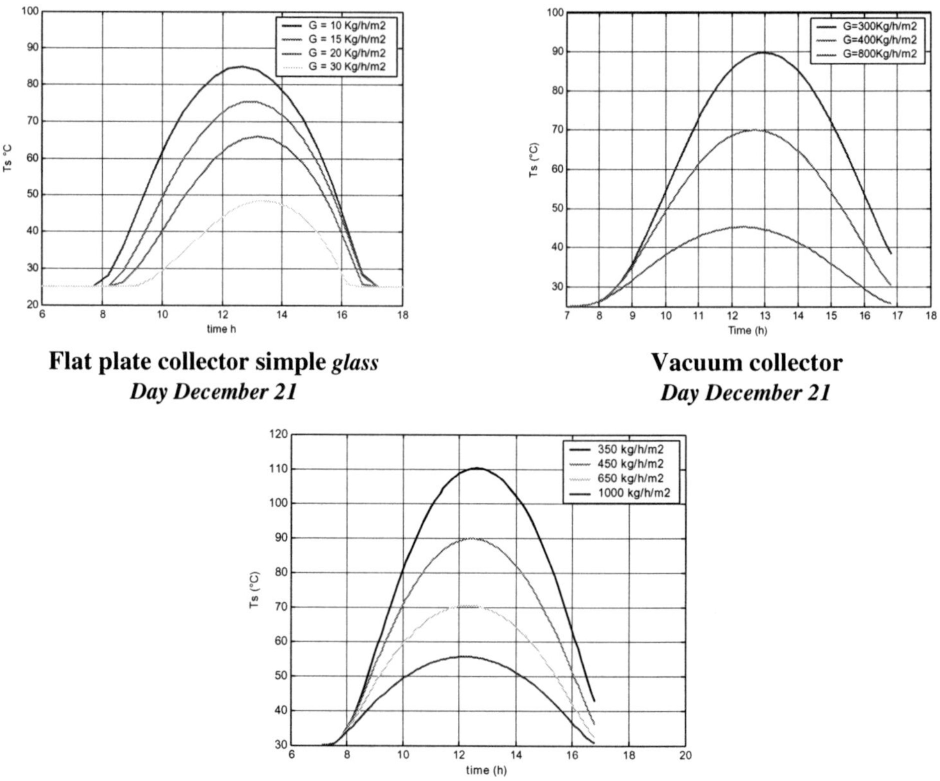

Flat plate collector simple *glass*
Day December 21

Vacuum collector
Day December 21

Cylindro parabolic Collector
Day December 21

Figure 6.11 Evolution of the temperature in the collectors

6.3.2 Experiments at lab-scale on configurations coupling solar collector and VMD

Experiments were carried out with the VMD system at lab scale (and with experimental simulation of solar pre-heating) for different operating conditions. These experiments were performed with synthetic seawater and also with actual seawater sampled from Mediterranean Sea.

Three different flatsheet polymeric membranes have been used: Fluoropore (3.26×10^{-6}s.mol$^{1/2}$.m^{-1}.kg$^{-1/2}$), Durapore (2.29×10^{-6}s.mol$^{1/2}$.m^{-1}.kg$^{-1/2}$) and GVS PVDF (2.27×10^{-6}s.mol$^{1/2}$.m^{-1}.kg$^{-1/2}$).

Interesting permeate fluxes (10–15 L/h/m^2) can be reached even for low permeable membrane for well-chosen operating conditions (high temperature and low permeate pressure). Importance of the feed seawater Reynolds number is limited since no concentration and temperature polarisation effects are observed. Experiments with real seawater have shown no decrease of the permeate flux due to an eventual fouling (organic, scaling, biofouling) over six hour experiments. Study of apparent Knudsen permeability has confirmed this observation. However, some salts precipitation occurred on membrane surface but its effect on permeate flux is nearly inexistent.

Compared to classic distillation desalination process, its advantage is that VMD operates at a lower temperature. Indeed, a 50°C temperature allows the VMD to reach high permeate flux whereas distillation process needs a minimum 80°C temperature or more. In comparison with conventional distillation, the main advantage of solar enhanced membrane distillation is the compactness of the membrane distillation process: for example, for a 1000 m^3 per day water production, membrane area is about 2777 m^2, i.e. a 2.7 m^2 area by water m^3 per day whereas multi effect distillation requires 2.5 to 4 m^2 exchange surface by water m^3 per day.

6.3.3 Design of a semi-industrial pilot plant

On the basis of the previous points, the principle of the flow sheet of the demonstration pilot plant to be installed in Tunisia was established according to configuration C which couples external thermal solar heating equipment and some membrane modules. The characteristics necessary to design the plant were defined: flow rates, temperature, constraints on membrane operation. Figure 6.12 shows the scheme of the designed plant.

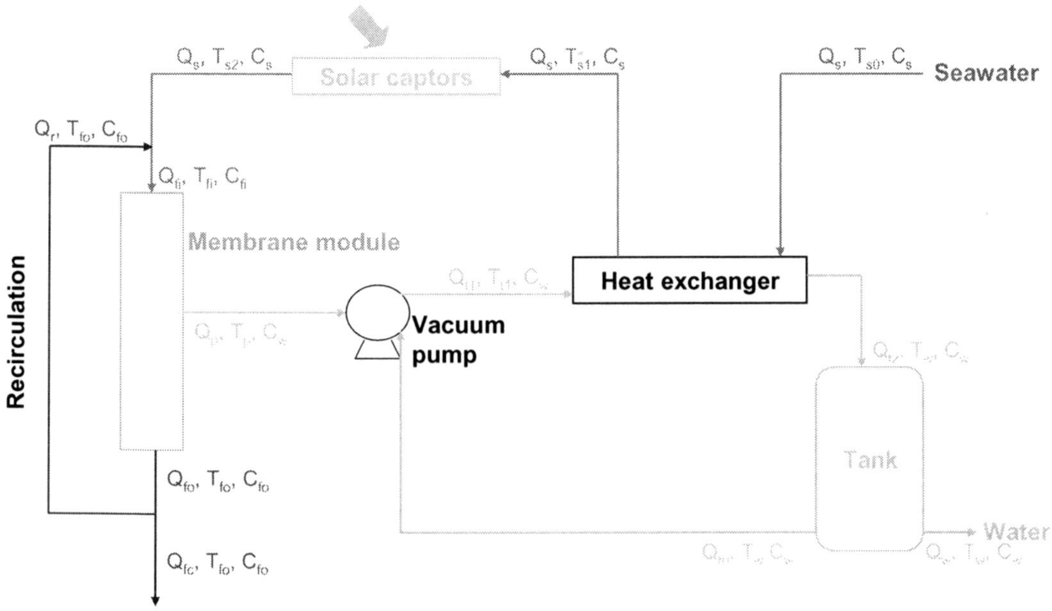

Figure 6.12 Scheme of the designed plant. This figure is available in colour at http://www.iwapublishing.com/template.cfm?name=isbn9781843393214

6.4 STUDY AND DEVELOPMENT OF SOLAR SYSTEMS COUPLED WITH MEMBRANE DISTILLATION

The development of a user-oriented software (which is a tool of assistance to chose the type of adequate solar collector) was finalized.

This software is based on two parts; the first one aims to simulate solar radiation (direct, diffuse and total) and the ambient temperature and the second the exit instantaneous temperature of each solar collector (flat plate, cylindro-parabolic, vacuum). An example of simulation is presented in the Figure 6.13.

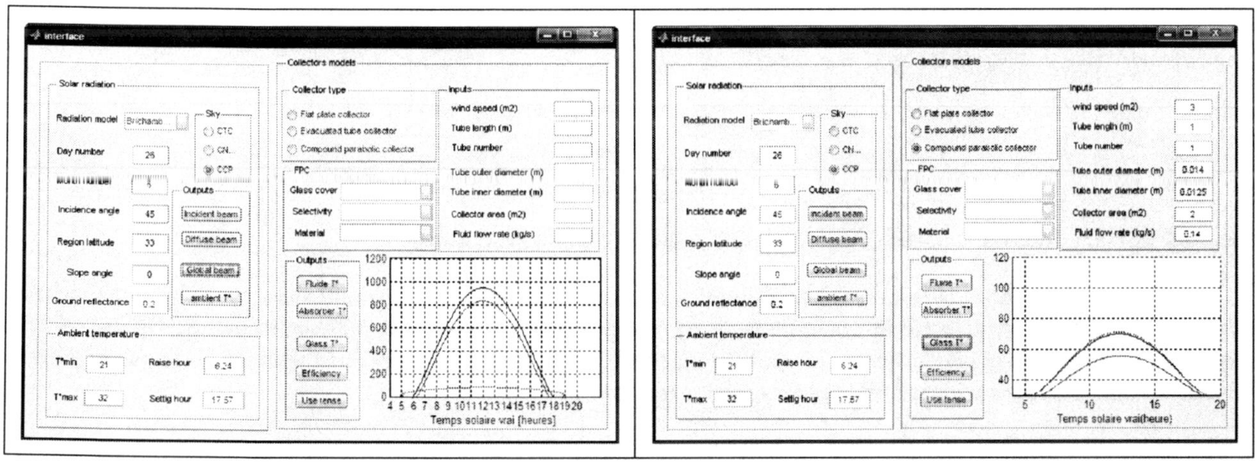

Figure 6.13 Example of typical information obtained during simulation runs. This figure is available in colour at http://www.iwapublishing.com/template.cfm?name=isbn9781843393214

6.4.1 Modelling and simulation of the two configurations of membranes modules

This study mainly focused on the integration of:

- a hollow fiber module in a thermal absorber of a cylindro-parabolic collector,
- a flat sheet membrane module in a flat plate solar collector.

In both cases, mathematical models were elaborated. The mass and energy balance in the various compartments made it possible to determine the interface temperature profile, the number of compartments necessary, the daily instantaneous and annual permeate flow rates. An example of result is showed in Figures 6.14–6.15. The model allows also a comparison between the integrated (configuration D) and not integrated (configuration C) configurations in the solar collector. Effect of concentrate recycling on heat recovery was also considered.

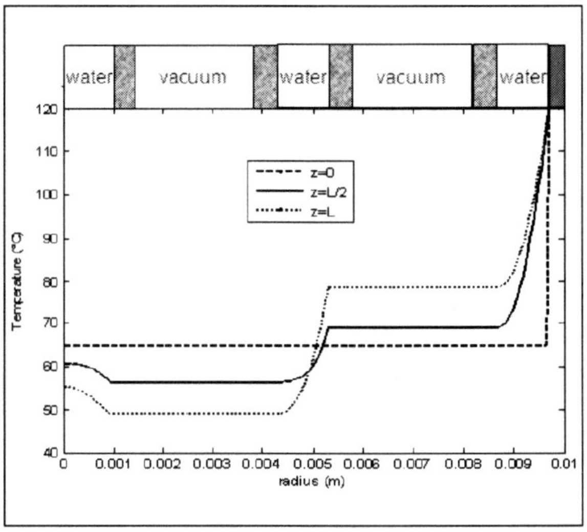

Figure 6.14 Temperature profiles. This figure is available in colour at http://www.iwapublishing.com/template.cfm?name=isbn9781843393214

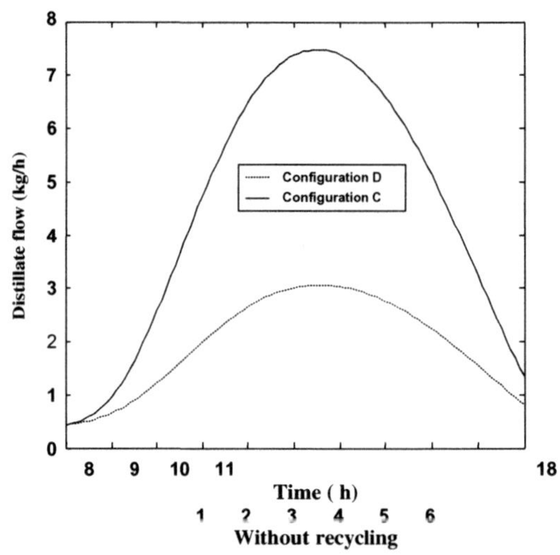

Figure 6.15 Instantaneous variation of the distillate flow for configurations C and D

According to simulations, whatever the configuration chosen (C or D), recycling could allow to enhance the quantity of distillate produced by a factor of about three in comparison with on operation without recycling. Operation with recycling improves energy effectiveness thus daily production. The distillate mass obtained with the hollow fiber module for the two configurations with and without recycling is reported in Table 6.3.

Table 6.3 Comparison between two modules configurations with recycling

Distillate mass obtained for an Hollow fiber module	Configuration C	Configuration D
without recycling	7.92 kg	10.39 kg
with recycling	19.8 kg	45.45 kg

Although the configuration D appears as more effective on the basis of these computations, but radial heat losses inside the module which will affect productivity have not been considered. This applicability of this configuration is probably limited to low diameter modules because oh radial heat profiles in the module and thus to small systems and low flow rates. Moreover its practical realization is very delicate. Configuration C was chosen for the following. The selected modules to be implemented on the pilot plant are commercial available hollow fiber modules.

6.4.2 Design and implementation of the demonstration pilot plant

The Figure 6.16 presents the flow sheet of VMD desalination demonstration plant and the various instrumentations of control and regulation.

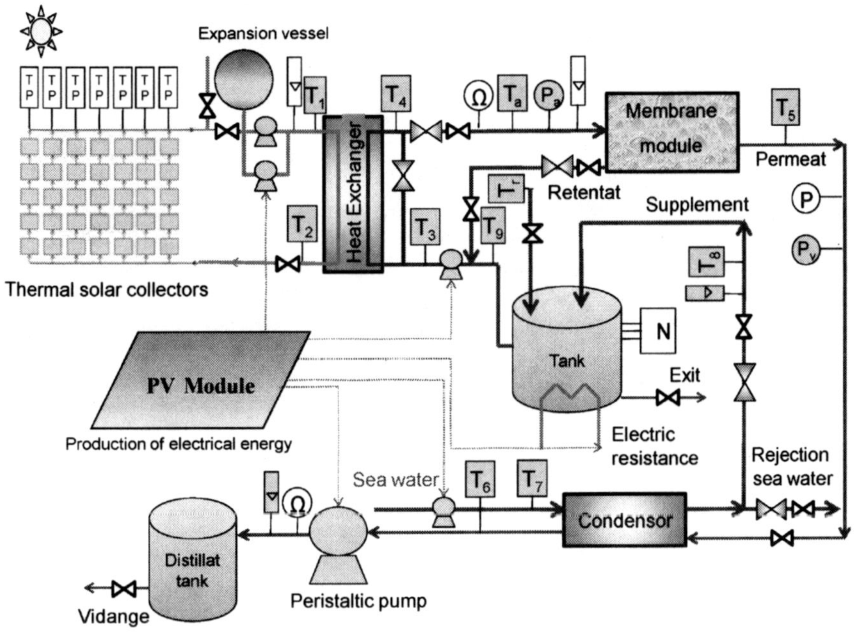

Figure 6.16 Diagram of the VMD desalination plant

The principal components of the vacuum solar membrane desalination demonstration plant are:

- a commercial PVDF hollow fiber module (UMP 3247 R): 806 fibers, 4 m^2 total membrane area.
- a system of thermal solar collectors comprising 7 lines of 5 collectors in series.
- a system of photovoltaic solar cells. This system is composed of 16 modules of cells LC 120 WC and 2 regulators voltage continuous 48 VDC 40A. An assembly of 8 solar tubular accumulators 12V/230 AH (4 lines of 2 accumulators) ensures the accumulation of the electric power produced. This installation is carried out of a power inverter 220/2 kW.
- Pumps for feed water and circulating heat water
- a pump for pumping the permeate, which can ensure a vacuum of 5000 Pa with 100 l/h flow rate
- an heat exchanger with titanium plates of 26 kW power, exchanger area 1,08 m^2 and 27 plates,
- a tubular condenser in titanium 60 kW power with 41 tubes, 7 mm internal diameter and 1 mm thickness,
  - Instrumentations for process control and regulation,
  - a 1 m^3 tank for fresh water
  - a mixing tank which makes it possible to mix the retentate outgoing membrane module and the supplement out of seawater
  - an electric resistance in the mixing tank thus allowing to heat it when the electric charge accumulated in the batteries exceeds the electric needs necessary for the pumps.

The demonstration pilot plant was installed in Tunisia in a plat-form in the village of orphaned children (S.O.S MAHRES), as shown in Figure 6.17. This is a non-governmental social and foster care of children without family support. It was created in 1949 to improve the living room conditions of children in distress. The S.O.S Children's Village of Mahres was created in March 2000. There are currently 13 family houses, playground, kids club, shops and services and domestic appliances. The capacity is 104 children with 8 children per family.

Figure 6.17 Demonstration solar/VMD plant installed in Tunisia

This choice was made for several let us reasons: firstly, this place offers great security for our facilities and at the same time it provides utilities needed to operate the unit as electricity. On the other hand, the water produced by the unit will be consumed by this small village. The village that is about 1500 m from the sea, now it was has survey that enables the delivery of brackish water. While this water is different from the seawater they wanted dealt with this project started goal eventually we edge build has pipe pumping seawater and after this we ensure the necessary funding for this operation.

Experimental tests were carried out, showing the effectiveness of the solar collectors and photovoltaic cells as well as the plates heat exchanger selected and allowing to estimate the quantity of water to be produced.

REFERENCES

Simone S., Figoli A., Criscuoli A., Carnevale M. C., Rosselli A. and Drioli E. (2010). Hollow fibre membranes prepared from PVDF/PVP blends: VMD tests. *Journal of Membrane Science*, **364**, 219–232.

Méricq J. P., Laborie S. and Cabassud C. (2009). Vacuum membrane distillation for an integrated seawater desalination process. *Desalination and Water Treatment*, **9**, 287–296.

Méricq J. P., Laborie S. and Cabassud C. (2010). Vacuum membrane distillation of seawater reverse osmosis brines. *Water Research*, **44**, 5260–5273.

Frikha N., Chaouachi B. and Gabsi S. (2009). Design of a semi industrial scale pilot plant for sea water desalination with a solar membrane distillation, Workshop MEDINA, Qingdao, China.

Chapter 7

Optimization and modelling of seawater and brackish water reverse osmosis desalination processes

S. Panglisch[1], M. Keller[2], C. Linder[3], G. Weihs-Fimbres[4], D. Wiley[4], S. Gaeta[5], E. Drioli[6,7] and A. Criscuoli[6]

[1] IWW (Germany)
[2] University of Duisburg-Essen (Germany)
[3] Ben Gurion University (Israel)
[4] University of New South Wales (Australia)
[5] GVS (Italy)
[6] ITM-CNR (Italy)
[7] University of Calabria (Italy)

7.1 INTRODUCTION

Seawater (SWRO) and brackish water reverse osmosis (BWRO) desalination plants are often very complex. They are also energy- and cost-intensive and above all, crucial to life support in several regions of the world. Consequently, desalination plants must meet high standards of performance, including optimality, cost effectiveness, reliability, and safety. Many of these criteria can be satisfied by optimized processes and/or improved design, operation and control. The optimization has to be seen against the background from the integrated approach. However, not only the desalination step itself has to be considered but also the other steps taking place in RO desalination like pre-treatment and post-treatment and their possible backlash to the core process have to be taken into account.

Removal of scaling ions by means of a nanofiltration (NF) pre-treatment stage is expected to provide drastic improvement in the efficiency of reverse osmosis and thus with a huge potential of cost savings. Turbidity, microorganisms and hardness can be strongly reduced, as well as the most part of multivalent ions. If NF is used monovalent species can be retained by 10–50% depending on the membrane properties. As consequence the osmotic pressure of the RO feed stream will decrease, thus allowing the unit to operate at high recovery factors (of about 50%) without scaling problems. However, costs have to be minimized beyond what is now possible. Consequently, one of the objectives of the presented work was to develop low cost and low fouling NF membranes for removing both scaling ions and organic foulants.

In desalination, the content of oxygen and carbon dioxide in the seawater considerably affects the performance and the material life of the desalination plants. Carbon dioxide also affects the pH and the conductivity of the water and could influence the salts precipitation. The content of these gases in the liquid stream can be optimized by using membrane contactors (MC). Thus, further objectives of the presented work were to investigate the potentialities of MC for controlling the water gas composition in desalination and to optimize them in terms of membrane properties and operating conditions.

An essential and economical method to optimize the single processes or the overall process is using modern computer-based simulation techniques. Modelling of RO and NF performance on the one hand side is beneficial to design and operation of membrane elements as well as of entire membrane desalination plants. On the other hand side modelling is also essential for comprehension of the fundamental processes which take place inside of a membrane element, which in turn is the basis for the development and validation of measures against fouling, of cleaning procedures, and of tools for the assessment of fouling potential. RO desalination processes make good candidates for using artificial neural networks (ANN) modelling due to their process complexity, nonlinear behaviour with many degrees of freedom, and the presence of uncertainty in the control environment. Further on, an important topic for the improvement of RO desalination processes is still the optimizing of the hydraulics in spiral wound membrane elements (SWM), which are predominantly used in RO and NF. In the last five to ten years many efforts were made to optimize the spacer geometry to enhance the mass transfer to hinder or avoid scaling and fouling processes and/or to decrease the pressure loss using computational fluid dynamics (CFD) modelling. Thus, the presented work also focused on the process modelling by using ANN and CFD.

7.2 OPTIMIZATION OF NF MEMBRANES USED IN THE PRE-TREATMENT OF MEMBRANE BASED DESALINATION

The limiting factors of SWRO are high concentrations of total dissolved solids (TDS) causing high osmotic pressure and the presence of scaling ions and the presence of suspended particles characterized as turbidity both of which can cause fouling. Because of the characteristics of the class of NF membranes characterized as water softening membranes and their intrinsic non-fouling properties as compared to SWRO membranes both scalants and foulants are readily removed from sea water and the total TDS is also significantly reduced. As a result the performance characteristics of the SWRO are significantly improved when it operates on this NF pre-treated SW stream. In fact numerous studies have shown multiple benefits of the combined NF/RO system such as higher recovery ratio (e.g., from 25 to 40% to as high as 60–70%), lower energy consumption, lower feed, and less waste concentrate. Another effort to decrease the cost is to recover valuable minerals from the NF and RO concentrate. The primary rational of increasing the recovery ratio of SWRO by the use of NF as a pre-treatment is in part to remove scalants such as $CaCO_3$, $CaSO_4$, $MgSO_4$, $Mg(OH)_2$, and this operation can be complemented by recovering these salts from the same concentrate especially such as gypsum and epsomite salts. Thus the optimized operation of the individual but integrated subsystems of a NF-SWRO unit with mineral precipitators and crystallizers can be used to supply not only potable drinking waters but also commercially valuable minerals at potentially cost effective operations.

It is reported, that the potential of using NF-RO systems for sea water desalination is in the overall product recovery due to the selectivity of NF which reduces scalants and organic foulants on the RO membrane, and reduction in chemical pre-treatment costs. Erikson *et al.* concluded that NF as pre-treatment for seawater RO could be economical only for cases were fouling (both by organics and inorganic scaling) is of importance and TDS levels of the sea water feed are high. In this case the intrinsically lower fouling of NF makes a NF-RO system more economical than a two pass RO. A NF-SWRO system to minimize the cost of the potable water should be designed to operate at a high recovery and high RO operating pressures. Consequently, the objective was to develop low cost and low fouling NF membranes for removing both scaling ions and organic foulants.

7.2.1 Membrane preparation

The membranes made were of composite structure comprising a nano-selective thin layer upon a UF support. The UF support in this case was commercially available 10 kDa PES from GE-Osmonics and the thin selective layer was made from different combinations of sulfonated polyphenylsulfone (SPPS), polyethyleneimine (PEI) and polyhydroxystyrene (PHS). The SPPS was prepared from commercially available polyphenylsulfone (PPS) which was sulfonated to different degrees. As NF membrane selectivity's to ions are known to vary as a function of anionic/cation charge ratio and degree of crosslinking, the membranes were made with different variations in the anionic charge (SPPS), cationic (PEI) charge and crosslinker PHS. Examples of the different selective layer compositions are given in Table 7.1 and vary between anionic (A1 and A2), amphoteric (A3/B1, B2, B3, B4) and primarily cationic A4.

The actual net ion exchange capacity of each membrane was measure and the measured ratio of cationic/anionic is similar to the calculated values (Table 7.2). The measured capacities are lower than the calculated ones since SPPS and PEI for a polyelectrolyte complex. This reaction was also observed in the solubility tests.

7.2.2 Membrane characterization

The pure water permeability L_p of the non crosslinked 'A' membrane series is represented in Table 7.3 along with the squares of the correlation coefficients.

Table 7.1 Variation in net ionic charge and different quantities of polystyrene hydroxy (PSH) in each membrane

Membrane	% SPPS	% PEI	% PHS	Positive to negative charge ratio	Calculated thickness (nm)
A1	100	–	–	0 (anionic)	200
A2	98	2	–	0.3 (anionic)	120
A3/B1	93.3	6.7	–	1 (amphoteric)	120
A4	80	20	–	3.5 (cationic)	200
B2	88.6	6.4	5	1 (amphoteric)	120
B3	84	6	10	1 (amphoteric)	120
B4	74.6	5.4	20	1 (amphoteric)	120

Table 7.2 Measured and calculated (in parentheses) charge capacities of the membranes A1, A2, A3 and A4 without PSH

Membrane	Negative charge (meq/g)	Positive charge (meq/g)	Positive to negative charge ratio
A1	1.46 (1.58)	0 (0)	0 (0)
A2	1.01 (1.55)	0.54 (0.44)	0.54 (0.3)
A3	1.25 (1.47)	1.33 (1.47)	1.06 (1.0)
A4	0.84 (1.26)	2.20 (4.40)	2.63 (3.0)

Table 7.3 Permeability L_p of membranes of series A and the squares of the correlation coefficients

Membrane	L_p (L/m^2 · h · bar)	R^2
A1	4.11	0.995
A2	6.36	1.000
A3	6.52	0.998
A4	3.73	0.995

There is a major difference between the water permeability of membranes A1 and A4 compared to membranes A2 and A3 that is likely due to difference in the thickness of the membranes. For technical reasons membranes A1 and A4 could not be prepared in thickness of 120 nm as A2 and A3 (Table 7.1). Therefore, these membranes were prepared with calculated thickness of 200 nm. Both A1 and A4 have similar fluxes even though the former is anionic and the latter cationic (Table 7.1), with respect to rejection however that have quite different selectivity's corresponding to their differences in charge. Thus the different mechanisms of selectivity's may be resolved independent of difference in permeability.

The L_p of the B membrane series as function of applied pressure is given in Table 7.4 with the squares of the correlation coefficients. The PHS functions as a crosslinker and the decrease in L_p of the membranes with the increasing addition of PHS (Table 7.1) can be rationalized in terms of increased degree of crosslinking.

Table 7.4 Permeability L_p of membranes in series B and the squares of the correlation coefficients

Membrane	L_p (L/m^2 · h · bar)	R^2
B1	6.52	0.998
B2	4.50	0.997
B3	2.85	0.993
B4	2.20	0.999

A transmission electron microscopy (TEM) of membrane A1 and B4 stained with uranyl acetate and sodium tungstate respectively, shows a composite structure with a thin selective layer of about 130 nm. Using both staining methods gave rise to clear image of the selective layer, which indicates the existence of both ionic and cationic groups in the membranes. The selective layer appears uniform with a straight smooth surface.

Membranes of series "A" were characterized with different salt solutions (Table 7.5) to determine the selectivity between monovalent and divalent ions and rejections to low molecular weight (LMW) organic solutes represented by a 5% sucrose solution. Sucrose has a molecular weight of 342 Da and a Stokes radius of 47Å and hence represents a relatively small size organic molecule.

The results for the membrane of series "A" (Table 7.6) with respect to organic solute rejection show the rejection of membrane A1 (anionic) is about 95% for all applied pressures. The other membranes increase in sucrose rejection at the higher pressures with the cationic membrane having the lowest rejection (<90%) at all pressures.

Table 7.5 Salt solutions tested

Solution number	[NaCl] (mM)	[$CaCl_2$] (mM)	[Na_2SO_4] (mM)
S1	20	–	–
S2	–	8	–
S3	–	–	8
M1	10	4	–
M2	10	–	4
M3	9.6	1.3	2.6

Table 7.6 Solute passage from 5% sucrose solution and relative standard deviation (in parentheses) in 10, 20 and 30 bar

	10bar	20bar	30bar
A1	5.8 (1.7)	4.6 (1.1)	5.6 (1.7)
A2	14.6 (8.9)	9.1 (5.3)	6.8 (3.4)
A3	13.3 (6.1)	6.4 (2.7)	3.0 (2.6)
A4	14.7 (5.4)	15.1 (3.3)	10.2 (0.2)

Figure 7.1 shows the behaviour of the ions in mixture M3, were four ions are present in the solution. Rejection properties appear to be dominated by the chloride ion which readily permeates the membrane due to its size. In addition two transport mechanisms can be seen in this case. The first is size exclusion and the 2nd is dielectric exclusion, since the divalent ions which are also the larger ions are more rejected than the monovalent ions. There is good separation between the mono and divalent ions in amphoteric membrane A3 which also has a high rejection to sucrose (97% at 30 bar-Table 7.7). This membrane can potentially be used in water softening processes.

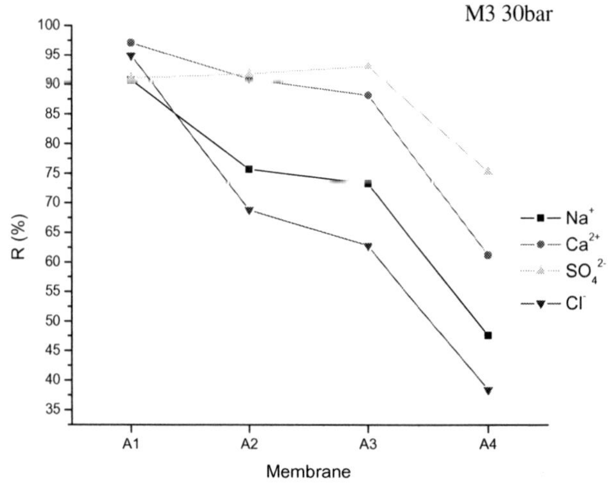

Figure 7.1 Rejection of ions in mixture solution versus membrane type. Solution M3 at 30 bar

As the membranes of series "B" (B1, 2, 3, 4 in Table 7.1) are based on the amphoteric membrane A3 (also called B1) with increasing amount of the crosslinker PSH, it was possible to study the effect of crosslinking on membrane solute selectivity. Table 7.7 shows the rejection of the respective membranes towards 5% sucrose solution. The high rejections of this compound in all membranes at 30 bars imply that the membranes are tight.

Table 7.7 Rejection of 5% sucrose solution and relative standard deviation in 10, 20 and 30bar

	10bar	20bar	30bar
B1	86.7 (6.1)	93.6 (2.7)	97.0 (2.6)
B2	84.2 (5.2)	93.1 (2.7)	96.6 (3.0)
B3	90.2 (2.0)	96.2 (3.6)	97.6 (2.9)
B4	91.1 (2.2)	96.2 (1.6)	96.8 (2.8)

The rejection of the ions in mixture solution M3 also correlates with the addition of PHS (see Figure 7.2). As shown below this result fits with the results of water permeability that decrease with addition of PHS. There is a larger change in the rejection from membrane B2 to B3 than from B3 to B4 or B1 to B2. That might imply that there is saturation in the change in the morphology caused by the PSH. The saturation in the degree of cross-linking is seen in this case as well. The declines seen in sulfate rejection from membrane B1 to B2 is minor and is within the range of error. As the amount of PHS in the membranes is growing the ions are less separated one from each other. Therefore, although the tighter membranes might have more stable structure, as will be discussed later, they are more susceptible to scaling of calcium sulfate.

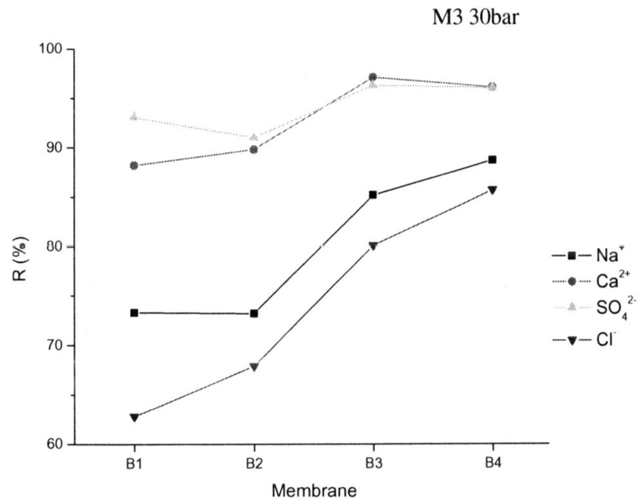

Figure 7.2 Rejection of ions in mixture solution versus membrane type. Solution M3, 30bar

The "A" series membranes were compared with commercially available NF membranes Dow Filmtech polyamide NF 270, NF 90 and Hydranautics sulfonated polysulfone NF 7450 (Table 7.8). The differences are:

- The A1 membrane is similar in rejection characteristics to the NF 90 which is a tight NF and both have a relatively high saline rejection. A1 however has a lower rejection to multivalent ions though still above 90%. This may be a positive factor as it would allow for a high recovery ratio of the NF element without a high passage of scalant ions and a high retention of salt (NaCl). Also the difference in surface properties
- A1 being an anionic polyphenylsulfone as compared to the NF 90 polyamide structure may give improved antifouling and cleaning properties. The A3 membrane is similar to the NF 270 membrane but with higher rejection to Na and Cl ions. Higher salt retention will lower the TDS to the RO unit increasing its recovery ratio. Also the difference in surface properties
- A2 being an amphoteric polyphenysulfone/PEI combination as compared to the NF 90 polyamide structure may give improved antifouling properties. Membrane A4 is similar in selectivity to NF 7450. The latter is not normally used as a pre-treatment to RO and is considered an open NF membrane. The membranes with increased crosslinking, B3 and B4 have fluxes which are considered too low to the cost effective for applications in sea water desalination.

Further on the membranes of series "A" were checked for antifouling properties in municipal waste streams. The composition of the municipal waste stream is characterized with respect to its conductivity (ms) and UV absorbance $UV_{254\,nm}$ for salt and organic content respectively. Example of representative values for conductivity and $UV_{254\,nm}$ are 1.052 mS/cm and 0.231 cm^{-1} respectively. The membranes were characterized in the municipal waste stream at constant flow with circulation under 10 atm pressure without volume concentration and cleaned periodically. When checked in these streams the membranes with selective layers containing PEI (A2, A3 and A4) lost 30 to 60% of their initial flux after 24 hours of operating indicating they were rapidly fouled. The membrane A1 with a selective layer of only SPPS maintained a constant Lp of 2 to 3 L/m^2/hr-bar which was considered as too low for the desired application. Based on these results the objective was the development of an NF membrane with low organic fouling while still maintaining the ion selectivity needed for softening.

Table 7.8 Comparison between developed and commercial membranes

Membrane rejection to ions in a mixture NaCl(9.6 mM), $CaCl_2$(1.3 mM), Na_2SO_4(2.6 mM)						
Ion	**A1**	**A3**	**A4**	**NF7450**	**NF270**	**NF 90**
$SO4^{2-}$	90	94	80	78	94	97
Ca^{2+}	95	88	65	77	86	99
Na^{+}	90	75	50	48	57	91
Cl^{-}95	63	38	41	47	89	
Flux Lp	4	6.5	7	7.6	8.2	9

−/+ Membrane charge ratio A1(0- anionic), A3 (1-amphoteric), A4 (3.5-cationic).

7.2.3 Development of NF membranes with low fouling properties

The approach for improving the antifouling properties of the membranes was the introduction of neutral hydrophilic polyvinyl alcohol (PVA) into the selective layer. Thus the selective layer was a combination of SPPS (IEC ~1.6 meq/gram) and PVA. We investigated different types of PVA. For examples:

- PVA with different ratios of hydroxyl to acetate groups characterized as polymer with 80%, 88% and 98+% hydrolysis of the acetate to hydroxyl.
- Different molecular weights form Mw of 9 K, 13-23 K, 31 K-50 K, 85 K-146 K.
- And ionomeric copolymers of polyvinyl alcohol with small amounts of co-ethylene (27 mole %).

After screening these PVA polymers we decided from both the performance and handling characteristics to concentrate on PVA of 98+% hydrolysis with a MW of 31 K-50 K.

A selective example of the different membranes formed is given in Table 7.9. The variables shown in this set of membranes are the concentration of PVA which varies from 20% (A) to 30% (B and C) w/w of the selective layer where the other component is SPPS, and the post curing step's time of 1 hr (A and C) and 1.5 hrs (B) at 110°C temperature. A control using polystyrene hydroxy (PStyOH) with SPPS is given in membrane D, which is similar to A1 discussed above. The membranes are conditioned overnight in 1000 ppm NaCl before testing.

Table 7.9 Preparation of hydrophilic membranes

Membrane	**Composition of selective layer**	**Curing temperature and time**
A	70% P.P.S 1.6 meq/gr, H form + 30% P.V.A 19/07/06	Oven: 1 h at 110°C
B(2)	80% P.P.S 1.6 meq/gr, H form + 20% P.V.A 19/07/06	Oven: 1.5 h at 110°C
C(1)	80% P.P.S 1.6 meq/gr, H form + 20% P.V.A 20/07/06	Oven: 1 h at 110°C
C(2)	Repeat of C(1)	
C(3)	Repeat of C(1)	
D	80% P.P.S 1.6 meq/gr, H form + 20% PStyOH 26/07/06	Oven: 1 h at 110°C

The membranes from Table 7.9 were characterized for L_p and rejection in pressure (dead-end) and flow cells (Table 7.10). Two commercial NF membranes NF 270 and NF 90 from Dow FilmTech were also tested for comparison. The results show that membrane C has the best potential as a sea water pre-treatment to RO with respect to good L_p, high organic (sucrose rejection) and a low mono salt rejection with a high divalent salt rejection. The results are similar to NF 270. Using a higher quantity of PVA (membrane A) gave higher L_p but a significantly lower rejection to organics, while increasing the curing time to 1.5 hours (membrane B) gave a lower L_p than C but a higher NaCl rejection indicating insufficient softening characteristics need for this application. When PVA is exchanged for PStyOH the resulting membrane (D) had much lower L_p and high rejection to NaCl as well as sucrose and sodium sulfate.

Table 7.10 Membrane Characterization in stirred and flow cells of membranes from Table 7.9

Series	L_p (L/m^2/h/bar)		%Rejection* Sucrose 5%	%Rejection* NaCl 20 mM	%Rejection* Na_2SO_4 8 mM
	stirred cells	flow cells			
A	8.3	7.87	70	60	98
B	5.6	4.93	90	81	98
C 1	7.5	8.4	85	70	97
C 2	8.8	6.8	86	68	99
C 3	9.3	8.9	90	66	98
D	2.11	2.9	96	93	99
NF 270	8.6	11	92	50	98
NF 90	12	9.5	98	91	99

**Rejection tests are carried out in flow cells at 15 atm.*

All the membranes made in that way were also checked for antifouling properties with the same conditions mentioned above. Table 7.10 presents the performance for the two commercially membranes NF 90 and NF 270, and membrane C (the developed membranes with the most promising characteristics of L_p, organic rejection and salt selectivity (softening)).

Further on, long term experiments (about 100 hours) in municipal waste stream after UF treatment were conducted with membrane of type C, NF 270 and NF 90, with respect to flux, ion/salt rejection and organic rejection respectively. The results show a constant flux for membrane C and for NF 270 with an L_p of ~6 or 10-12 L/m^2/hr-bar respectively, an overall salt rejection of 60% or 40% respectively and an organic rejection of about 99% for both.

By contrast NF 90 shows a rapidly dropping flux from 9 falling to less than 5 after 100 hours of operation even with periodic cleaning. The overall salt rejection is 95% with an organic rejection of about ~100%.

7.2.4 Conclusions and outlook

Promising membranes where developed based on selective layers of sulfonated polyphenylsulfone with polyvinyl alcohol, and compared to commercial NF (NF 90 and NF 270) in fouling municipal waste streams had improved performance characteristics. These membranes have a useful balance of salt rejections and separation of monovalent from divalent salts, a good flux and low fouling. These are characteristics needed for NF treatment of sea water feed streams prior to desalination by RO in order to achieve a higher recovery ration for NF-RO systems compared to RO-RO or single pass RO. In the future these membranes will be tested in sea and brackish water, and be further developed according to their performance in these tests.

7.3 USE OF MEMBRANE CONTACTORS FOR CONTROLLING THE WATER GAS COMPOSITION

7.3.1 Introduction

By considering that the final objective of the MEDINA project was to realize an integrated membrane system, membrane contactors (MC) tests with common and new membranes have been performed at different points of an integrated plant: on the water fed to the NF pre-treatment unit, on the permeates coming from the NF and the RO units and on the RO brine (see Figure 7.3).

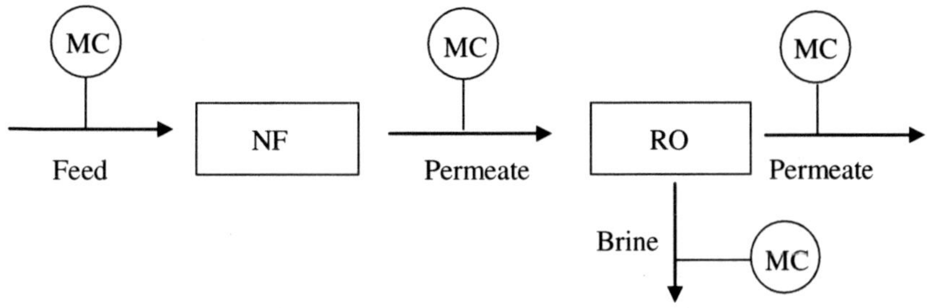

Figure 7.3 Membrane contactors used in different points of the integrated plant

The main objectives were to investigate the potentialities of MC for controlling the water gas composition in desalination and to optimize them in terms of membrane properties and operating conditions. In particular, the efficiency of the system for dissolved oxygen removal and pH control was studied. The use of MC was proposed as a way for reducing the chemical consumption inside the plant (normally, the pH control is, in fact, obtained by properly dosing acidic (H_2SO_4) or basic (NaOH) compounds), together with the corrosion and oxidation issues linked to the presence of oxygen (Criscuoli *et al.*, 2008 and Criscuoli *et al.*, 2010).

7.3.2 Membrane characterization

The new membranes used for the tests were provided by GVS (M09G0020-M09nt, made of PVDF, M09G0020E-M09t, which is the M09G0020 membrane, superficially treated, acrylic-based M07R0020, and M050020, made of PTFE). Further on a commercial polypropylene membrane was tested. All membranes have been characterized in terms of Scanning Electron Microscopy (SEM), Contact Angle Measurements (CAM) and water uptake.

Table 7.11 reports the results of the CAM measurements with distilled water for the two sides of each membrane. All of them are hydrophobic on both sides, but, excluding the polypropylene membrane, one side always presents a higher value of hydrophobicity. Therefore, for the membrane contactors tests, the side more hydrophobic has been putted in contact with the liquid stream.

Table 7.11 Contact angle values for the different membranes

M09G0020	M07R0020	M05E0020	M09G0020E	Commercial-PP02
side a/side b 144/138	side a/side b 140/130	side a/side b 145/100	side a/side b 147/140	side a/side b 134/134

Table 7.12 reports the results of the water uptake tests for the different membranes and liquid compositions. The M05 membrane presents the highest values of water uptake, followed by the M07.

Table 7.12 Water uptake results (%) for the different membranes and liquid compositions

Liquid	M09G0020	M07R0020	M05E0020	M09G0020E	Commercial-PP02
Distilled water	0	0	47	0	0
RO permeate	0	2.74	48	0	0
NF permeate	0	4.39	49	0	0
NF feed	0	4.95	55	0	0
RO brine	0	7.5	87	0	0

7.3.3 Results and discussion

Tests with distilled water and synthetic solutions, prepared in lab and containing only inorganic salts, have been carried out. Moreover, the effect of the presence of humic acid (HA) as organic species has been studied and experiments on real water samples provided by partners BGU and VEOLIA have been also performed.

Concerning the effect of the temperature, liquid flow-rate and gas flow-rate on the oxygen removal, it was found that higher removals can be reached at higher temperatures (due to the reduction of the oxygen solubility) and at higher liquid flow-rates (due to the reduction of the mass transport resistances at the liquid side). No significant effects have been observed by acting on the gas flow-rates.

Figure 7.4 shows the comparison of the membranes in terms of oxygen removal, for different liquid compositions. For all membranes there is an increase of the oxygen removal with the liquid concentration, due to the "salting-out" effect, with exception for the M05. The M05 has the highest oxygen removal with distilled water as feed but, after tests on the NF permeate, a decrease in the oxygen removal is registered.

This behaviour can be related to a partial closure of pores, as it was noticed by comparing the SEM picture of the M05 membrane before and after the tests. However, by washing the membrane with distilled water, the original performance was restored.

Figure 7.5 shows the trend of the oxygen content into the liquid with time for the different membranes.

A comparison between the oxygen removals achieved by using carbon dioxide and nitrogen as stripping gases is reported in Figure 7.6 for the M07 membrane. No differences are observed, then, depending on the water pH desired, one of the two gases can be chosen, leading to the same oxygen removals.

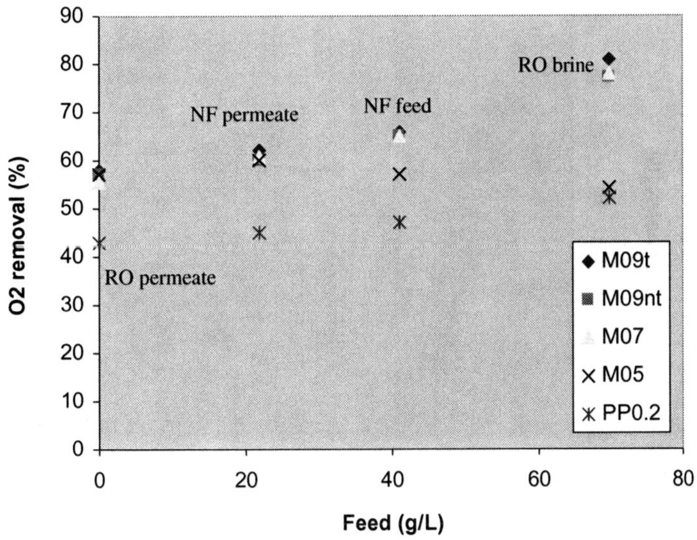

Figure 7.4 Membranes comparison for the different feeds (25°C, Qliq = 200 L/h, QCO_2 = 20 L/h, no HA, duration of test, 1 hour)

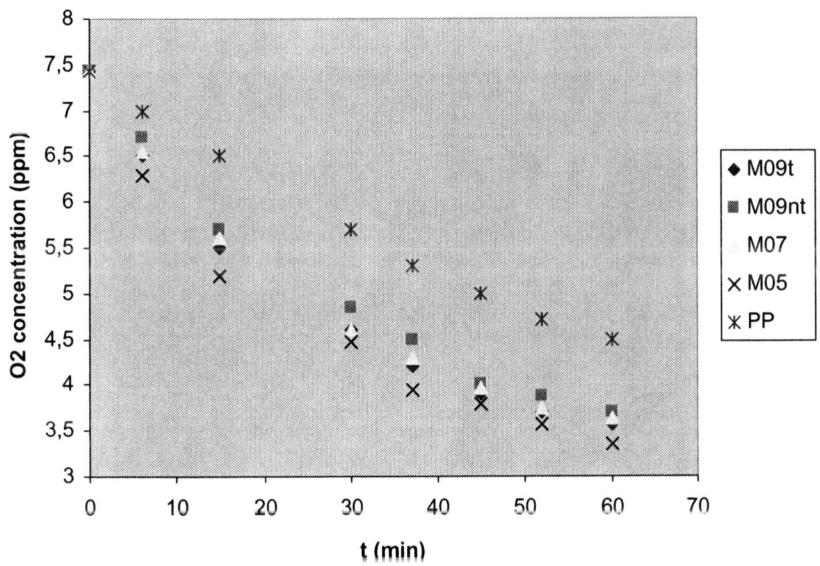

Figure 7.5 Trend of the oxygen content into the liquid with time for the different membranes (H_2Od, 25°C, Q_{liq} = 200 L/h, Q_{CO2} = 20 L/h, no HA, duration of test, 1 hour)

Figure 7.7 shows the reduction of pH achieved with the membranes at different liquid compositions, when carbon dioxide is used as stripping gas. The carbon dioxide solubility reduces when the content of salts increases and increases at higher feed pH. These two effects are balanced starting from the NF permeate and, therefore, no further variations are registered with the feed and the RO brine. All membranes behave similarly, with the exception of M05 that shows a slightly reduction of the pH variations after the NF permeate, for the same reasons observed for the oxygen removal trend.

The effect of nitrogen as stripping gas on the liquid pH is reported in Figure 7.8. When acidic liquids are fed to the system, the use of nitrogen leads to an increase of pH (up to neutral values), while maintaining the initial pH when neutral liquids are treated.

The comparison between the removals achieved with and without humic acid in the liquid is shown in Figure 7.9. For a humic acid content of 1.4 ppm, no big variations are observed with respect to the values achieved with a liquid

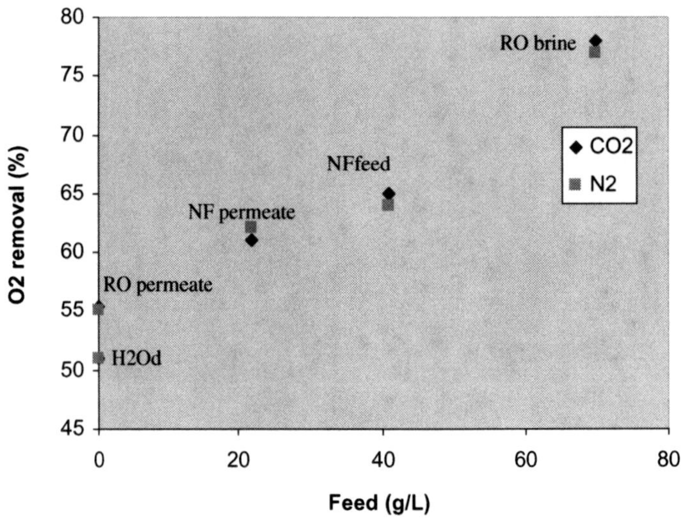

Figure 7.6 M07. Effect of the gas strip on the oxygen removal for different feeds. (25°C, $QCO_2 = QN_2 = 20$ L/h, Qliq = 200 L/h, no HA, duration of test, 1 hour)

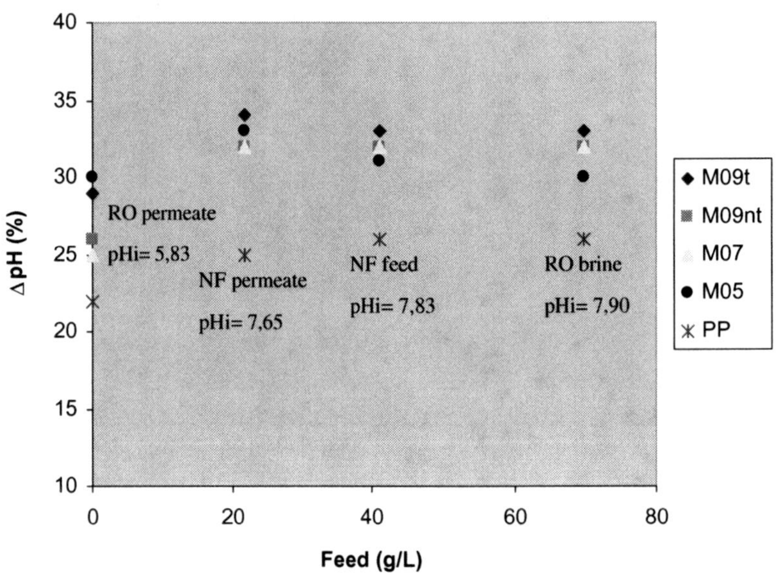

Figure 7.7 Reduction of the liquid pH for the different membranes at various feeds (25°C, $QCO_2 = 20$ L/h, Qliq = 200 L/h, no HA, duration of test, 1 hour)

organic-free. A small reduction is registered for humic acid contents of 2.7 ppm. This result can be attributed to the fouling of the membrane surface.

The efficiency of membrane contactors was, finally, investigated on real waters, with and without organics and antiscalants, provided by BGU and VEOLIA. Table 7.13 and Table 7.14 report the main composition of the tested waters. Besides them, tests were carried out also on seawater and UF permeate samples provided by VEOLIA.

Figures 7.10 and 7.11 summarize the obtained oxygen removals and pH reductions, respectively. The membrane contactor performed similarly for all tested waters and the final results did not deviate from those obtained during tests on synthetic solutions, confirming the potentiality of the proposed technique.

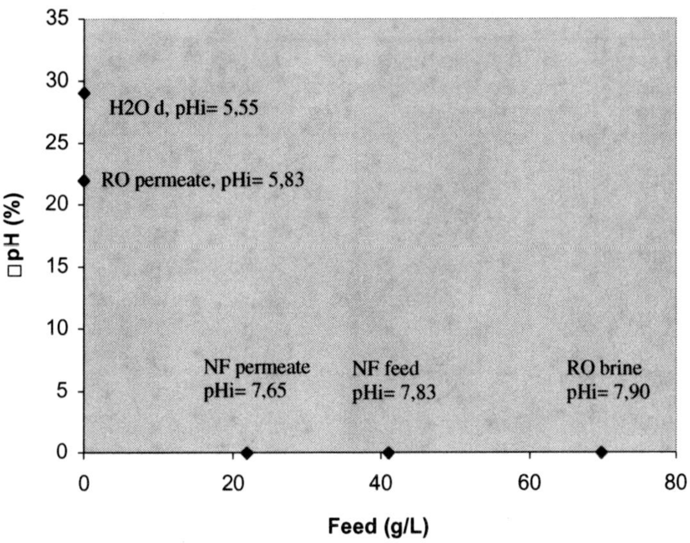

Figure 7.8 Increase of the liquid pH at various feeds. (25°C, QN_2 = 20 L/h, Qliq = 200 L/h, no HA, duration of test, 1 hour)

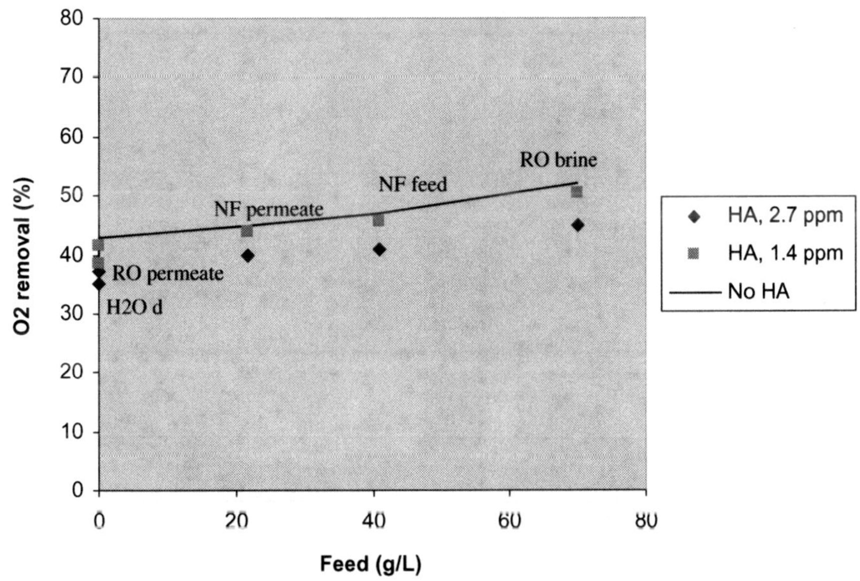

Figure 7.9 Comparison of the oxygen removals achieved with and without humic acid for different feeds (25°C, QCO_2 = 20 L/h, Qliq = 200 L/h, duration of test, 1 hour)

7.3.4 Conclusions and outlook

Based on the achieved results, it can be concluded that membrane contactors can be effectively applied for the control of the dissolved oxygen and of the water pH in a desalination plant without use of chemicals. All tested membranes kept their performance also when in contact with high concentrated solutions, except the M05. The best operating conditions in terms of liquid and gas flow rates have been identified and the efficiency of membrane contactors was proven also on real waters. As future research, an economical analysis of the proposed system will be at the basis of its possible implementation in desalination plants.

Table 7.13 Real water compositions (BGU) used for the membrane contactors tests (ppm)

	BGU_1: no org.; no antiscalant; 75% recovery	BGU_2: with org. and hexametaphosphate; 75% recovery	BGU_3: with org. and organophosphonate; 75% recovery	BGU_4: with org and organophosphonate; 88% recovery
Mg	3020	2897,2	5143.5	5130.75
Ca	1370	1026,4	946.5	786.38
Na	24080	24168	40861.5	38345.25
K	756	678,8	958.5	1107
SO4	9407.074	12511	16375.5	12103.37
Si	26.2	41.48	36	36
Sr	80.5	53	60.825	82.5
B	13.1	13.04	13.65	9.45
Ba	0.263	0.928	/	/
Fe	5.24	/	/	/
Ni	0.213	/	/	/
Zn	0.91	/	/	/
Li	0.78	11.6	18.4	/
TDS	/	90	239.78	142
EC (mS/cm)	109	78.7	142.4	158

Table 7.14 Real water compositions (VEOLIA) used for the membrane contactors tests (ppm)

	VEOLIA_RO brine: NOM, 1 C ppm; no antiscalant 20% recovery
Mg	1762
Ca	562
Na	15172
K	543
SO4	3700
SiO2	0.62
Sr	10
B	6.5
CO3	26
HCO3	214
Cl	27146

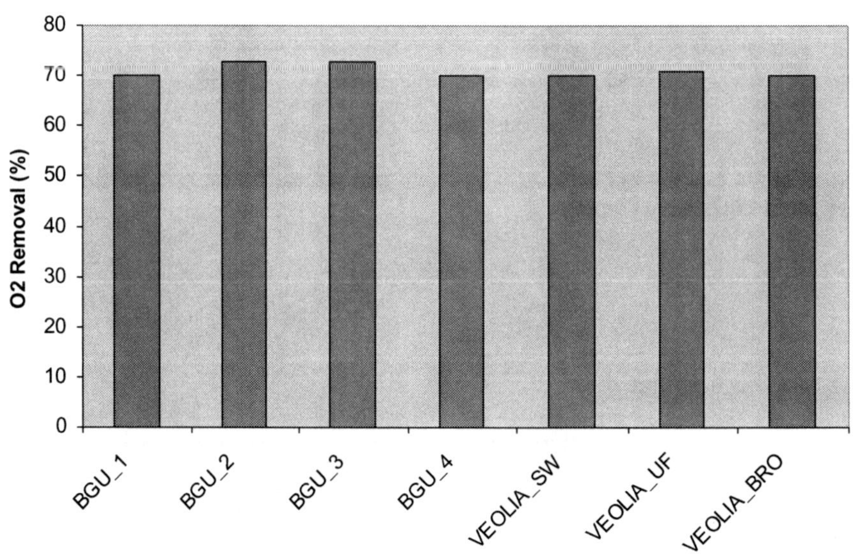

Figure 7.10 Oxygen removal for real waters

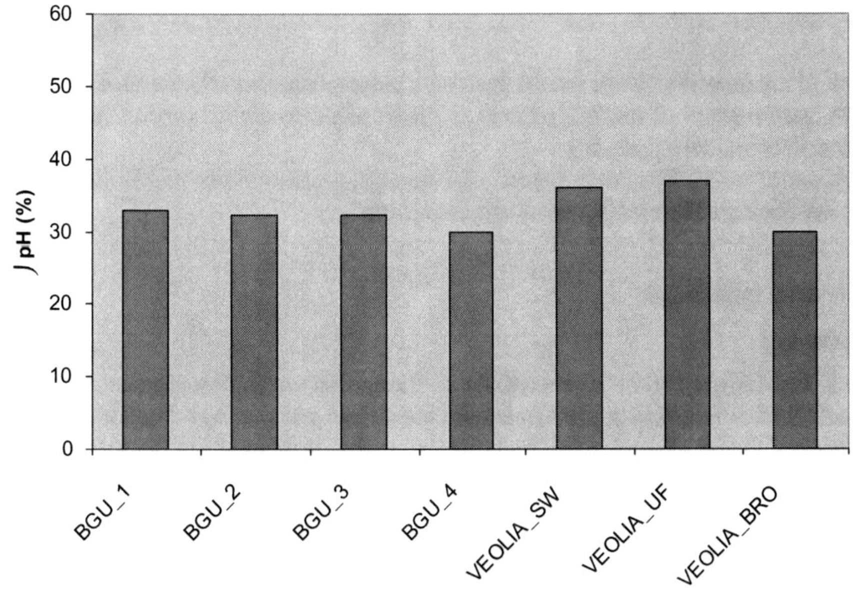

Figure 7.11 pH reduction for real waters

7.4 COMPREHENSIVE MODELLING OF THE RO DESALINATION PROCESS AND OPTIMIZING HYDRAULICS IN SPIRAL WOUND ELEMENTS

7.4.1 Introduction

During the last years, process modelling by the application of artificial neural networks (ANN) has gained acceptance in the field water treatment and membrane filtration. The primary advantage of ANN-modelling over theoretical approaches is the fact that it does not require any governing equations to model even complex multivariate processes. ANN have been successfully employed to predict various aspects of membrane performance including permeate flux, flux decline, and membrane rejection (Niemi *et al.*, 1995 and Curcio *et al.*, 2006) as well as the development of transmembrane pressure (Delgrange, 1998 and Strugholtz *et al.*, 2008) and membrane fouling effects (Delgrange *et al.*, 1998 and Shetty *et al.*, 2003). Also the applicability of ANN for prediction of membrane permeability using submerged ultrafiltration membrane capillaries has been shown (Kabsch-Korbutowicz *et al.*, 2009). Therefore it might also be possible to build an ANN model predicting the filtration performance in spiral wound modules derived from online data and laboratory measurements.

Unfortunately the strong reliance on specific training data can also be considered as the biggest drawback of the method of applying ANN for process modelling. Since ANN are solely extracting process information from the provided training data set, the predictions given by the generated ANN can only be considered reliable within the variation range of the variables applied during the training phase. Moreover the training data only represents the individual situation valid for the training data generating installation, thus, making the ANN model highly site specific. Since on-site data acquisition and processing is a complex and time consuming process it is of high interest to establish a method that makes the information provided by a neural net model applicable to other systems comprising for example a different module design.

Computational Fluid Dynamics (CFD) is a technique that allows simulation and subsequent analysis of fluid systems by solving conservation equations for mass, momentum and energy using numerical methods. Its results have become accurate and dependable enough to enable its application in areas such as aerospace engineering (Versteeg *et al.*, 1995). CFD has recently also become a more widely used tool in the field of membrane science (Ghidossi *et al.*, 2006), with more and more research groups utilising this technique in order to assist the design process, improve the performance, and gain insights into the phenomena taking place inside membrane modules.

The use of CFD for predicting flow and mass transfer conditions inside membrane modules reduces time, costs and risks associated with repeated experimental runs. Other advantages include its non-intrusiveness and the ease of manipulation and variation of flow conditions. However, the use of CFD as an analysis tool also presents disadvantages. One of the main difficulties most CFD studies face, particularly in the field of membrane science, is the lack of experimental data appropriate for validation purposes. In addition, the computational cost for the simulation of a full-scale membrane model is prohibitive at the current time. For this reason, most CFD membrane studies have focused on a small repeating section of the SWM module, and have assumed 2D flow.

A comprehensive literature research about using ANN and CFD modelling of membrane processes is given in deliverable D7.3.1.

The first approach of the presented work was to develop a comprehensive software tool on the basis of ANN, which is able to predict the performance of the RO process in terms of permeability dependent on the feed water quality, operation conditions and the module geometry.

The 2nd approach was to use CFD to study spacer mass transfer, pressure drops and fouling tendency in spiral wound elements, as well as developing and assessing novel spacer designs.

7.4.2 Materials and methods

7.4.2.1 ANN modelling

A laboratory test unit was designed and constructed due to the special needs. Since an extensive batch of training data including a broad variety of different pilot plant operation conditions and raw water qualities is needed for the training of the ANN, the pilot-plant design allowed a facilitated access to the membrane and the spacer channels, thus providing a high variability in terms of module geometry. The layout of the pilot-plant set-up is given in Figure 7.12.

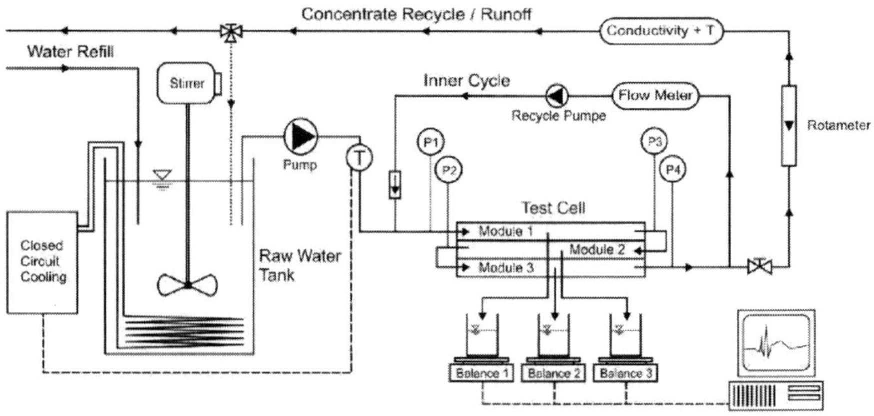

Figure 7.12 Process flowsheet of the modified pilot-plant set-up. Online sensor equipment is represented as cycles with "P" for pressure and "T" for temperature sensor

During all experimental studies sodium alginate has been used as a model substance to simulate organic and biofouling effects within the membrane system. Since alginate is a natural polysaccharide that is commonly found in gel matrices encapsulating microbial colonies in water systems, today alginate solutions are widely accepted and used to simulate Extracellular Polymeric Substance (EPS) induced organic- and biofouling mechanisms in membrane water treatment processes (Ye *et al.*, 2005 and Galle *et al.*, 2005). To provide a high variability in training data, a set of different operational conditions and raw water qualities were applied during the conducted experiments. Since due to the pilot plant specification no continuous variation of the raw water quality was possible the discrete levels given in Table 7.15 were chosen for the effectively variable parameters.

Table 7.15 Variation ranges for actively variable input parameters

	Feed-Spacer [mm]	Cross flow [l/h]	$CaCl_2$ [mmol/l]	NaCl [mmol/l]	Alginate [mg/l]
Exp.	0.7	40	0	0	0
Set 1			34	85.5	50
			68	171	100
Exp.	0.7	10	34	–	0
Set 2		20			
	0.85	40			50
Exp.	No Spacer	20	0	0	0
Set 3		30		6.25	10
	0.85	40	3.4	12.5	20
		50		25.5	40
		60		85.5	

Table 7.16 shows the input parameters applied for ANN training. The input parameters were chosen to represent the influences of water quality and module geometry volume flow on the membrane performance. Since current studies on the field of biofouling of spiral wound modules performed by Vrouwenfelder *et al.* revealed that a loss of filtration performance is commonly resulting from fouling phenomena induced or enhanced by the geometry of the feed channel and the applied spacer, which is indicated by a loss of pressure within the feed channel (Vrouwenvelder *et al.*, 2009 and Vrouwenvelder *et al.*, 2009), the feed channel pressure drop was chosen as an additional input variable to represent the extend of organic substance accumulation within the feed channel. Since the flux of all experiments was supposed to be constant with a certain range of variation the membrane permeability was selected to represent the filtration performance as the output of all established ANN.

Table 7.16 ANN input parameters and variation limits

	Unit	min	max	Measure
Water quality parameters				
Sodium chloride concentration	mol/l	0	0,086	preset
Calcium dichloride concentration	mol/l	0	0,0034	preset
Sodium alginate concentration	g/l	0	0,04	preset
Conductivity concentrate	μS/cm	0,6	200	online
Process parameters				
Volumetric feed flow rate	l/h	10	65	online
Feed channel pressure drop	bar	0,023	0,6	online
Mass transfer coefficient	m/s	Depending on calculation	calculated	

One of the mayor disadvantages of the application of ANN for the optimization and/or control of processes is its strong site specificity. In terms of membrane filtration this means that a neural network that is trained using operational data from a specific site can not be applied to a system comprising different module geometry. To provide an input parameter that represents the effect of the model geometry on the filtration performance a parameter is needed that incorporates a given continuous input parameter into a mechanistical equation that, based on an accepted theoretical basis, combines operational influences and the effects evolving from module geometry, into one new combined input variable. Thereby the respective equation is used to transform the continuous input into an effective parameter that makes the knowledge on the continuous variable that is enclosed in the structure of the trained neural network applicable to other membrane systems comprising different module geometries. Thus, the network should lose some of its side specificity and gains independence of module geometry.

Combining geometric, operational, and raw water influences, in this work the mass transfer coefficient (k; calculated in two different ways) was chosen to represent the desired comprehensive input for the ANN training. Due to neural net theory, only the relative progression of an input parameter and not its absolute value is affecting the developing neural net structure during the training phase. This means that as long as the neural net training is solely performed with data obtained from a single membrane system with constant geometric specifications the new input parameter k is basically representing the influence of the effective feed-channel flow rate on the filtration performance throughout the training. As soon as the model is readily trained it can be applied to membrane systems that comprise different module geometries. Since at this point the model is already trained, an absolute change of one of the input parameters will induce an immediate response of the neural net that is reacting on a virtual change of feed-channel flow velocity even if the measured volumetric flow rate is kept constant for both systems.

The incorporation of the effective feed channel velocity by applying the mass transfer coefficient in stead of (or in addition to) the measured volumetric flow rate during ANN training, thus, makes the neural net unsusceptible against changes of the module geometry and less side specific. In the same way the incorporation of an effective diffusion coefficient might create an even more site independent network. That, however, could not be tested within the present study. Based on the experimental data five independent ANN were trained. The neural net input parameters of all ANN are shown in Table 7.17.

To incorporate the system response on the given raw water quality and flow conditions into the input parameter set, the feed channel pressure drop (FCPD) was chosen to represent this factor. Since the FCPD changes according to a change of friction, contributed by the accumulation of organic fouling material within the flow channel, as well as on a changed flux, as a result of a varied osmotic pressure, the FCPD seems to apply as the optimal system response input parameter.

After training all generated ANN were tested by performing a validation with data that was not included in the training data set. For the primary validation a total of 10% of the available training data was randomly selected and separated from training data before the training was started. By later on applying the trained ANN on the validation data, the deviation between the predicted and the actually measured values provide a good estimation of the quality of the neural network as well as of its interpolation and extrapolation capacity (Geraldes *et al.*, 2006). Using this effect the transferability for each of the established ANN was evaluated by performing a 2nd validation using data from a separate set of experiments where identical volume flow rates and raw water conditions as included in the training data of the respectively validated ANN where applied but no feed channel spacers where used. Thus, the absolute deviation between the predicted and the measured values represents a direct measure for the module geometry independence of the evaluated ANN.

Table 7.17 Input Parameters of different neural networks for permeability prediction

	ANN No.				
	1	2	3	4	5
Input Parameters	Sodium chloride concentration	Sodium chloride concentration	Sodium chloride concentration	Sodium chloride concentration	Sodium chloride concentration
	Calcium dichloride concentration	Calcium dichloride concentration	Calcium dichloride concentration	Calcium dichloride concentration	Calcium dichloride concentration
	Alginate concentration	Alginate concentration	Alginate concentration	Alginate concentration	Alginate concentration
	Conductivity of concentrate	Conductivity of concentrate	Conductivity of concentrate	Conductivity of concentrate	Conductivity of concentrate
	Vol. Feed Flow Rate	–	–	Vol. Feed Flow Rate	Vol. Feed Flow Rate
	Feed channel pressure drop	Feed channel pressure drop	Feed channel pressure drop	Feed channel pressure drop	Feed channel pressure drop
	–	Mass transfer coefficient (k1)	Mass transfer coefficient (k2)	Mass transfer coefficient (k1)	Mass transfer coefficient (k2)

7.4.3 CFD modelling

The CFD model used simulated the steady fluid flow with mass transfer in three-dimensional narrow channels containing a novel non-woven 3-layer spacer with a filament diameter to channel height ratio of 0.6, and a mesh length to channel height ratio of 4, positioned at 0° and 90° against the bulk flow direction (Figure 7.13).

A solute with a Schmidt number of 600 (similar to NaCl) dissolving from the wall and hydraulic Reynolds numbers up to 200 were considered.

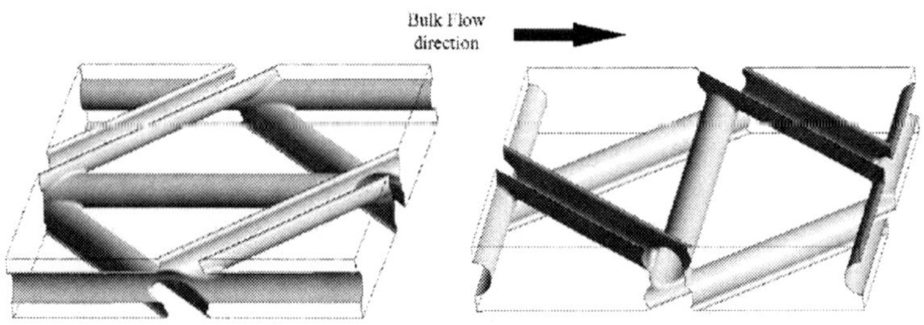

Figure 7.13 3-Layer spacer meshes modelled: A3LS-0° (left) and A3LS-90° (right)

Spacer performance at the different orientations was evaluated via a basic permeate processing cost analysis. An approximation of total costs was obtained by analyzing the direct costs for the production of permeate, without taking into account pre-treatment costs, which are independent of the spacer geometry in the SWM units. If the average permeate flux ($J_{sln,avg}$) is known, the capital cost for the membrane unit per cubic meter of permeate flow produced (C_c) was calculated using the following expression:

$$C_c = \frac{C_m F_a \rho}{t_{op} J_{sln,avg}} \tag{7.1}$$

where C_m is the cost per square meter of membrane equipment, F_a is the amortization factor, ρ is the solution density and top is the operating time. Operating costs (C_{op}) are proportional to the pumping energy (Table 7.18). Hence, the operating cost per cubic meter of permeate flow is then given by:

$$C_{op} = \frac{Q_{in}\Delta p_{ch}C_e}{\eta_{pump}J_{sln,avg}w_{ch}L} \tag{7.2}$$

where C_e is the energy cost, L is the membrane module length, w_{ch} is the width of the channel, η_{pump} is the pump efficiency, and Δp_{ch} is the pressure drop along the membrane module. Finally, the total processing cost per cubic meter of permeate flow produced (C_{tot}) was calculated as the sum of the operating and capital costs:

$$C_{tot} = C_{op} + C_c \tag{7.3}$$

Table 7.18 Case study parameters for cost analysis of CFD modelling of a membrane channel

Module length (L)	90 cm
Channel width (w_{ch})	90 cm
Salt rejection	99.6%
Feed mass fraction ($\omega_{b,in}$)	0.025
Inlet transmembrane pressure ($\Delta p_{tm,in}$)	60 atm
Reflection coefficient (σ)	1
Osmotic pressure coefficient (φ)	8.051×10^7 Pa
Membrane permeability (L_p)	3.94×10^{-6} $m\ s^{-1}\ atm^{-1}$
Channel height (h_{ch})	0.001 m
Energy cost (C_e)	\$0.10 kWh^{-1}
Membrane cost (C_m)	\$100 m^{-2}
Amortization factor (F_a)	0.4 yr^{-1}
Pump efficiency (η_{pump})	0.6
Operation time (t_{op})	8000 $hr\ yr^{-1}$

Although the CFD simulations did not calculate permeate flux directly, it was obtained using the following relationship, derived from the mass balance of solute at the membrane surface:

$$J_{sln} = \frac{1}{2}\left(\frac{TMP}{\mu R_m} + \rho k_{mt}\right) - \sqrt{\left[\frac{1}{2}\left(\rho k_{mt} - \frac{TMP}{\mu R_m}\right)\right]^2 + \frac{\rho k_{mt}\sigma\varphi(\omega_b - \omega_p)}{\mu R_m}} \tag{7.4}$$

where ω_b and ω_p are the solute mass fractions at the feed and permeate respectively, k_{mt} is the mass transfer coefficient, is the solution viscosity, R_m is the membrane resistance, ρ is the solution density, σ is the reflection coefficient, and ω is the osmotic pressure coefficient. This equation can be used to calculate the local permeate flux in terms of the operating parameters and the mass transfer coefficient. Since the mass transfer coefficient obtained from the CFD simulation results was obtained using impermeable wall conditions, the mass transfer coefficients for permeable wall conditions was calculated using the approach of (Geraldes *et al.*, 2006).

7.4.4 Results and discussion

7.4.4.1 ANN modelling

When experiments where performed as depicted in Table 7.15, it became obvious that, while the crossflow velocity could be properly adjusted and maintained by the employed set-up, it was hardly possible to completely decouple the applied flux and pressure. Due to the lack of a volume regulated control valve at the concentrate run off, constant flux conditions could not be guaranteed throughout the experiments. Thus, the generated conditions were not sufficient to evaluate membrane flux dependent mass transfer processes.

As described earlier ANN learn by recognizing patterns in the information provided with the training data. Thus, a good model is mirroring the real dependencies between the chosen input parameters and their influence on the target variable. Concluding from this, a networks analysis may provide information on the impact of neural net input parameters on the process.

By performing a sensitivity analysis, a certain input parameter is evaluated with regard to its influence on the modelling result. In this study the sensitivity analysis is realized by a so called iteration procedure, whereas the sensitivities are calculated from the reaction of the output parameter on defined changes (±0.5% of the value range) of each individual input parameter. Doing so, the average influence can be obtained for each input parameter. Thus, the sensitivity analysis provides quantitative information on the proportional contribution of each input onto the output factor as well as on the quality of the impact, respectively the way (positive or negative) in which the input contributes to the prediction of the output.

A sensitivity analysis was performed for each of the five generated neural networks, respectively for the corresponding input variables given in Table 7.17. From the sensitivity analysis the most negative impact has the alginate concentration. If the concentration is increased, the output or permeability respectively will decrease. The increase of the volume flow or the cross-flow velocity respectively has a small positive impact on the permeability. Why the concentration of both Ca and Na has a positive impact on the permeability on the one hand side and the conductivity on the other hand side has a negative impact is still unclear. Possibly, in the ANN structure the conductivity considers the osmotic pressure and the concentration of Na and Ca considers a positive impact on the permeation resistance of the alginate layer. This has to be proven by further data. It can further be seen that the relative impact of the cross-flow velocity is not completely accounted for by the mass transfer coefficient. Otherwise the impact of the flow rate should be close to zero in the respective ANN where the mass transfer is considered as well. Nevertheless, the positive impact of the cross-flow velocity on the permeability is obvious in all models. As these were trained using the same data basis, this result shows that the impact of flow velocity is coherently weighted by all 5 generated networks.

When the models were validated it showed that all established ANN basically posses the same quality when validated with a 10% fraction of the training data. Since the separation of the validation data fraction was automatically performed and data was randomly selected from the training data set, this result indicates good homogeneity of the training data set.

However, when the established networks where applied to data sets that where generated when no spacer was applied to the feed flow channel (external data), the respectively changed flow conditions within the feed channel resulted into significantly higher deviations between the predicted and the measured permeabilities. The validation results depicted in Table 7.19 further illustrate that while the validation with training data delivered similar results for all tested ANN, the validation with external data yielded higher standard deviations among the absolute validation results. Since the training data base as well as the influencing factors represented by the input variable set was the same for all developed ANN, this result indicates that the incorporation of deterministic filtration concepts into ANN input variables can positively affect the adaptability of the established neural networks. The validation results also reveal that validation quality can be further increased when the online measured input (flow velocity) and the corresponding deterministically specified input (mass transfer).

Table 7.19 Transferability test

ANN No.	Additional input variable(s)	Mean deviation: Prediction – Measured value [% of value range]	
		Validation with 10% of the training data set => same module geometry (spacer applied)	Validation with external data => changed module geometry (no spacer applied)
1	vol flow rate	4.31	17.69
2	k1	4.33	16.72
3	k2	4.35	15.47
4	vol flow rate; k1	4.44	15.25
5	vol flow rate; k2	4.35	12.88
	Standard deviation	0.05	1.81

Concluding from the provided results it can be summarized that the incorporation of theoretical filtration models by specification of continuously measured input parameters is generally applicable and might positively affect the adaptability of neural networks. Nevertheless, the implementation of deterministic models implies applicability of the theoretical concept on the specific situation. Thus, the adaptive capacity of neural networks established by the described method can only be expanded on the application range of the underlying theoretical model, which is mostly restricted to certain specific boundary conditions. Considering this constriction a suitable application for the developed method may be found in the field of piloting and up scaling applications, where certain boundary conditions are maintained while hydrodynamic conditions vary for example according to changed module dimensions.

7.4.4.2 CFD modelling

Comparison of permeate processing costs in the Reynolds number range of 50 to 200 suggests that although the 90° orientation of the 3-layer spacer results in a 150% to 200% higher channel pressure drop respectively, lower processing costs were achieved due to a 55% to 14% increase in mass transfer and therefore 24% to 7.6% lower membrane area requirements respectively (cf. Figure 7.14). Moreover, the costs achieved by the 90° orientation in the laminar flow regime were lower than those achieved with common 2-layer ladder or diamond type spacers. Further reductions in processing costs may be achievable by varying the geometric ratios of the 3-layer spacer mesh, as the optimal ratio for multi-layer spacers may differ from traditional 2-layer spacers.

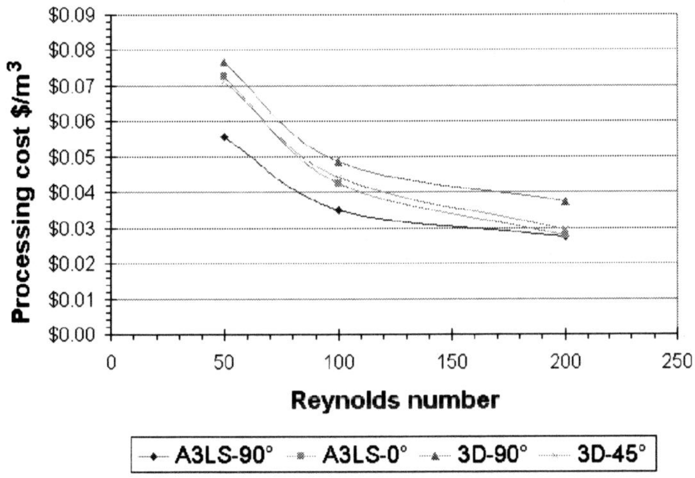

Figure 7.14 Comparison of processing costs per cubic metre of permeate for the two 3-layer spacer orientations considered, and for two typical ladder type spacers positioned at 45° and 90° against the bulk flow

The shape of extrusion profile of middle spacer layer also lends itself to optimisation, given that previous research shows that form drag in the bulk flow does not improve mass transfer. In order to analyse the effect of the shape and orientation against the bulk flow of the middle spacer layer, various configurations incorporating elliptical filaments as the middle spacer layers were simulated and analysed (Figure 7.15). Due to the higher number of degrees of freedom in 3D spacer geometries, only 2D flow was considered for this study.

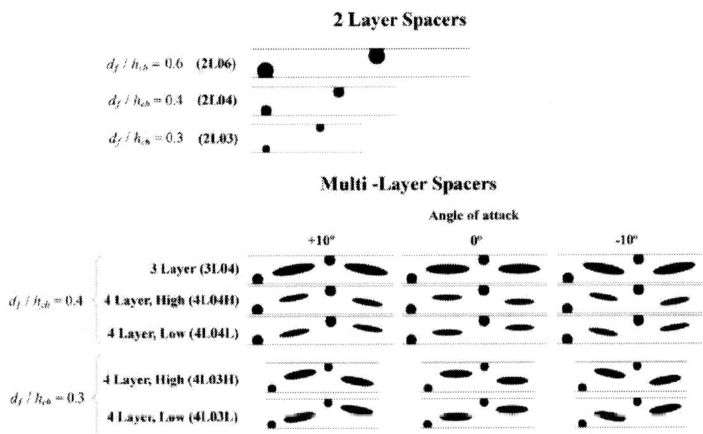

Figure 7.15 Schematic of the unit cells for the different spacer configurations analyzed

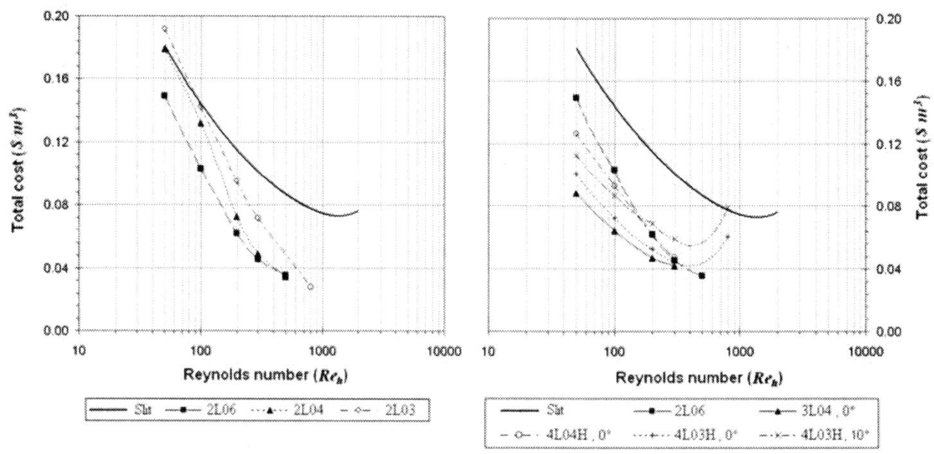

Figure 7.16 Dependence of total processing cost per cubic meter of permeate on hydraulic Reynolds number, for empty, 2-, 3- and 4-layer spacer-filled channels. The graph on the left shows effect of filament diameter on cost. The graph on the right compares costs for selected multi-layer spacers against empty channel and 2L06 spacer. A membrane cost of $100 per square meter was assumed

A basic permeate processing cost analysis was performed for these spacer configurations using the CFD simulation results (Figure 7.16). One of the main components of processing cost which was not incorporated into the economic analysis carried out in this work was that of fouling and cleaning. Since this cost is usually proportional to the membrane area, it was therefore assumed to be included in the membrane cost. If fouling and cleaning costs are significant, the analysis presented in this work suggests that increasing the operating Reynolds number will reduce total processing costs. The direct effect of multiple spacer layers on fouling was not studied in this work. However, based on the study of (Schwinge *et al.*, 2004) and on the results presented here, it is expected that the higher local velocities and shear rates at the membrane wall will reduce the extent of fouling.

The multi-layer configurations with elliptical filaments showed potential for increased productivity due to mass transfer enhancement. However, lower permeate processing costs were obtained by using a simple 2D zigzag spacer (2L06). The middle layers of elliptical filaments in multi-layer spacer configurations promoted mass transfer enhancement when compared to 2-layer spacer filled channels with the same filament diameter to channel height ratios. However, stagnant fluid regions near the membrane walls resulted in lower mass transfer for the multi-layer spacer configurations than for the 2L06 and 2L04 spacers. Variation of the angle of attack of the elliptical filaments showed potential to increase mass transfer enhancement at the same or lower energy losses by optimizing the profile of the middle spacer layer, such that little or no recirculation regions are formed on its downstream side. In addition, multi-layer spacer showed potential for reducing processing costs in high membrane cost scenarios.

Although negative attack angles usually reduced energy losses, they failed to direct flow towards the membrane wall, and thus resulted in lower mass transfer and hence higher operating costs per unit volume of permeate (Figure 7.17). On the other hand, positive angles of attack generally resulted in lower total permeate processing costs, due to increased mass transfer. Thus, mass transfer performance was the main factor in determining the total processing costs for the conditions of this study. As such, improvements in mass transfer enhancement have the potential to reduce both capital and operating processing costs per unit volume of permeate produced. This is possible even if energy losses are increased, since larger permeate fluxes will reduce the operating cost per unit volume of permeate.

With regards to the relative positioning of circular and elliptical filaments in 4-layer spacers, it was found that the "low" (–L) configuration resulted in higher mass transfer enhancement, and hence lower permeate processing costs. 4L spacers in the L configuration, coupled with positive angles of attack for the elliptical filaments were found to be the most effective in disrupting flow and redirecting low concentration fluid towards the membrane walls. However, further studies are required in order to optimise the hydrodynamic profile and angle of attack of the submerged layers, and thus reduce the energy losses incurred by the addition of further flow obstacles. Variation of the mesh length to channel height ratio should also be explored.

7.4.5 Conclusions and outlook

The substantial effort that has been invested into the field of modelling mass transfer and hydrodynamics in spacer filled spiral-wound membrane modules shows that the mutual relations of the involved variables are that complex and situation specific that an assessment of meaningful correlations by means of conventional data evaluation based on mechanistical modelling approaches is highly time consuming and labour-intensive. In contrast, the evaluation of correlations by applying ANN is fast and relatively simple in its application. Since neural networks analysis is reducing the time expenses for the evaluation of correlations implied in the given training data set as well as reducing the bias emerging due to the personal perception of the analyst, it provides a good base for further investigations.

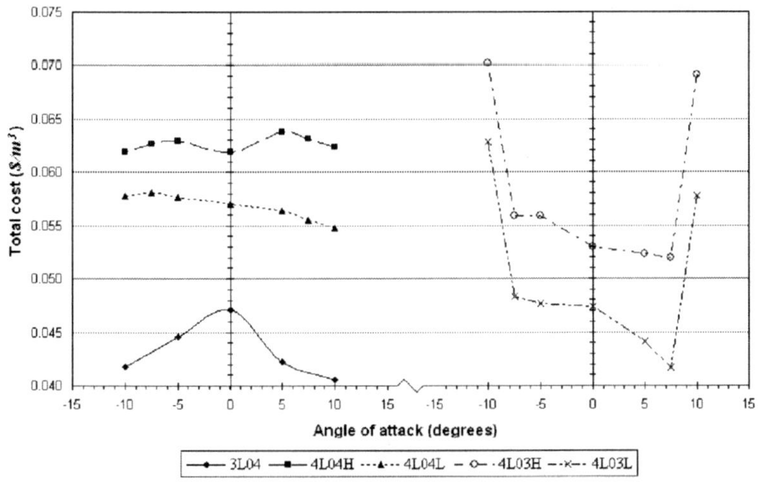

Figure 7.17 Dependence of total processing cost per cubic meter of permeate on angle of attack of elliptical filaments, for multi-layer spacer-filled channels at a hydraulic Reynolds number of 200. A membrane cost of $100 per square meter was assumed

However, the idea of generating a comprehensive ANN model that is capable of predicting the filtration performance of a given spiral-wound module independent of its geometry and applied spacer types presumes an exhaustive library of training data that comprises a huge variety in operation conditions, raw water compositions and their respective combinations. Since during this study severe problems in generating reproducible training data were encountered even for the use of model fouling substances and idealized laboratory operation conditions, it is doubtful if such a huge and comprehensive data set can be established by applying an arguable work and time expenditure. Nevertheless the results showed that the implementation of mechanistical correlations can be applied as ANN input, thus, increasing the predictive quality and strengthen the transferability of the network.

In future studies it might be interesting to investigate the use of empirical ANN knowledge to validate or complete mechanistical correlations, for example by the application of genetical algorithms.

A more suitable application for CFD is during the design stage of membrane processes, such as the design of spacer meshes. However, one of the main difficulties most CFD studies face, particularly in the field of membrane science, is the lack of experimental data appropriate for validation purposes. Future studies utilizing CFD models should address the issue of validation. For this purpose, Particle Image Velocimetry (PIV) is a developing technology for acquiring data that might be compared to data obtained from CFD simulations. Permeation and fouling phenomena should also be incorporated in more detailed design studies of spacer geometries for improving membrane system performance while keeping energy losses at a minimum. In addition, several of the assumptions made in the CFD model should be relaxed in subsequent studies, such as constant fluid properties. This should enable the design of a new generation of spacer meshes with the ability to improve mass transfer enhancement while keeping energy losses at a minimum, thus improving the performance of membrane systems.

7.5 SUMMARY

The presented work was focused on the research and development of new processes and tools to optimize membrane based seawater and brackish water desalination. Novel Nanofiltration (NF) membranes with good flux and softening characteristics as well as lower fouling characteristics compared to commercial NF were developed. In the future the membranes will be tested for sea and brackish water, and be further developed. Theoretical and experimental studies on membrane contactors (MC) showed that MC can be effectively applied for the control of the dissolved oxygen and of the water pH in a desalination plant without use of chemicals. The efficiency of MC was proven on real waters and by considering as case study their application for the pH adjustment in the pre-treatment line of a desalination plant when coagulation is used. Computational Fluid Dynamics (CFD) was used the optimise characteristics of the spacer in spiral wound elements in terms of mass transport enhancement, fouling and scaling mitigation, and reduction of energy consumption. Spacer performance for the different configurations was evaluated via a basic permeate processing cost analysis. By these calculations the spacer configuration could be optimised which resulted in higher mass transfer enhancement, and hence lower permeate processing costs. Based on experimental data sets five independent artificial neural networks (ANN) were trained and afterwards validated. The structure of the ANN was modified by using the equation for the determination of the mass transfer coefficient as alternative input parameter in order to increase the predictive quality and strengthen the transferability. However, the study further showed that even for the use of model fouling substances and idealized laboratory operation conditions it is hardly practicable to establish a comprehensive data set necessary for training the modified ANN by applying an arguable work and time expenditure.

REFERENCES

ATLAN-tec GmbH, Handbook for NeuroModel Release 2.3, Willich, GER, 2007.

Babovic V. and Minns A. W. (1994). Proc.; 1st Int. Conf. on Hydroinforamtics, Balkema, Rotterdam, The Netherlands.

Balster J., Pünt I., Stamatialis D. F. and Wessling M. (2006). Multi-layer spacer geometries with improved mass transport. *Journal of Membrane Science*, **282**(1–2), 351–361.

Bergles A. E. (1997). Heat transfer enhancement – The encouragement and accommodation of high heat fluxes. *Journal of Heat Transfer*, **119**(1), 8–19.

Bongards M., Hilmer T. and Ebel A. (2005). Computational intelligence-based optimisation of wastewater treatment plants. *Water Science and Technology*, **52**(12), 99–104.

Chang Y. C. and Myerson A. S. (1985). The diffusivity of potassium chloride and sodium chloride in concentrated, saturated, and supersaturated aqueous solutions. *AIChE Journal*, **31**(6), 890–894.

Criscuoli A., Carnevale M. C., Mahmoudi H., Gaeta S., Lentini F., Reggiani S. and Drioli E. (July 12-18 2008). "Oxygen and carbon dioxide control by membrane contactors in desalination". *International Conference on Membranes, ICOM 2008*, Honolulu, Hawaii (USA).

Criscuoli A., Carnevale M. C., Mahmoudi H., Gaeta S., Lentini F. and Drioli E. Membrane contactors for the oxygen and pH control in desalination *(Submitted to Journal of Membrane Science, 2010)*.

Curcio S., Calabro V. and Iorio G. (2006). Reduction and control of flux decline in cross-flow membrane processes modeled by artificial neural networks. *Journal of Membrane Science*, **286**(1–2), 125–132.

Delgrange N., Cabassud C., Cabassud M., Durand-Bourlier L. and Laine J. M. (1998). Neural networks for prediction of ultrafiltration transmembrane pressure – application to drinking water production. *Journal of Membrane Science*, **150**(1), 111–123.

Delgrange N., Cabassud C., Cabassud M., Durand-Bourlier L. and Laine J. M. (1998). Modelling of ultrafiltration fouling by neural network. *Desalination*, **118**(1–3), 213–227.

Emmerichs N. (2004). *Untersuchung der Wechselwirkungen von Mangan- und Calciumionen mit Alginat von Algen und von verschiedenen mucoiden Stämmen des Bakteriums Pseudomonas aeruginosa*. Dissertation, Duisburg: University of Duisburg-Essen.

Galle C. O. (2005). *Alginat- und Alginat/Protein-Systeme als Modelle der EPS-Matrix von Biofilmen – NMR-Untersuchungen zur Wasserdiffusion*. Dissertation, Duisburg: University of Duisburg and Essen.

Garson G. D. (1998). *Neural Networks – An introductory Guide for Social Scientists*. London: SAGE Publications Ltd.

Geraldes V. and Afonso M. D. (2006). Generalized mass-transfer correction factor for NF and RO. *AIChE Journal*, **52**(10), 3353–3362.

Geraldes V. and de Pinho M. N. (2006). Mass transfer coefficient determination method for high-recovery pressure-driven membrane modules. *Desalination*, **195**(1–3), 69–77.

Ghidossi R., Veyret D. and Moulin P. (2006). Computational fluid dynamics applied to membranes: State of the art and opportunities. *Chemical Engineering and Processing: Process Intensification*, **45**(6), 437–454.

Grant G. T., Morris E. R., Rees D. A., C., S. P. J. and Thorn D. (1973). Biological Interactions between Polysaccharides and Divalent Cations: The Egg-Box Model. *FEBS Letters*, **32**(1), 195–198.

Gumrah F., Oz B., Guler B. and Evin S. (2000). The application of artificial neural networks for the prediction of water quality of polluted aquifer, *Water Air and Soil Pollution*, **119**(1–4), 275–294.

Holtan S. (2006). *Structural tailoring of alginates – A study of the mode of action of mannuronan C-5 epimerases*, Doctoral thesis, Trondheim: Norwegian University of Science and Technology.

Izydorczyk M., Cui S. W. and Wang Q. (2005). Polysaccaride Gums: Structures, Functional Properties, and Applications. In: S. W. Cui (Ed.), *Food Carbohydrates – Chemistry, Physical Properties, and Applications*. Taylor & Francis, Boca Raton, FL, pp. 293–295.

Kabsch-Korbutowicz M. and Kutylowska M. (2009). Neural network approach for prediction of submerged membrane permeability in surface water coagulation/ultrafiltration process. *Chemical and Process Engineering*, **30**(2), 307–318.

Li F., Meindersma W., De Haan A. B. and Reith T. (2005). Novel spacers for mass transfer enhancement in membrane separations. *Journal of Membrane Science*, **253**(1–2), 1–12.

Maier H. R. and Dandy G. C. (1996). The use of artificial neural networks for the prediction of water quality parameters. *Water Resources Research*, **32**(4), 1013–1022.

Mirsepassi A., Cathers B. and Dharmappa H. B. (1995). Application of artificial neural networks to the real time operation of water treatment plants, at IEEE International Conference on Neural Networks – Conference Proceedings.

Niemi H., Bulsari A. and Palosaari S. (1995). Simulation of membrane separation by neural networks. *Journal of Membrane Science*, **102**, 185–191.

Rojas R. (1996). *Theorie der neuronalen Netze : eine systematische Einführung*. Berlin: Springer, 4.

Rumelhart D. E. and McClelland J. L. (1986). *Parallel Distributed Processing: Explorations in the Microstructure of Cognition. Band I: Foundations*. Cambridge, MA, USA: MIT Press.
Schock G. and Miquel A. (1987). Mass transfer and pressure loss in spiral wound modules. *Desalination*, **64**(C), 339–352.
Schwinge J., Wiley D. E. and Fane A. G. (2004). Novel spacer design improves observed flux. *Journal of Membrane Science*, **229** (1–2), 53–61.
Shetty G. R. and Chellam S. (2003). Predicting membrane fouling during municipal drinking water NF using artificial neural networks. *Journal of Membrane Science*, **217**(1–2), 69–86.
Smidsrød O. (1970). Solution properties of alginate. *Carbohydr. Res.*, **13**, 359–372.
Strugholtz S. (2009). Neuronale Netze als Grundlage für die Analyse und Optimierung von Flockungsprozessen in der Praxis der Trinkwasseraufbereitung PhD Thesis. Duisburg: University of Duisburg-Essen.
Strugholtz S., Panglisch S., Gebhardt J. and Gimbel R. (2008). Neural Networks and genetic algorithms in membrane technology modelling. *Journal of Water Supply: Research and Technology – AQUA*, **57**(1), 23–34.
VDI-Gesellschaft Verfahrenstechnik und Chemieingenieurwesen (Hrsg.), Kapitel: Stoffwerte von Wasser (Dba), VDI-Wärmeatlas, Springer-Verlag, 2006.
Versteeg H. K. and Malalasekera W. (1995). *An Introduction to Computational Fluid Dynamics: The Finite Volume Method*. London: Prentice Hall.
Vrouwenvelder J. S., Graf von der Schulenburg D. A., Kruithof J. C., Johns M. L. and van Loosdrecht M. C. M. (2009). Biofouling of spiral-wound NF and RO membranes: A feed spacer problem. *Water Research*, **43**(3), 583–594.
Vrouwenvelder J. S., van Paassen J. A. M., Kruithof J. C. and van Loosdrecht M. C. M. (2009). Sensitive pressure drop measurements of individual lead membrane elements for accurate early biofouling detection. *Journal of Membrane Science*, **338**(1–2), 92–99.
Ye Y., Le Clech P., Chen V., Fane A. G. and Jefferson B. (2005). Fouling mechanisms of alginate solutions as model extracellular polymeric substances. *Desalination*, **175**, 7–20.
Zell A. (2003). *Simulation neuronaler Netze*. München: Oldenbourg, 4.
Zhang H., Zheng H., Zhang Q., Wang J. and Konno M. (1998). The interaction of sodium alginate with univalent cations. *Biopolymers*, **46**(6), 395–402.

Chapter 8

Integrated system configuration

F. Macedonio[1,2] and E. Drioli[1,2]
[1]University of Calabria (Italy)
[2]ITM-CNR (Italy)

In this chapter the most interesting results carried out in Work Package 8 (WP8) of MEDINA project are summarized. The WP8 leader has been Prof. Enrico Drioli, whereas the contributors have been all the research partners due to the fact that the results achieved in all the previous WPs were used for implementing the integration of the different membrane units in this WP. In order to reach its aims, WP8 was structured into the following five complementary Work Tasks:

- ***WT 8.1: Critical State of the Art of the Membrane-based Desalination Technologies***, in which a critical analysis on the current desalination technologies has been carried out in order to individuate the main items which have to be improved and to identify possible solutions.
- ***WT 8.2: Study of different integrated systems***. In this work task, the different membrane systems studied, as single units, in the previous WPs were integrated each other in order to analyse how both the single and the overall performance changes because of the integration.
- ***WT 8.3: Optimisation of the integrated system configuration***. For each flow-sheet an optimization was carried out in terms of fresh water recovery, energy consumption, crystal recovery and quality, brine reduction.
- ***WT 8.4: Economic evaluation of the integrated membrane system***. In this work task an economic evaluation was carried out in order to define (i) the different associated costs and (ii) if the optimal integrated configuration in terms of overall performance is also the most convenient from an economic point of view.
- ***WT 8.5: Quantitative indicators***. Quantitative indicators (such as, water recovery factor, brine reduction, permeate quality improvement and cost) were elaborated and used for comparing the performance of the different integrated membrane systems analyzed.

In the following, the research activities carried out in the 39 months and the final results are reported.

8.1 INTRODUCTION

The prospect of severe water shortage, which restricts future development and may cause environment damage, stresses the need for new strategies in the field of water resources and water supply management. Among water treatment technologies, desalination is increasingly thought as the answer to the world water crisis in the 21st century.

The modern era of desalination is considered to have begun in the second half of the 20th century with the development of distillation processes. By the late 1960s, interest in a new desalination process using semi-permeable membranes gathered momentum. The process, known as Reverse Osmosis (RO), had some potential advantages over distillation. Recent advances in membrane technology and performance have led to a its broad application, and RO systems now represent the fastest growing segment of the desalination market, with more than 50% of the world's desalination capacity and whose installations account for close to 80% of all desalination facilities. The growth in membrane applications has resulted in a decreasing cost of the desalination facilities, with the consequence that the unit cost of the desalted water from membrane plants has been also lowered (since one or more dollar per cubic meter until about 0.50 $\$/m^3$). Nevertheless, despite the great success, some critical problems still remain open, such as: further reduction of desalted water cost, improvement in water quality, enhancement of recovery factors, minimization of brine disposal impact. In particular, the latter represents one of the big problems in desalination and the primary limitation to widespread application of RO inland brackish water desalination.

Objective of MEDINA project was *to implement an integrated approach to the design of advanced membrane desalination systems*. It means to couple to RO desalination process different membrane operations. This strategy represents an attractive opportunity because of the synergic effects that can be reached, the simplicity of these units, the possibility of advanced levels of automation and remote control, and for the improvement of the overall efficiency of the desalination processes. Work package 8 (WP8) of MEDINA project was the WP aimed to use the results achieved in the others WPs as guidelines for implementing the integration of the different membrane units in reverse osmosis pre-treatment and post-treatment stages. In order to reach this aim, the following specific objectives have been identified:

1) to provide a critical state of the art of the existing desalination technologies (WT 8.1);
2) to develop and optimize an integrated membrane system, assuming that this strategy improves significantly the efficiency of membrane desalination technology (WT 8.2, WT 8.3);
3) to evaluate and compare the costs associated to the integrated membrane system with respect to conventional technologies (WT 8.4, WT 8.5).

In the following sections the results and the objectives reached in the various work tasks of WP8 of MEDINA project are presented.

8.2 WT 8.1: CRITICAL STATE OF THE ART OF THE DESALINATION TECHNOLOGIES

A critical analysis on the current desalination technologies is of fundamental importance in order (i) to individuate the main items which have to be developed and/or improved, and (ii) to identify possible strategies for reliable solutions.

Historically, desalination industry started in the late 1950s, when the first Multiple Effect Distillation (MED) plants were installed. Thermal desalination technology (essentially MED and Multistage Flash distillation (MSF)), although energetically expensive, grew exponentially in Middle East and South Mediterranean countries due to their large fossil fuel availability.

In the 1960s, the first protocols for the preparation of asymmetric cellulose acetate membranes were elaborated by Loeb and Sourirajan. Since then, the growth of desalination industry has been driven by the falling cost of membrane processes and by the decreasing electric energy consumption in membrane plants.

At the present status, expectations for substantial improvements of thermal desalination technologies are substantially low.

Reliable proposed actions are limited to: (i) increasing the production capacity of MED and MSF, and (ii) hybridization with membrane processes. In fact, feed water pretreatment carried out through the operation of a hybrid NF-MSF or NF-MED system, with the filtration unit used to remove double-charged-ion groups such as Mg^{2+}, Ca^{2+} and SO_4^{2-}, allows to increase the operating temperatures to values above 110°C without risk of scaling and, therefore, with the possibility to increase the performance ratio of water production to steam consumption. Among conventional thermal desalination processes, Vapour Compression (VC) is less energy-consuming with respect to MSF and MED. However, it is a practical option only for small unit capacities.

On the opposite side, despite the exponential growth of RO desalination of the last two decades, the perspectives for membrane technology advancement are much more promising and stimulating than that of thermal processes. For example, at present the most critical issues to face seem membrane fouling and brine disposal. Fouling affects membrane lifetime, with important relapse on membrane replacement rates that usually range between 5 and 20 per cent depending on the nature of feed water and on the pretreatment procedures. A significant membrane fouling reduction can be achieved replacing conventional pre-treatment (that still represent the preferred approach because of a lower investment cost) with

MF/UF, where the higher investment cost is almost completely compensated by costs reduction in the successive RO stage, by enabling higher permeate flux and longer membrane lifetime, and by dropping chemicals consumption in the disinfection, coagulation and dechlorination processes.

For what concerns brine disposal problem, the increasing number and capacity of installed desalination plants exacerbated the problem of the negative environmental impacts of high concentrated streams that desalination facilities often discharge into surface waters or into the ocean. Cost effective and environmentally sensitive concentrate management is today recognized as a significant hurdle to extensive implementation of desalination technologies.

Also from this point of view membrane engineering can help to improve current desalination processes through Membrane Distillation/Crystallization operations. These technologies offer a powerful tool for inter-phase mass transfer driven by partial pressure gradients, not limited by concentration polarization. An innovative evaporation process that can be used for reducing brine volumes is also Wind-Aided Intensified eVaporation (WAIV) technique.

In principle, by using Membrane Distillation/Membrane Crystallization for the treatment of NF/RO retentate streams of the desalination plants, water recovery factors up to 90% can be reached, approaching the concept of Zero Liquid Discharge; in addition, the ability to concentrate RO or NF brines above supersaturation limit might offer the opportunity to recovery salts dissolved in seawater (sodium chloride, epsomite etc.) as crystalline product. Therefore, among the technological developments of the desalination processes, the adoption of Membrane Distillation/Membrane Crystallization might represent a solution to brine disposal problem and, as a consequence, to the environmental impact of the current desalination systems.

However, for a real implementation of the sustainability concept for desalination processes, traditional economic indicators referring to investments and profits need to be completed by new technical, social and environmental metrics, mainly related to the health and safety of labour, benefits and troubles for humans and environment etc. These new metrics have to take into account the size, the weight (especially for installations off-shore or for plants built in remote areas), the flexibility (variations of productivity and/or product of an existing operating unit) and the modularity (for an easy scale-up) of various plants (Criscuoli *et al.*, 2007). According to the logic of MEDINA project, the achievement of this goal involves improving the interchanges between the various areas of research: deep knowledge and better control of fouling, clear and optimized processing protocols, development of innovative membrane units, new comparative indexes and metrics, accurate modelling and optimization.

The critical state of the art of the existing desalination technologies has been accomplished thank to a multi- and cross-disciplinary team involving all partners, offering a wide and well recognized expertise in various fields. The achieved information were collected in the deliverable D 8.1 "Report on the critical analysis of the desalination technology" (see Appendix 1).

8.3 WT 8.2 AND WT 8.3: STUDY AND OPTIMIZATION OF DIFFERENT INTEGRATED SYSTEMS

In order to reach the objective 2 ("*to develop and optimize an integrated membrane system*") the different membrane units, investigated in the previous work packages, have been integrated each other in order to analyse how the single and the overall performance change because of the synergic interactions. Experimental data obtained from the different lab scale plants (reported in the different work packages/chapters involved, i.e. WP2-WP5-WP6-WP7) have been used to individuate the best operative conditions to be validated on the integrated lab-scale plant. For selected flow-sheets, optimization was carried out in terms of fresh water recovery, energy consumption, crystal recovery and quality, brine reduction.

The lab-scale integrated membrane-based desalination prototype shown in Figure 8.1 was built for carrying out the experimental tests.

The lab-scale plant consists of different membrane units: MBR or MF or UF and NF have been introduced for feed water pre-treatment and load reduction to the following RO unit; MD and/or MCr operate on NF and/or RO brine as post treatment (intended as brine management) in order to increase the quantity of desalted fresh water produced combined to solid salts production; finally membrane contactors (MCs) have been also tested in different points of the integrated plant (i.e., on the water fed to the nanofiltration unit, on the permeates coming from the nanofiltration and the reverse osmosis units and on the reverse osmosis brine) for the control of the oxygen content and of the pH of the water in a desalination plant. The different units have been designed so that, although consisting in separate equipments, each step is completely compatible according with the proposed scheme in terms of feed water composition, feed flow rate, permeate flow rate, and other operative conditions. Experiments were carried out on the single units considering the feed for each step having the composition of the permeate/brine of the stage immediately before, this simulating the overall process.

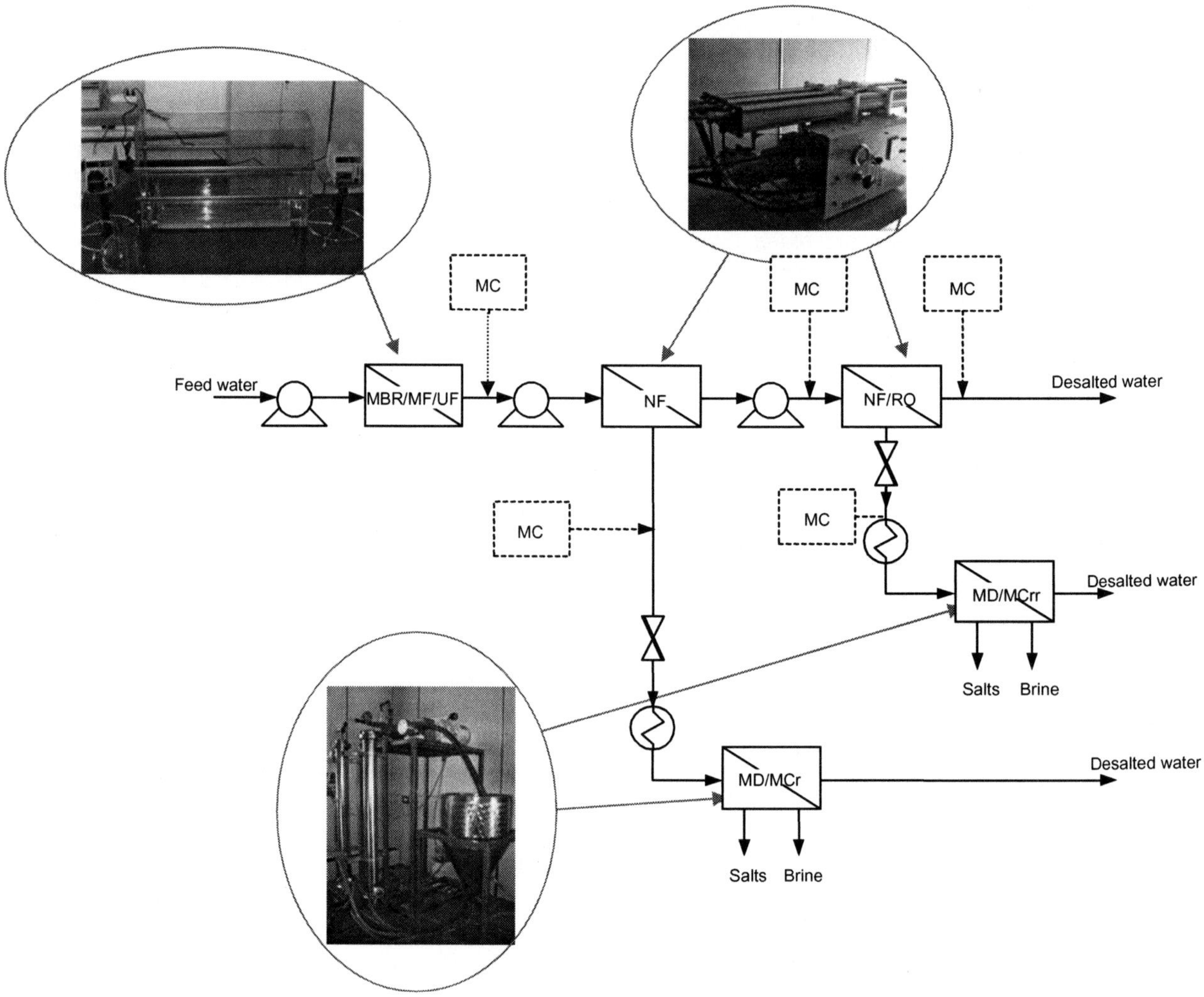

Figure 8.1 Integrated lab-scale membrane desalination system

The details of the carried out experimental tests on the single unit are reported in the previous WP/chapters together with the optimization of the operative conditions. In particular, the main achieved results useful for the modelling of the integrated system are as follows:

- pre-treatment via MBR is an effective tool for reducing organic content from seawater thus decreasing fouling potential and improving water's quality (removal of TOC and EPS) for further process (see Chapter 2). Moreover, MBR eliminates those molecules that affect/hinder crystallization kinetics if RO/NF brine are further concentrated through membrane crystallizers. However, supplementary economical and implemental examinations are necessary in order to decide if this treatment can be adopted as an efficient and suitable alternative pre-treatment.
- MC can be effectively applied for the control of the dissolved oxygen and of the water pH in a desalination plant without use of chemicals (see Chapter 7). All the tested membranes kept their performance also when in contact with high concentrated solutions. However, also in this case like for MBR, an economical evaluation is essential for a MC possible implementation in future desalination plants.
- MD/MCr/WAIV can be efficiently used downstream RO and/or NF retentate, as post-treatment stages, for increasing the overall recovery factor, thus reducing the volume of concentrated streams usually discharged by the desalination plants and, in the case of MCr, recovering the dissolved salts in the form of high-quality crystals (see Chapter 5). As stated in Chapter 5, MD and/or MCr can treat the highly concentrated solutions without suffering the large drop in permeability observed in other membrane processes (such as the pressure-driven membrane processes) due to the fact that they are temperature driven membrane processes. Therefore, the effect of concentration polarization (i.e. formation of a boundary layer on the feed membrane surface) is very small compared to that of temperature polarization. As a consequence, MD and MCr allow to produce fresh water also from highly concentrated feeds (such as the brine streams) with which RO cannot operate due to the osmotic phenomena.

- Membrane pretreatment (MF/UF) provides a better SDI abatement than granular media filtration (see Chapter 2). As reported in many of the deliverables of MEDINA project (see Appendix I), in more and more projects the use of membrane pre-treatment (such as MF, UF and NF) prior to RO stage is being considered as an alternative to conventional pre-treatment in particular due to the decreasing membrane costs and to the declining raw water quality: MF is an obvious technique for the removal of suspended solids and for lowering the silt density index (SDI). By replacing MF by UF further improvements of the RO feedwater can be obtained due to the removal of dissolved macromolecules, colloids and smaller bacteria. Therefore, the MF/UF introduction provides an NF/RO feedwater of good quality in view of lower capital and operating cost. A feed of high-quality means a reduction of membrane fouling with consequent extension of the NF/RO membranes life time and reduction of their maintenance and replacement costs.

If NF is used as RO softening, turbidity, microorganisms, hardness and the most part of multivalent ions are strongly reduced whereas monovalent species are retained by 10–50% depending on the membrane properties. As a consequence, the osmotic pressure of the NF permeate (that is the RO feed stream) is decreased and RO unit can operate at higher recovery factors without scaling problems. In conclusion, the selection of an appropriate membrane-based pre-treatment process leads to the minimisation of membrane fouling, chemical consumption and operating costs.

8.3.1 Integrated membrane processes: systems' design and analysis

Since various combinations of membrane operations are possible, depending by site-specific water quality, performance requirements and, as a consequence, by the energetic consumption and economical evaluations of the desalination plant, different flow sheets (FSs) for seawater desalination have been analyzed and compared. In each of them, except for the first (FS1) constituted only by the RO unit, different membrane operations have been integrated: (i) RO operates on NF permeate in FS2; (ii) both MF and NF have been introduced for feed water pre-treatment and load reduction to the following reverse osmosis unit in FS3. In other four flow sheets, at the basis process represented by FS3, a precipitator and a membrane crystallizer (MCr) have been added on NF and/or RO brine as post-treatment step according to the results achieved in Chapter 5/WP5: MCr operates on NF brine in FS4, on RO brine in FS5 (Figure 8.2), both on RO and NF brine in FS6. In the last flow sheet (FS7), MCr has been introduced on NF brine while membrane distillation (MD) operates on RO brine. In each MD/MCr process, the fresh water coming out from the RO has been used as cold water stream.

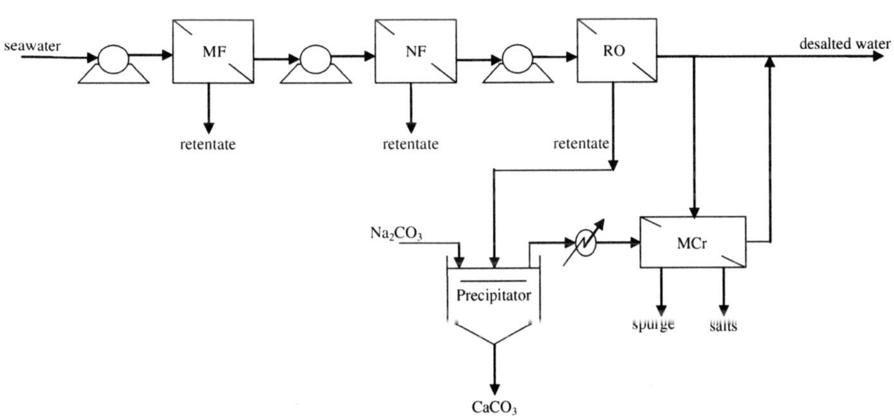

Figure 8.2 Flow sheet 5 (FS5): MF/NF/RO and MCr on RO retentate

The proposed desalination systems have been compared on the basis of their energetic requirement, water cost, amount of discharged brine, fresh water and salts produced. Moreover, in Chapter 9/WP9, some of the studied systems were evaluated also in terms of exergetic efficiency and entropic losses.

For an objective comparison of the proposed flow sheets, the following assumptions were considered for all the analysed flow sheets:

- *Seawater*. The synthetic seawater with the compositions shown in Table 8.1 was chosen as stream to be fed to the various processes on the basis of the experience and knowledge acquired in Chapter 1/WP1;
- *Membrane modules*. The FILMTEC SW30HR LE-400 (one of the commercial membrane module chosen by MEDINA Consortium) was utilized as RO membrane. The RO rejections values and recovery ratio are reported in Table 8.2 together with those related to the used NF membrane (the Dow SR90 modules). The MEMCOR 20M10 characterized by a recovery factor equal to 94.7% was considered as MF membrane.

Table 8.1 Inorganic seawater composition

Inorganic composition [mg/L]	
Na+	10,500
K^+	380
Ca^{2+}	400
Mg^{2+}	1,350
Cl^-	19,000
SO_4^{2-}	2,700
HCO_3^-	142
CO_3^{2-}	3.5
Br^-	65

Table 8.2 NF/RO rejection values and recovery ratios

Ion	RO rejection values [%]	NF rejection values [%]
Cl^-	99.6	16.6
Na^+	99.6	6.3
SO_4^{2-}	99.6	98.9
Mg^{2+}	99.6	94.8
Ca^{2+}	99.6	80.7
HCO_3^-	99.6	33.5
K^+	99.6	7.4
CO_3^{2-}	99.6	70.6
Br^+	99.6	10.0
Recovery ratio	45%	75%
Pressure	55.8 bar	20 bar

Considering the same feed flow rate for all the proposed flow sheets (1051 m^3/h), Figures 8.3–8.4 and Tables 8.3–8.4 show the results of the analysis in terms of recovery factor, quantity of produced salts, energy consumption, brine/permeate flow rate and concentration.

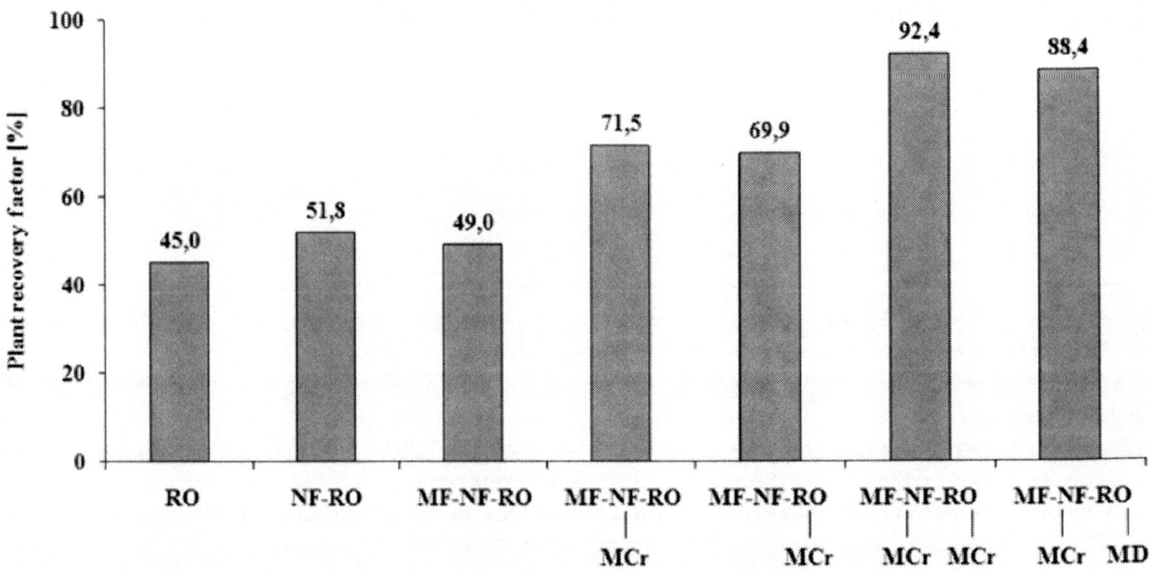

Figure 8.3 Plant global recovery factor in the analysed flow sheets

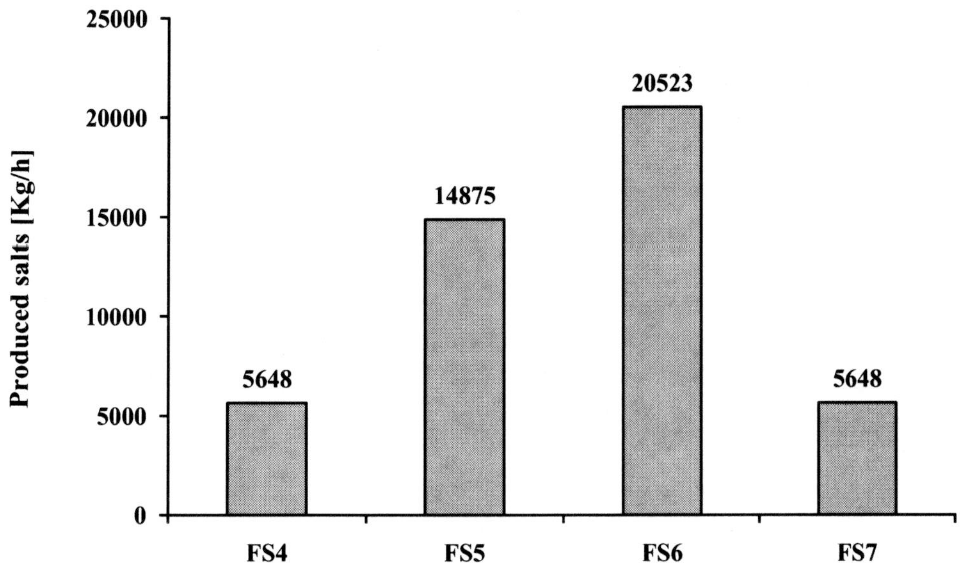

Figure 8.4 Quantity of salts produced in the analysed flow sheets

Table 8.3 Energy consumption

Flow sheet	FS1	FS2	FS3	FS4	FS5	FS6	FS7
Energy consumption [KWh/m^3]	3,76	3,58	3,64	19,37	18,30	26,91	28,11
Energy consumption with EERD [KWh/m^3]	2,09	2,97	3,03	18,95	17,87	26,59	27,77
Energy consumption with ERD and TERD [KWh/m^3]	2,09	2,97	3,03	2,08	2,13	1,61	1,68

EERD= Electrical Energy Recovery Device.
TERD= Thermal Energy Recovery Device.

Table 8.4 Product characteristics for the analyzed flow sheets

Flow sheet	FS1	FS2	FS3	FS4	FS5	FS6	FS7
Brine flow rate [m^3/h]	578,1	506,7	534,0	297, 8	314,7	78,56	120,4
Brine concentration [g/L]	62,69	71,54	67,69	105,9	68,08	214,2	259,41
Fresh water flow rate [m^3/h]	473,1	544,0	515,6	751,8	734,8	971,0	929,6
Fresh water concentration [g/L]	0,138	0,105	0,105	0,072	0,074	0,056	0,058

By introducing a MCr unit on NF brine (FS4), the quantity of fresh water produced is higher with respect to when it is on RO brine (FS5) because MCr works at the same degree of efficiency but, in FS5, on a lower brine flow rate. The introduction of a membrane crystallizer unit on both retentate streams increases plant recovery factor so much that it can reach 92,4% in FS6, value higher than that of a conventional membrane desalination process (such as a typical RO unit – about 45%) and higher than that of a typical MSF (about 10–20%). With respect to FS6, in FS7 the quantity of produced fresh water decreases because MD has a lower recovery factor (77%) with respect to MCr.

For what concerns salts, the highest production is when the membrane crystallizer unit operates on both retentate streams (FS6).

From an energetic point of view (see Table 8.3), the FS1, FS2 and FS3 processes use only electrical energy. The introduction of MCr and/or MD (in FS4, FS5, FS6 and FS7) introduces a thermal energy requirement, due to the retentate flow rate which has to be heated and which increases the global energy demand: in FS5, where the RO brine has to be further concentrated in a MCr unit, the thermal energy necessary is reduced with respect to the other systems because a lower flow rate has to be heated; in FS6, where both retentate streams have to be heated, the total energy consumption is the highest. The same thing occurs in FS7. However, if thermal energy is available in the plant, than the energy requirements of the integrated systems with MCr/MD decrease, reaching competitive values with those of the other desalination processes.

From this preliminary analysis, FS6 (that means MCr operating on both retentate streams) is the desalination integrated membrane based system that allows better to exploit brine added value, not only increasing plant recovery factor, but also extracting the salts naturally present in the concentrated streams of the desalination plants and decreasing brine disposal problem and its negative environmental impacts.

8.3.2 Integrated membrane processes: systems' modelling

The various membrane unit operations constituting the membrane-based desalination system have been modelled through the recourse of a "block structure". The chosen software has been "Simulink", an extension of MATLAB, a programming language specifically designed for simulating dynamical systems using standard block diagram notation. The aim of the simulation has been to facilitate the manipulation of the different membrane operations in the building of the integrated desalination system, and to analyze and foresee the performance of different and possible integrated desalination systems.

Blocks are, in fact, conceptually simple objects, easy to integrate each other, hiding the complexity of their implementation.

Blocks representing membrane unit operations (e.g. NF, RO, MCr) as well as the service equipments (e.g. pumps, valves, heaters) have been created. Examples of the "extended structure" of some modelled blocks are reported in Tables 8.5 and 8.6.

Table 8.5 Feed seawater block

FEED SEAWATER BLOCK	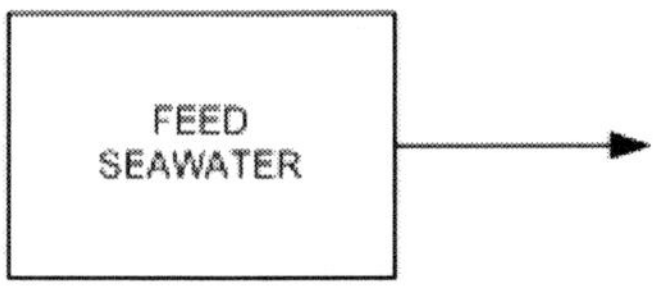
EXPANDED "FEED SEAWATER BLOCK"	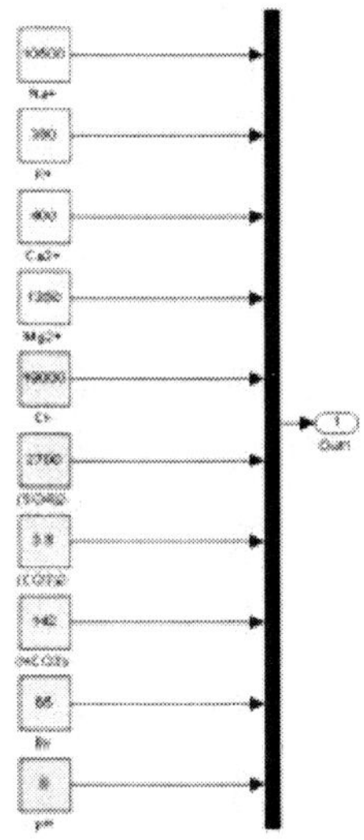
Note:	The block is simple and contains information about the ionic composition of seawater. It is used as input for specific analysis tool (TDS, etc).

When the single blocks are built, the integration of the different units can be easily realized. For example, in the integrated MF/NF/RO desalination system (FS3), the composition of the NF and RO retentate streams have been determined coupling NF and RO blocks. This means that, in the integrated flow sheet NF + RO, the OUTLET values of the "NF" block are the INLET values of the "RO" block.

The results achieved through the computer simulation have been compared with those experimentally obtained in order to evaluate the performance of the computer simulation and to estimate the error between the theoretical and experimental results. An example can be found in Table 8.7 (*case study: NF brine*) that reports the experimentally achieved recovery factor of the MCr lab plant and the recovery factor of the MCr in the proposed FS4, the latter obtained through computer simulation when FS4 and the lab plant produce the same amount of salts and use the same feed seawater (Table 8.1). The comparison of the calculated with the experimentally determined MCr recovery factor shows a good agreement, with error less than 6%.

Table 8.6 Feed seawater block

PUMP BLOCK

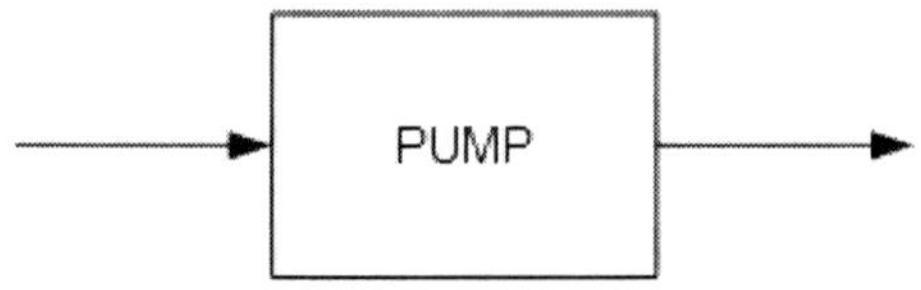

EXPANDED "PUMP BLOCK"

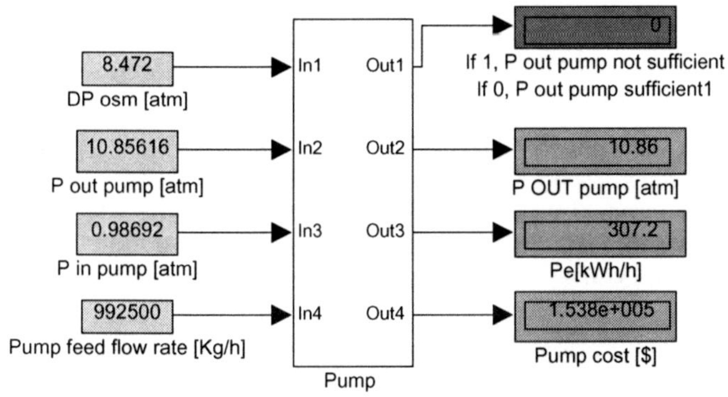

Note:

The INPUT values of the "PUMP" block are highlighted in the green cells and are as follows:

i. pressure [atm] of the stream to be pressurized;
ii. pump feed flow rate [kg/h];
iii. pressure of the stream to be fed to the following membrane module [atm];
iv. osmotic pressure difference pressure of the stream to be fed to the following membrane module [atm];
v. pump efficiency; Marshall & Swift index (M&S) for the calculation of the pump cost;

The "PUMP" block furnishes, as OUTPUT, the following information:

i. the indication (in the red cell) if the chosen pressure for the stream to be fed to the following membrane module is sufficient for carrying out the separation;
ii. the electrical power consumption Pe [KWh/h];
iii. the pump cost [$].

Table 8.7 MCr recovery factors in the *case study: NF brine*

Retentate flow rate [L/h]	Retentate temperature [°C]	Produced salts [g/L]	Experimentally determined MCr recovery factor [%]	MCr recovery factor in FS1 as achieved from computer simulation [%]	Error [%]
120	39 ± 1	6.16	76.71	87.13	6.36
150	39 ± 1	12.22	79.83	88.81	5.32
200	39 ± 1	17.05	80.84	90.15	5.45
250	39 ± 1	21.90	81.22	91.49	5.95

8.4 WT 8.4: ECONOMIC EVALUATION OF THE INTEGRATED MEMBRANE SYSTEMS

The essential elements for the economic evaluation of the desalination plants may be categorized as *direct capital costs, indirect capital costs* and *annual operating costs*:

- *direct capital costs* are the costs associated for purchasing the major items for plant construction and they include:
  - land,
  - building and construction,
  - process equipment,
  - auxiliary equipment;
- *indirect capital costs* are normally expressed as a percentage of the total direct capital costs and they include:
  - insurance,
  - freight,
  - construction overhead,
  - owner's costs,
  - contingency costs;

- *annual operating costs* are the costs incurred after plant commissioning and during actual operation and they include:
  - energy cost,
  - steam cost,
  - labour cost,
  - membrane replacement costs (for membrane-based process),
  - maintenance and spare parts costs,
  - chemicals,
  - amortization or fixed charges,
  - brine disposal costs.

In order to reach the objective 3 ("*to evaluate and compare the costs associated to the integrated membrane system with respect to conventional technologies*"), an economic evaluation has been carried out for all the investigated flow-sheets in order to determine the desalted water cost and the gain for the salts sale, and to indicate if the optimal integrated configuration in terms of overall performance is also the most convenient from an economic point of view. All calculations were based on recent economic data obtained from field data and literature and using the equations reported in *Chapter 5 – Section 4.*

First of all, in order to check if the utilized economic approach is reasonable and realistic, the desalted water cost of a desalination plant constituted by MF/NF/RO (similar to FS3 but working at different RO pressure inlet) has been calculated and compared with the fresh water cost of an already existent membrane desalination plant. Table 8.8 shows the achieved results for the MF/NF/RO system.

Table 8.8 Summary of annual cost data for the MF/NF/RO desalination plant

Cost item	Without Energy recovery devices (ERD)	With Pelton turbine as ERD
Direct capital cost [$]	2,227,000	2,447,000
Plant capacity [m^3/d]	12,370	12,370
Specific consumption of electric power [KWh/m^3]	4.23	3.47
Annual fixed charges [$/yr]	67,230	81,530
Indirect capital cost [$/yr]	6,723	8,153
Annual steam cost [$/yr]	–	–
Annual electric power cost [$/yr][1]	1,549,000	1,270,000
Annual cost for chemicals [$/yr]	73,170	73,170
Annual brine disposal cost [$/yr]	6,315	6,315
Annual labour cost [$/yr]	121,900	121,900
Annual maintenance and spare parts cost [$/yr]	134,100	134,100
Annual membrane replacement cost [$/yr][2]	119,400	119,400
Total annual cost [$/yr]	2,078,000	1,814,000
Unit product cost [$/m^3]	0.51	0.45

[1] *For the comparison with available actual field data data, a pressure equal to 6.9 MPa at the RO inlet has been considered.*
[2] *A membrane life of 10 years has been considered.*

The desalted water cost in the most part of recently building SWRO desalination plants varies in the range 0.50 ÷ 0.70 $/m^3, not too much different from the one calculated (0.51$/m^3) in the analyzed system. The accuracy of the used equations and assumptions can be compared with the data reported in Table 8.9, related to the water desalination cost for an RO plant estimated by Borsani and Rebagli (2005) and to the Ashkelon seawater RO desalination plant data reported by Dreizin (2006).

Table 8.9 Comparison between water price of Ashkelon RO plant and the one calculated by Borsani and Rebagliati (2005)

Cost item	Ashkelon [$/m^3]	Borsani and Rebagliati [$/m^3]
Base fixed price	0.311	0.22
Energy	0.134	0.148
Chemical	0.21	0.078
Post-treatment	0.009	
Filters	0.005	/
Membranes	0.28	/
Others	0.17	/
Total water price	0.525	0.446

One cannot be surprised of the higher water cost reported in Table 8.9 for Ashkelon plant, because it is characteristic of a RO desalination plant operating with high salinity feedwater.

The results achieved for some of the studied membrane-based desalination systems are reported in Table 8.10.

Table 8.10 Desalted water unit cost comparison

	Fresh water cost* [$/m³]	Fresh water cost with EERD* [$/m³]	Fresh water cost with EERD and TERD* [$/m³]	Fresh water cost considering the gain for the salt sale [$/m³]
FS3	0.459	0.407		
FS4	0.675	0.639	0.513	0.256
FS5	0.596	0.560	0.441	−0.017
FS6	0.730	0.703	0.515	−0.058
FS7	0.739	0.710	0.514	0.400

EERD = Electrical Energy Recovery Device.
TERD = Thermal Energy Recovery Device.
**Potable water cost calculated not considering the gain for the salts sale.*

In all the examined flow sheets, the fresh water cost is lower than that of thermal desalination processes (about 0.9 ÷ 1.4 $/m³ for MSF (Ettouney *et al.*, 2002 and Fritzmann, 2007) and 0.7–1.0 for MED/TVC (Semiat, 2000)) and ranges from 0.41 $/m³ for FS3 with Pelton turbine like Electrical Energy Recovery system, to 0.74 $/m³ for FS7 if the gain for the salts sale is not considered. The higher water cost in the integrated system with the MCr is due to the thermal demand of the MCr unit needed for heating the brine. If thermal energy is available in the plant, the water cost is reduced to about 0.44–0.52 $/m³. However, it should be pointed out that in the integrated system with a MCr the quantity of produced salts is high enough that the gain for the salts sale covers more than entirely the cost of the desalination process, particularly in the system with MCr operation on both NF and RO retentate stream (FS6). Therefore, the FS6 integrated desalination process becomes very attractive also from an economical point of view.

8.5 WT 8.5: QUANTITATIVE INDICATORS

Apart from water recovery factor, energy consumption, water cost, amount of discharged brine and salts produced, the proposed membrane-based desalination systems have been evaluated also in terms of exergy (Chapter 9). Moreover other quantitative indicators have been elaborated and used for comparing the performance of the different analyzed membrane desalination systems and in particular:

a. brine reduction (Figure 8.5) and
b. permeate quality improvement (Figure 8.6).

"*Brine reduction*" is a parameter indicating how much the analysed membrane-based desalination system discharges less brine with respect to a conventional desalination plant constituted only by RO (i.e. FS1). "*Permeate quality improvement*" is a parameter indicating how much the desalted water produced from the integrated membrane-based desalination system is less concentrated than that of a conventional desalination process constituted only by RO (i.e. FS1). In both cases, the highest brine reduction and permeate quality improvement were achieved in FS6.

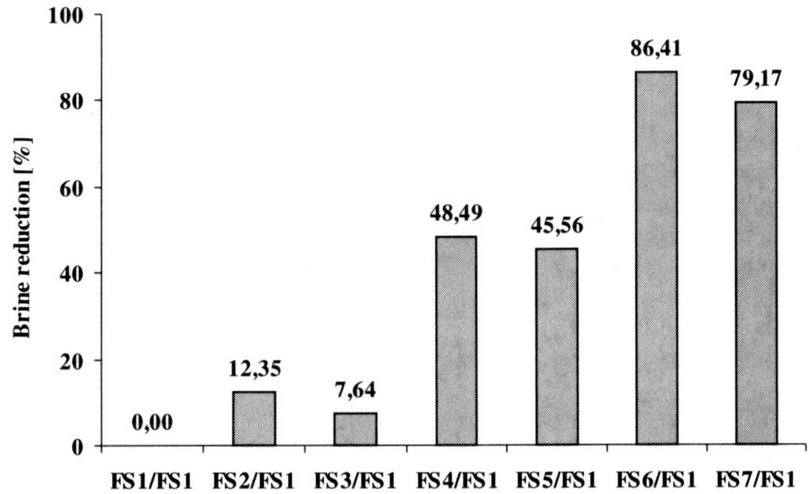

Figure 8.5 Brine reduction (estimated with respect to FS1)

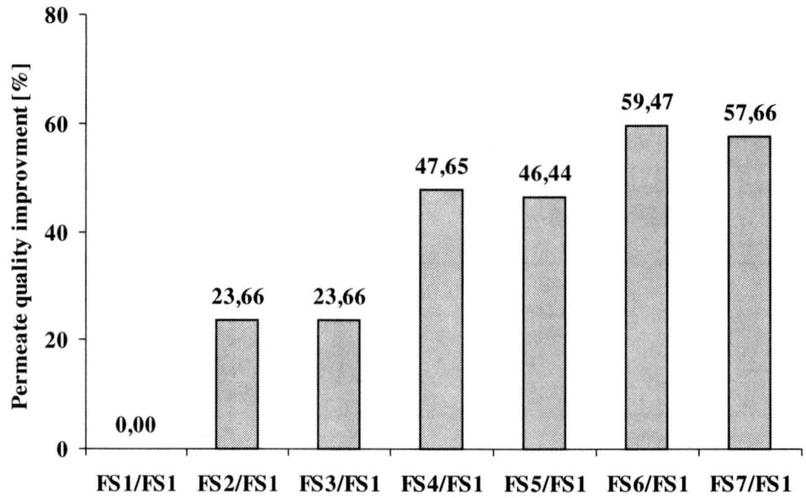

Figure 8.6 Permeate quality improvement (estimated with respect to FS1)

8.6 CONCLUSIONS

One of the objectives of MEDINA project was to determine the more appropriate integrated membrane system for water desalination. In order to reach this target, the results achieved in all the project work packages (in those related to the RO pre-treatment and post-treatment steps as well as in the work aimed at the reduction of membrane fouling) have been taken into account for the development of the integrated membrane system.

First of all a *critical* state of the art was accomplished in order to individuate the advantages and drawbacks of the current desalination technologies, and to find possible strategies for the improvement of the membrane-based desalination systems. Then, a lab-scale integrated membrane system has been realized in order to analyse how the single and the overall performance change because of the synergic interactions. Finally, different membrane-based systems have been modelled and compared in terms of water recovery factor, permeate quality, brine concentration and amount, energy consumption and cost. The modelling of the integrated systems in the specified conditions showed that membrane based desalination systems coupled with MCr unit have the important advantage to increase plant recovery factor thus reducing the economic and environmental problems related to the brine disposal. In fact, the introduction of a membrane crystallizer unit on NF and RO retentate streams of a desalination plant constituted by MF/NF/RO, increases plant recovery factor so much that it can reach 92,4%, value higher than that of a conventional membrane desalination process (such as a typical RO unit – about 45%) and higher than that of a typical MSF.

The results achieved through the computer simulation have been compared with those experimentally obtained in order to evaluate the performance of the computer simulation and to estimate the error between the theoretical and experimental results. The comparison showed a good agreement, in particular for what concerns the values of the water recovery factor.

Moreover, the comparison of the results achieved for the different flow sheets showed that, among the desalination process with MCr unit, FS6 (which means the system with MCr operation on NF and RO retentate streams) is that to prefer when thermal energy is available in the plant *or* the gain for the salts sale is considered because it is characterized by the highest recovery factor (92.4%), the lowest amount of drained off retentate stream, the lowest specific energy consumption (Table 8.3) and desalted water cost (Table 8.10).

In conclusion, the choice of the most convenient and suitable membrane desalination system depends by many parameters, first of all by seawater composition and by the possibility to use low-grade or waste heat streams, as well as alternative energy sources (solar, wind or geothermal) for a cost and energy efficient desalination system. What is undoubtedly is that *integrated membrane desalination systems* offer a reliable solution to the water shortage problem well approaching the concept of "zero-liquid-discharge", "total raw materials utilization" and "low energy consumption".

REFERENCES

Borsani R. and Rebagliati S. (2005). Fundamentals and costing of MSF desalination plants and comparison with other technologies, *Desalination*, **182**, 29–37.

Criscuoli A. and Drioli E. (2007). New metrics for evaluating the performance of membrane operations in the logic of process intensification. *Ind. Eng. Chem. Res.*, **46**, 2268–2271.

Curcio E., Criscuoli A. and Drioli E. (2001). Membrane crystallizers. *Ind. Eng. Chem. Res.*, **40**, 2679–2684.

Dreizin Y. (2006). Ashkelon seawater desalination project—off-taker's self costs, supplied water costs, total costs and benefits, *Desalination*, **190**, 104–116.
Ettouney H. M., El-Dessouky H. T., Faibish R. S. and Gowin P. J., *Evaluating the Economics of Desalination*, CEP Magazine, (December 2002), pp. 32–39, www.cepmagazine.org
Fritzmann C., Löwenberg J., Wintgens T. and Melin T. (2007). State-of-the-art of reverse osmosis desalination. *Desalination*, **216**, 1–76.
Greenlee L. F., Lawler D. F., Freeman B. D., Marrot B. and Moulin P. (2009). Reverse osmosis desalination: water sources, technology, and today's challenges, *Water Research,* **43**, 2317–2348.
Macedonio F. and Drioli E. (2010). Hydrophobic membranes for salts recovery from desalination plants. *Desalination and Water Treatment*, **18** 224–234.
Macedonio F., Drioli E., Curcio E. and Di Profio G. (2009). Experimental and economical evaluation of a membrane crystallizer plant. *Desalination and Water Treatment*, **9** 49–53.
Semiat R. (2000). Desalination – present and Future. *Invited article for IWRA'21, "Water International*", **25**, 54–65.
Semiat R. (2008). Energy demands in desalination processes. *ES&T.*, **42**(22), 8193–8201.

Chapter 9

Environmental impact assessment (EIA) and life cycle analysis (LCA) of membrane-based desalination plants

S. Lattemann[1], M. Kennedy[2], G.Amy[2], J. C. Schippers[2], E. Drioli [3,4,5] and A. Criscuoli[5]

[1]ICBM (Germany)
[2]UNESCO-IHE (The Netherlands)
[3]University of Calabria (Italy)
[4]Hanyang University (S. Korea)
[5]ITM-CNR (Italy)

9.1 INTRODUCTION

The first paper which noted that the discharges of desalination plants may pose a risk to the marine environment appeared in 1979. It called for a thorough investigation of both the physical and biological components of the environment prior to construction and on a regular basis during operation (Winters *et al.*, 1979). However, it took until the 1990's before the scientific interest in the environmental concerns of desalination plants and the number of publications increased (e.g., Abdel-Jawad and Al-Tabtabaei, 1999 and Hoepner, 1999). The first comprehensive review and evaluation of desalination plant effluents and their potential impacts on the marine environment followed in 2001 (Lattemann and Höpner, 2003). It concluded that more experimental data is needed, including field investigations, laboratory toxicity and modeling studies. No more than a dozen of these experimental studies have been published since. An explanation may be that project developers are statutorily not required in many countries to conduct extensive experimental studies, or to provide access to the results to a wider public. In 2004, the World Health Organization (WHO) identified a clear public health and environmental protection argument to develop a guidance document on desalination, which also addressed environmental concerns. The results and recommendations concerning the environment were later published by the United Nations Environment Programme (UNEP, 2008).

Despite this progress, the environmental impacts of desalination plants still remain the subject of considerable debate. A report of the U.S. National Research Council concluded in 2008 that there is still "a surprising paucity of useful experimental data, either from laboratory tests or from field monitoring, to assess these impacts" (NRC, 2008). Two long-term goals for further research identified by the NRC report were to develop monitoring and assessment protocols for evaluating the potential ecological impacts of surface water concentrate discharge, and to carry out site-specific assessments of the impacts of source water withdrawals and concentrate management.

Simultaneously, the European Community advanced research into the sustainable development and use of desalination technologies by funding the MEDINA project within the Sixth Research Framework. One goal of the project was to address the environmental concerns of desalination technologies by developing strategies for the environmental impact assessment

 Membrane-Based Desalination: An Integrated Approach (MEDINA). Edited by Enrico Drioli, Alessandra Criscuoli and Francesca Macedonio. ISBN: 9781843393214. Published by IWA Publishing, London, UK.

(EIA) of desalination plants and for reducing the overall impacts of desalination plants on the environment. This chapter gives an account of the results of work package 9 of the MEDINA project, which had the following objectives:

- to provide a systematic analysis and documentation of all relevant impacts of desalination plants, with emphasis on concentrate and chemical discharges into the sea, and energy use of the process;
- to elaborate and validate a methodology and criteria for conducting project-specific environmental impact assessment (EIA) studies, which are specific to the prerequisites and requirements of membrane desalination plants and include monitoring and assessment protocols;
- to provide guidance on process design and site-selection by adopting the concept of best available techniques (BAT) to desalination technologies in order to mitigate potential impacts on the environment; and
- to develop a decision-support system based on multi-criteria analysis for comparing desalination technologies and for decision making in EIAs.

Special emphasis was given to the process of seawater reverse osmosis (SWRO). Although thermal distillation plants still account for 61% of the global seawater desalination capacity, they are mainly restricted to the sea areas of the Arabian Gulf and the Red Sea, which account for 44% and 13% of the global desalination capacity, respectively. In these sea regions, distillation plants have been listed as a main source of land-based pollution by the UNEP regional seas reports. SWRO, which is the predominant process in most countries outside the Middle East, including Southern Europe and the Mediterranean Sea region, has a huge development potential in the face of an ongoing urbanization and industrialization of many coastal area. The question is, however, if it can pose a sustainable solution to water shortages in coastal areas.

A recent review in the Journal Nature described desalination as a water treatment technology that is often "chemically, energetically and operationally intensive, focused on large systems", and thus requires "considerable infusion of capital, engineering expertise and infrastructure". The costs as well as the environmental concerns are still an impediment to the widespread use of desalination technologies today (Shannon *et al.*, 2008). At the same time, some SWRO projects in California and Australia made headlines claiming that desalination is a "green" technology and that project developers are working towards "sustainability" (Crisp, 2006; Crisp *et al.*, 2009; Port *et al.*, 2009 and Voutchkov, 2008). The seemingly contradictory statements are indicative of the current debate on the extent in which desalination plants will actually affect the environment. As desalination capacities and in particular SWRO capacities are expected to grow rapidly in the future, a critical examination and appraisal of the resource consumption and environmental impacts of desalination technologies was necessary, followed by the identification and development of measures to increase the sustainability of desalination projects if necessary.

9.2 ENVIRONMENTAL IMPACTS

The main environmental concerns of new projects usually revolve around a few key issues (Figure 9.1). Specific to all desalination processes is the discharge of a concentrate, which may contain residual chemicals from pretreatment and cleaning solutions, and which may impair coastal water quality and ecosystems. When evaluating these impacts, one has to distinguish between the salt, which is a natural component, and the chemicals and their reaction products. Correspondingly, different standards should apply for the disposal of the concentrate and disposal of pretreatment and cleaning wastes.

9.2.1 Concentrate disposal

A commonly raised concern is that concentrate disposal may increase the overall salinity of an enclosed sea region. Considering the Arabian Gulf as an example with the world's highest density of desalination plants in a semi-enclosed body of water, the probability of such an effect seems to be rather low. Desalination only accounts for a fraction – an estimated 0.05% – of the annual water losses in the Gulf as compared to natural evaporation – which accounts for about 5.7% of the losses. Moreover, evaporation is a main driving force behind water mass circulations in the Gulf and the formation of a deep water current which exports higher salinity water from the Gulf and results in a net inflow of lower salinity water from the open ocean. As a result, the seawater in the Gulf is assumed to be exchanged every 3 to 5 years, which should effectively counteract any long-term increase in salinity. In the absence of a field monitoring programme, however, any statement on the long-term impacts of desalination plants on the Gulf's marine ecosystems remains speculative.

What has been ascertained through several toxicity studies, however, is that *locally* elevated salinity levels caused by desalination plant discharges can be potentially harmful to marine organisms depending on the species' sensitivity and life cycle stage. The concentration of salts and the exposure time to the discharge is the problem rather than the salt itself. Discharge into the sea is an adequate means of concentrate disposal if dilution to ambient levels is achieved within a very short distance from the outfall and if sensitive ecosystems are not impacted by the dispersing plume. A suitable approach to minimize the impacts of the concentrate is to establish a regulatory mixing zone around the outfall in which dilution to a level close to ambient is achieved by technical means (multi-port diffusers) and natural processes (strong currents).

Mixing zone regulations combine parameters that define the spatial extent of the allowable mixing zone with water quality standards that apply at the edge of this mixing zone. As it is generally difficult to develop a universal set of mixing

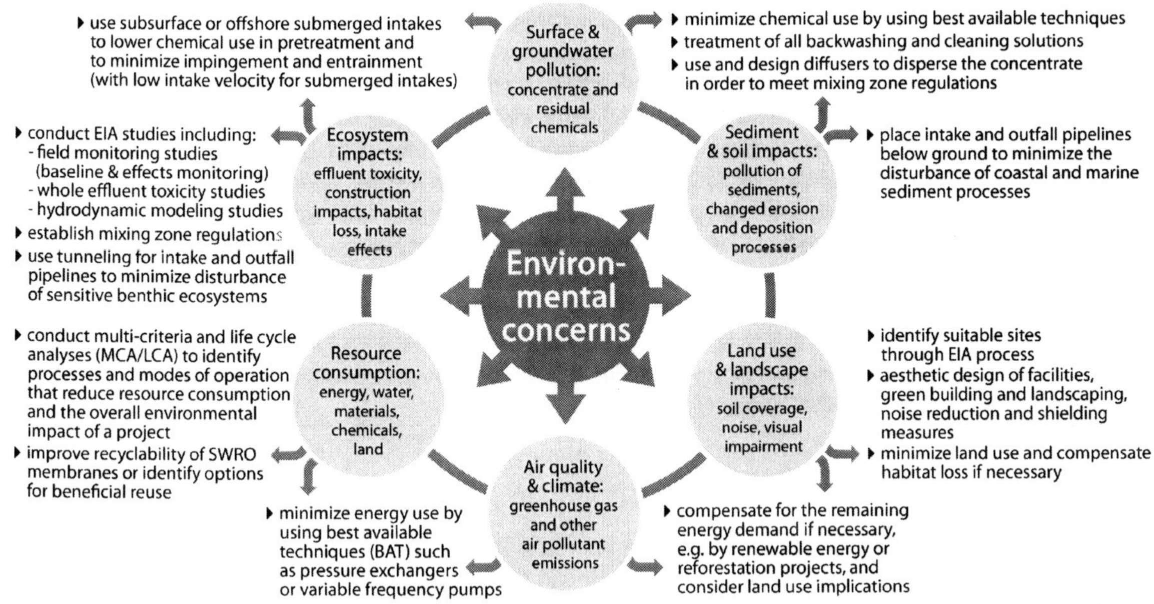

Figure 9.1 Key environmental concerns of seawater desalination projects and measures for impact mitigation (Lattemann, 2010)

zone regulations and standards that apply equally to the wide range of marine ecosystems, a common approach is to tailor regulations and standards for local conditions. Moreover, single standards for each physical and chemical stressor (e.g., salinity and residual chlorine) do not take potentially synergetic effects into account. The basic concept of the tailor-made approach is to derive a single threshold or trigger value from a suite of bioassays that use the whole effluent of a given discharger to measure the acute and chronic toxicity to different local marine species representing different taxonomic and trophic levels (WET tests).

In a first step, the no observed effect concentration (NOEC) or EC_{10} value (statistically calculated concentration that causes a 10% effect) is established for each species. From this data set, a threshold value is calculated that indicates the safe dilution for a given percentage of species. A water quality standard would combine the dilution ratio (e.g., 1:50 dilution, e.g. corresponding to a salinity of 40 g/L) with the species protection level (e.g., a level which protects 95% of the marine species from experiencing a sub-chronic effect of greater than 10%). Depending on the ecosystem, a higher or lower species protection level may apply. In Australia, for example, a level of 95% applies to slightly to moderately disturbed ecosystems and 99% to ecosystems with a high conservation value. The return water dilution usually has to be met at the edge of the mixing zone in a given percentage of time (e.g., 90%) to avert negative effects. This is because it is assumed that marine organisms can temporarily cope with increased salinity levels. The approach reflects that toxicity is a function of species sensitivity, concentration and exposure time. To conclude, narrow mixing zone regulations should be established site-specifically which specify:

- the *spatial extent* of the mixing zone depending on the local conditions (open coast, bay, etc.) and the marine ecosystems in the vicinity of the outfall. The *best location and design of the outfall* to meet mixing zone regulations should be determined through field studies in combination with hydrodynamic modelling studies that investigate the mixing conditions under different ambient, including quiescent conditions.
- the *dilution rate* that must be met at the edge of this mixing zone in a given percentage of time. The safe dilution ratio to protect a given percentage of the local species can be determined by WET tests.

9.2.2 Pretreatment and cleaning chemicals

Contrary to the concentrate, impact mitigation for pretreatment and cleaning chemicals should emphasize avoidance and treatment rather than discharge and dilution by:

- avoidance and minimization of chemical use in pretreatment and cleaning by using best available techniques (BAT);
- substitution of harmful chemicals by less harmful compounds, e.g., by using chlorine dioxide instead of chlorine for the control of marine growth, or acid instead of antiscalants for the control of scaling;
- treatment, where possible, such as intermittent wastes from filter backwash, ultrafiltration (UF) or SWRO cleaning operations. Untreated discharge of these waste waters into the sea cannot be considered BAT.

The most preferred intake and pretreatment alternative for a SWRO plant from an environmental perspective would be a subsurface intake. A subsurface intake can potentially provide a consistently high feed water quality with advantages for pretreatment, cleaning and membrane life, hence considerably reducing chemicals, materials and energy use over the life-time of a project. Moreover, a subsurface intake eliminates impingement and entrainment effects of marine organisms. It also minimizes land use and landscape impacts as structures are embedded in the offshore or beach sediments, and as the engineered pretreatment system following the natural intake system usually has a smaller "footprint" in terms of area size. A constraint of vertical beachwells is the limited production capacity, so that their use is restricted to smaller SWRO plants. Horizontal drains in the offshore sediments may be a promising alternative to beachwells for large SWRO plants, however, experience and performance data of horizontal drains is still very limited. A constraint of all subsurface intakes is the possible appearance of iron (II) and manganese (II) in anoxic/anaerobic well water, which may precipitate when oxidized, and which may necessitate additional media filtration after the subsurface intake.

An acceptable alternative where a subsurface intake is not possible due to geological, environmental or operational constraints is a submerged intake in deep water in an offshore location. It should have a large surface area resulting in low flow velocities (passive screen) of less than 0.15 m/s, velocity caps, and fine-mesh screens which can be backwashed with air. A suitable alternative to conventional pretreatment, which is often needed for surface water and even submerged intakes, may be UF with a low chemical approach.

9.2.3 Energy use

SWRO plants can achieve a specific energy demand of less than 2.5 kWh per cubic meter of desalinated water and a total energy demand <3.5 kWh/m^3 under favorable conditions (i.e., a salinity <35 g/L, a temperature >15°C, and a low fouling potential of the feed water) and by using state of the art equipment (i.e., isobaric energy recovery systems, variable frequency pumps and drives, etc.). The total energy demand is generally between 4 and 5 kWh/m^3 if one includes the pretreatment of the feedwater and other auxiliary equipment in the plant, the transfer of the water, and the materials and construction stages.

Carbon dioxide emissions can be estimated with a high degree of certainty as they mainly depend on the carbon content of the fuel and the energy mix of the electricity grid. For example, the global warming potential of the Australian SWRO projects (Perth, Melbourne, Sydney, Gold Coast) ranges between 2.3 and 7.8 kg CO_2-equivalents[1] per cubic meter of desalinated water. The lower value is for the lowest recorded specific energy demand (2.3 kWh/m^3), whereas the upper value reflects the full life cycle cost of desalinated water including electricity for desalination (4.2 kWh/m^3), water transfer, emissions attributed to the transportation of workforce, offsite waste decomposition, chemical and material transportation and use during construction and operation of the project.

On the one hand, SWRO plants require much less energy per cubic meter product water than thermal distillation plants if one takes the thermal energy requirements of distillation plants into account (the energy demand of distillation plants ranges between 145–390 MJ/m^3 in addition to an electricity demand of 1.5–5 kWh/m^3 depending on the process (NRC, 2008). On the other hand, SWRO is much more energy-intensive than conventional treatment of local ground and surface waters with an energy demand of 0.2–0.4 kWh/m^3. In locations where the water has to be transported over long distances, the relative energy increase caused by a seawater desalination plant may be marginal or desalination may also be the more energy-efficient option. Compared to other activities and amenities of modern lifestyles, such as air conditioning or heating, desalinated water is not an overly energy-expensive product.

In the end, it depends on the perception and definition of "significance" and on local circumstances whether or not energy use of desalination is considered as a significant factor. In many countries, policy initiatives and stricter technological standards are being introduced in order to reduce energy consumption and increase energy-use efficiency in all sectors of use. The increasing use of energy-intensive desalination technologies counteracts these efforts to some extent. Technological standards for desalination processes are missing and should be introduced in the form of BAT standards in order to facilitate the selection of resource-saving technologies at the project level. Furthermore, a decision may be taken on the national or project level to compensate for the energy use of desalination. For example, carbon dioxide emissions resulting from energy use of all major Australian SWRO projects are being compensated by renewable energy sources.

9.2.4 Other environmental concerns

The remaining environmental concerns of desalination projects (Figure 9.1) show certain parallels to other coastal development projects. They include:

- impingement and entrainment effects, which also occur at coastal power plant intakes and which can be avoided by using subsurface intakes, or minimized through proper facility design, using a combination of screens and velocity caps, intake site location in offshore submerged locations, and operation using low intake velocities of <0.15 m/s;

[1] CO_2-e estimate the global warming potential of all climate change gases by transforming the non CO_2 emissions into an equivalent amount of CO_2 emissions that would have the same global warming potential.

- impacts on sediments and benthic ecosystems which may be similar to the construction of harbors and jetties and which may be minimized by placing pipelines below ground and the use of tunneling techniques;
- construction, land use and landscape impacts in the terrestrial site of the project, which are intrinsic to any development project, and which can be minimized by identifying suitable sites with preference on existing industrial sites, facility design and landscaping measures; and
- resource consumption including material, chemical, water and energy use, which can be minimized by the use of BAT.

Although SWRO plants consist of numerous membrane elements which have to be replaced every few years, material use only plays a minor role in the overall resource consumption of desalination projects. A life cycle assessment found that material use and disposal has little weight (10%) compared to plant operation (90%) due to the high energy demand of all desalination processes (Raluy *et al.*, 2006; Raluy *et al.*, 2005 and Raluy *et al.*, 2004). Assessments of the Sydney and Melbourne desalination projects arrived at similar conclusions, that is, the total project emissions of CO_2 and other greenhouse gases associated with the materials and construction stages were <5% whereas operation accounted for 95% of all project-related emissions.

9.3 ENVIRONMENTAL IMPACT ASSESSMENT (EIA)

Impact mitigation measures are usually required for all impacts which are found to be significant for a given project. The significance of impacts depends on the size, process and location of a project and on the environmental characteristics of the project site. Moreover, a universally valid standard for significance does not exist. What is considered as significant depends on a society's values, on environmental regulations, on economic potentials, and the availability of water supply alternatives. For these reasons, a final evaluation of the environmental impacts of desalination projects and the identification of the best practicable environmental option is only possible at a project- and site-specific level through an EIA. A second main objective of the MEDINA project was therefore to develop an EIA approach applicable to large SWRO projects and including an environmental monitoring framework (monitoring and assessment protocols).

9.3.1 EIA process

The primary goals of an EIA are to provide information on the environmental consequences of a project for decision making, and to promote environmentally sound and sustainable development through the identification of appropriate alternatives and mitigation measures. EIAs are usually not limited to environmental aspects, but where appropriate, also address public health and socio-economic concerns. Public participation is therefore an integral element of EIAs in many legislative systems. As a result, EIAs are often multi-stage, multi-disciplinary, and multi-participatory processes.

A ten step EIA process for desalination projects is proposed (Figure 9.2 and Figure 9.3). The pre- or initial EIA phase includes the steps of screening and scoping of the project, in which a decision is taken on whether or not an EIA is required for a particular project, and in which the scope, contents and methodologies of the EIA and expert studies are specified in the terms of reference (ToR). A reference list for preparing the ToR has been provided as part of the MEDINA project, which may also serve as a "blueprint" for preparing the EIA report. During the main EIA phase, a detailed description of the preferred project configuration including site and process alternatives is required, and other statutory permits applicable to the project should be identified and obtained. The scientific studies and analyses are also conducted during this phase including baseline studies, the prediction and evaluation of impacts, and the identification of alternatives and impact mitigation measures. The final EIA phase involves decision making and a review of the EIA process. An environmental management plan is often established at this stage, which specifies the monitoring requirements during the installation and operation of the plant.

In principle, EIAs for large desalination projects will not differ in terms of complexity and level of detail from those of other water supply infrastructure projects. Depending on the proposed project, it is generally the responsibility of the competent authorities to individually define the need, scope and complexity requirements of each EIA study. The results of the MEDINA project can be used as a guidance in this process. When dealing with a larger number of desalination proposals, a collaborative effort between the main government agencies and participatory groups should be initiated to elaborate a national EIA guideline for desalination projects. It would facilitate the EIA process by establishing equal standards for the environmental studies to be undertaken and the information to be submitted as part of the EIA for each individual project in the future. Moreover, as a number of agencies usually have permitting authority over the project, a "lead agency" should be designated to coordinate the process by involving other agencies and by informing the project proponent about permitting requirements.

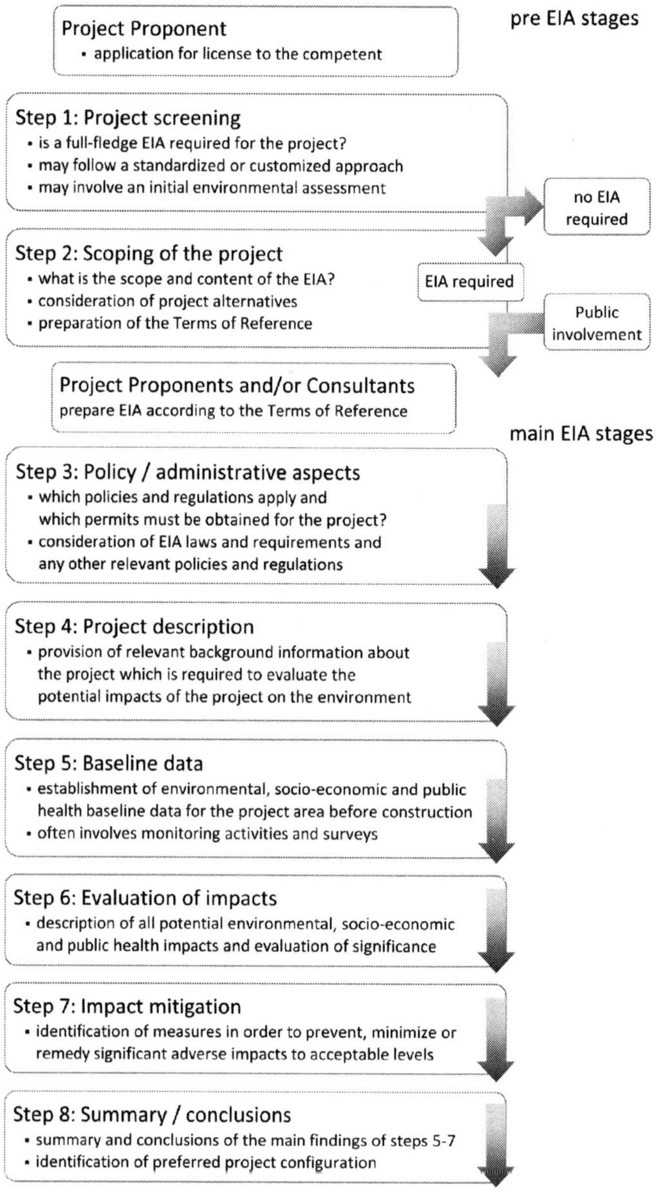

Figure 9.2 10-step EIA process (Lattemann, 2010)

To this day, only a limited number of EIA studies for desalination projects have been carried out and published. Most of them are from Australia and the United States. In the EU, the EIA Directive[2] regulates which project categories have to be made subject to an EIA by member states. It lists major water supply projects such as groundwater abstraction schemes, dams, and works for the transfer of water resources between river basins, but not water treatment installations and desalination plants. The list should be expanded to include desalination projects. EIAs for desalination plants from other parts of the world, particularly the Middle East, are scarcely available. The reason may be that EIA studies are considered to be the intellectual property of the project proponent. This is contradictory to the notions of transparency and public participation, and EIAs should be made available to a wider public audience.

9.3.2 Environmental monitoring

As EIA studies make predictions about the expected impacts of a desalination project on the environment, it is necessary to validate the accuracy of these predictions against observations during project implementation and operation. The longest

[2]Directive 85/337/EEC on the assessment of the effects of certain public and private projects on the environment, amended by Directive 97/11/EC.

effects monitoring programmes for a few SWRO projects worldwide have just accumulated two to three years of cohesive monitoring data. Although this may allow for some conclusions regarding those particular projects, it is too early to use these results as conclusive evidence concerning the long-term impacts of desalination plants as a whole. A review of the existing monitoring studies showed that most were either limited in scope – addressing only one aspect such as salinity; short-term – without a continuous baseline and effects monitoring; and localized – without adequate spatial replication.

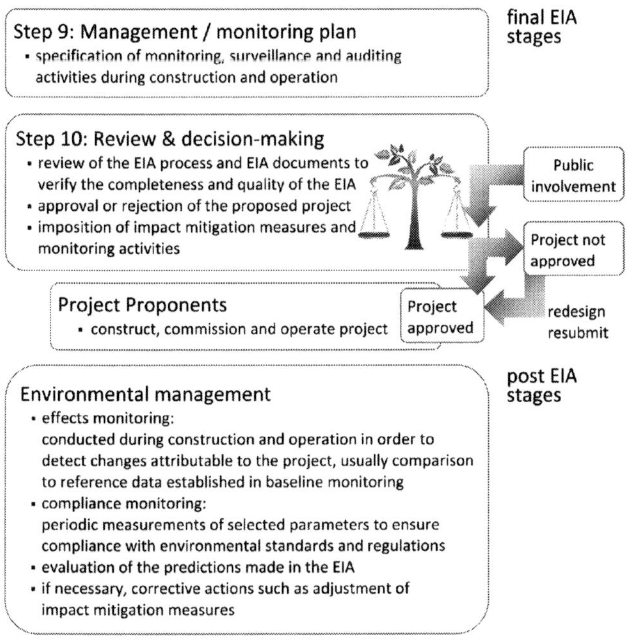

Figure 9.3 10-step EIA process (cont.) (Lattemann, 2010)

The requirements of a holistic marine monitoring programme for desalination projects have been elaborated in the MEDINA project (Lattemann, 2010). The core of the problem is to design a monitoring programme that provides sufficient statistical power to be able to distinguish the effects of the desalination project from natural processes. For this reason, sufficient spatial and temporal replication is needed in field monitoring studies. Field monitoring for desalination projects should include the project (impact) site and several control sites which adequately represent the habitats of the impact site. To capture the temporal variance at these sites, paired sampling should be carried out at several times before, during and after project implementation (BACIPS approach). Baseline monitoring for major development projects is usually carried out over a period of two years before project start-up and effects monitoring for two to three years during commissioning and operation. A monitoring framework for desalination projects should furthermore integrate field monitoring with hydrodynamic modelling and bioassay studies, preferably using whole effluent toxicity tests.

9.4 BEST AVAILABLE TECHNIQUES (BAT)

While EIA aims at minimizing impacts at a site- and project-specific level, BAT is a technology-based approach. The concept of BAT has been established by different legislative systems, e.g., in Europe and the United States[3], and has been applied to similar industries and applications, such as power plants and seawater cooling systems for power plants. A BAT reference for water treatment and desalination plants is lacking so far and should be established to guide practitioners, consultants and decision makers in selecting adequate techniques and modes of operation to minimize impacts on the environment. For example, the U.S. Environmental Protection Agency has now initiated a new rulemaking to address the direct discharge of drinking water treatment residuals to surface water, such as aluminum or iron salts, organic matter, polymers, *desalination concentrates* or other residuals (EPA, 2009). The EPA's "effluent guidelines" for categories of dischargers are technology-based and have the objective to establish a basic national standard for all facilities within a category using BAT.

[3]EC Directive on Integrated Pollution Prevention and Control (IPPC), the Conventions for the Protection of the Marine Environment of the North-East Atlantic (OSPAR), of the Baltic Sea Area (HELCOM), and the Protocol for the Protection against Pollution from Land-Based Sources of the Mediterranean Action Plan (LBS protocol), the U.S. Clean Air Act and Clean Water Act.

BAT is defined as state of the art processes, facilities or methods of operation which indicate the practical suitability for limiting discharges, emissions and waste, and for reducing the impact on the environment as a whole. The term "technique" includes both the technology used and the way in which the installation is designed, built, maintained, operated and dismantled. The techniques that are considered BAT should be economically and technically feasible, should be used or should have been tried out on an industrial scale, and should take technological advances in scientific knowledge into account. Special consideration in the development of BAT is typically given to the consumption of raw materials, water, energy, less hazardous substances, and the possibility for recovery and recycling of any resources used or wastes generated.

According to the general concept and definition of BAT, it is proposed to consider the following order of measures when determining individual BAT solutions for desalination projects:

- selection of the desalination process with the highest energy efficiency;
- optimization of energy and water use efficiency of that process;
- lowering the chemical use of that process by:
  - reducing the occurrence of fouling and corrosion through process design (i.e., intake design and location) and thus minimizing cleaning and pretreatment requirements;
  - giving preference to no or low chemical use/waste designs;
  - substitution of harmful substances with less harmful substances;
  - optimizing the application and dosage of pretreatment chemicals based on pilot testing and/or monitoring of the feedwater quality;
  - treatment of wastes before discharge/disposal;
- selection of manufacturing materials that can be reused or recycled, and identification of appropriate waste disposal options at the end of their useful life.

A first analysis to identify general BAT solutions for seawater desalination plants has been conducted in the MEDINA project. If it would be possible to choose freely between the different process alternatives, leaving out technical, economical and site-specific environmental limitations and taking only environmental benefits into account, the most preferred design would be a SWRO plant with a multi-port diffuser in a suitable oceanic site (cf. Section 9.2.1) and a subsurface intake (cf. Section 9.2.2).

9.5 DECISION SUPPORT SYSTEM

EIAs typically generate large volumes of complex information which often exceed the capacities of decision makers to integrate and process all this information in decision making. Moreover, different decision makers and stakeholders often have conflicting preferences about a project so that the process of decision making in an EIA can be described as a conflict analysis between different value judgments. The process can be facilitated by a formalized decision support tool such as multi-criteria analysis (MCA), which allows the integration of multiple quantitative and qualitative criteria and different value judgments in decision making. One objective of the MEDINA project was to explore the usefulness of MCA as a decision support tool for the EIA process of new desalination projects and to implement MCA in a case study.

9.5.1 Case study

The case study compared different intake and pretreatment systems for SWRO plants, which can range from minimalist to operationally intensive depending on local circumstances, the latter consuming considerable amounts of chemicals, materials, energy, land, and water. The overall environmental impact of a desalination process can therefore by minimized by selecting the best pretreatment option in a given site. The methodology follows the MCA approach described in (Lattemann, 2010), using the software tool DEFINITE 3.1 (Janssen *et al.*, 2003). The approach involved three main steps:

- structuring the decision problem, i.e. establishing a complete set of operational alternatives (Figure 9.4) and evaluation criteria;
- scoring the alternatives on all criteria (Table 9.1) and attachments of weights to the criteria to reflect four different perspectives, which were gathered by means of a questionnaire (Figure 9.5, pie diagram);
- ranking of the alternatives (Figure 9.5, bar diagram) and a sensitivity analysis to investigate the robustness of the ranking results.

The following conclusions were drawn from the MCA and sensitivity analysis:

- Beachwells were found to be dominant over the other alternatives, that is, beachwells scored better or at least as well as the other alternatives on all criteria (Table 9.1). Beachwells therefore also ranked first in the MCA, irrespective of the weights that were attached to the criteria to represent different perspectives.
- The MCA showed that the value judgments were generally similar in the two groups that represented plant operators and companies on the one hand, and in the two groups that represented university and environmental

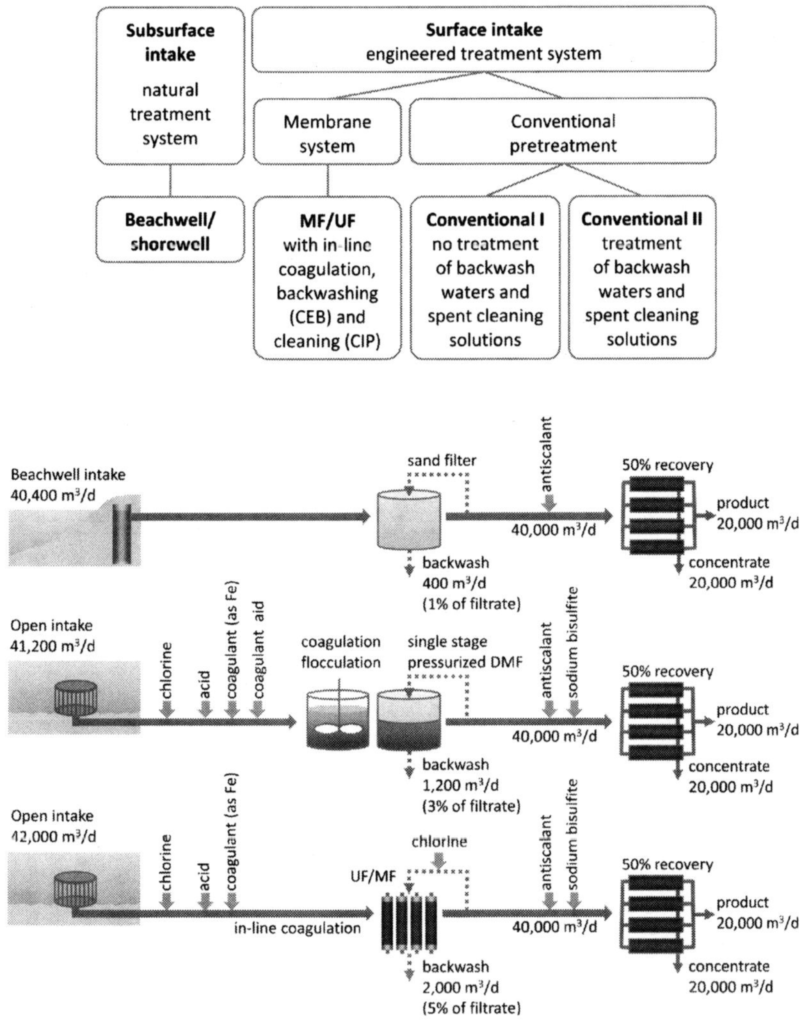

Figure 9.4 Intake and pretreatment alternatives considered in the MCA (top), assuming the process flow specifications shown below for a beachwell intake, an open intake with conventional pretreatment and ultrafiltration/microfiltration (UF/MF) (Lattemann, 2010)

backgrounds on the other hand. While plant operators gave highest priority to the criteria "water quality" and "costs", the university and environmental groups gave highest priority to the criterion "ecological impact". Chemical and energy use varied in importance between ranks two to four in all groups.

- The MCA ranking showed that conventional pretreatment II (with sludge treatment) was the second best alternative according to the value judgments of the expert groups. The preference over conventional pretreatment I was more distinct in the university and environmental groups (>0.1 total score difference) than in the plant operators group, in which conventional II ranked second by a narrow margin (0.02 difference), and in the company group, in which both conventional pretreatments had equal preference.
- UF ranked at the last position in all groups, although the difference to the next best alternative, i.e., conventional I, was small.
- A sensitivity analysis of *weights* showed that conventional pretreatment I would rank before conventional II if the weights of the criteria "cost" or "energy" were to be increased to 50%. This can be attributed to the assumption in this study that sludge treatment causes a 10% increase in cost and energy use. UF would rank second if the weights of the criteria chemical or land use were to be increased to 50%.
- A sensitivity analysis of *scores*, assuming 50% and 100% score uncertainty, showed that beachwells had a high probability to rank first. Either conventional pretreatment I or II was likely to rank second. UF pretreatment had a high probability of ranking on third or fourth position. In order to reverse a ranking, either discharge or intake effects had to be eliminated, or the scores for the various aspects of resource consumption had to be reduced by about 20–60%. A decisive criterion was energy use. If reduced by 11% or less, it brought about a change in the ranking between the alternatives conventional pretreatment I and II, which can again be attributed to the assumed higher energy use for conventional pretreatment II.

Table 9.1 MCA reference values used to calculate the scores for a 20,000 m^3/d SWRO plant, assuming the flows and chemical dosing points as shown in Figure 9.4. For details concerning the reference values, see (Lattemann, 2010)

Criterion	⊖/⊕ scale	Beachwell selected reference value	calc. score for 20,000 m^3/d	Conventional pretreatment I[a] selected reference value	calc. score for 20,000 m^3/d	UF pretreatment selected reference value	calc. score for 20,000 m^3/d
Chlorine (feed water)	⊖ ratio	– water from beachwells is not chlorinated	–	**1 mg/l** if continuous chlorine use: 0.5-1 mg/l residual usually maintained throughout pretreatment line	**41 kg/d**	**1 mg/l** if continuous chlorine use: 0.5-1 mg/l residual usually maintained throughout pretreatment line	**42 kg/d**
Chlorine (back-wash)	⊖ ratio	– does not apply	–	– does not apply	–	**100 mg/l 1x/d** varied between 5-20 mg/l 1x/15-60 min. up to 500 mg/l 1x/d	**4 kg/d**
Sodium bisulfite	⊖ ratio	– no dechlorination required	–	**3 mg/l** dosing rate usually three times the residual chlorine concentration	**120 kg/d**	**3 mg/l** same assumption as for Conv. I	**120 kg/d**
Coagulant	⊖ ratio	– not used after beachwells	–	**1.2 mg/l Fe** (≡3 mg/l $FeCl_3$, 40%) values were 0.5-3 mg/l as Fe and <1-18 mg/l as $FeCl_3$ or $FeSO_4$	**49 kg/d**	**0.4 ppm** 0.1-0.6 mg/l Fe (≡0.3-1.5 mg/l $FeCl_3$, 40%) in full scale plants, 0.2-4 mg/l as Fe in pilot plants	**17 kg/d**
Coagulant aid	⊖ ratio	– not used after beachwells	–	**0.85 mg/l** values were 0.2-1.5 mg/l	**35 kg/d**	– typically not used	–
Acid (H_2SO_4)[b]	⊖ ratio	– no pH adjustment required (no coagulation step)	–	**30 mg/l** values were: 22-55 mg/l, 20-25 mg/l 30-100 mg/l	**1236 kg/d**	**10 mg/l** no doses available, 2 full scale plants used acid in addition to Fe, which is assumed to be for coagulation and not for scale control, as both plants also used antiscalant, selected value is 1/3 of H_2SO_4 dose in Conv. I	**420 kg/d**
Anti-scalant[c]	⊖ ratio	**2 mg/l** values were 0.9-1.2 mg/l for polymer antiscalants and 8 mg/l for phosphate antiscalant	**80 kg/d**	**2 mg/l** values were 1-2 mg/l or 1-4 mg/l if one counts an earlier literature review and company information	**80 kg/d**	**2 mg/l** probably similar to Conv. I; 1 pilot plant reported 3 mg/l, a literature review reported 1-3 mg/l	**80 kg/d**

[a]This column indicates whether the criterion is a "cost" (?) or "benefit" (+) criterion, and the scale of the criterion (ratio or binary).
[b]For the alternative conventional II (treatment of backwash waters) a 10% higher energy use, cost and land use was assumed.

Criterion	⊖/⊕ scale	Beachwell selected reference value	calc. score for 20,000 m^3/d	Conventional pretreatment I[a] selected reference value	calc. score for 20,000 m^3/d	UF pretreatment selected reference value	calc. score for 20,000 m^3/d
Land use for pre-treatment	⊖ ratio	**0.8 m^2 per 1000 m^3/d** inst. capacity 4.7 m^2 was given for a 6,000 m^3/d plant with a vertical beachwell	**16 m^2**	**38 m^2/1000 m^3/d** value selected from a range of 35-40 m^3/d	**Conv. I: 760 m^2** **Conv. II: 836 m^2**	**27 m^2/1000 m^3/d** ≡30% reduction from Conv. I, 25% 33% reduction reported	**540 m^2**
Energy use for pre-treatment	⊖ ratio	**0.1 kWh/m^3** of pretreated water for a vertical beach well incl. sand filter	**4000 kWh/d**	**0.15 kWh/m^3** of pretreated water for intake + 1 stage pressurized DMF, reported values were: 0.035 kWh/m^3 RO permeate ≈0.018 kWh/m^3 pretreated water for single GF excl. intake 0.35 kWh/m^3 RO permeate ≈0.16 kWh/m^3 pretreated water for 2 stage DMF incl. intake 0.12 kWh/m^3 pretreated water for single stage pressure filter	**Conv. I: 6000 kWh/d** **Conv. II: 6600 kWh/d**	**0.15 kWh/m^3** of filtrate values ranged between 0.05-0.2 kWh/m^3 filtrate	**6000 kWh/d**
Water use for BW, CEB, CIP	⊖ ratio	**1%** of pretreated water assuming a sand filter backwashed 1x every 3 days	**400 m^3/d**	**3%** of pretreated water assuming a 1 stage DMF backwashed 1x every day; 2-3% of the filtrate capacity per filtration stage used for backwashing	**1200 m^3/d**	**5%** of filtrate assuming a BW/CEB 24-48x/d and CIP 1-2x/a	**2000 m^3/d**
Costs	⊖ ratio	no data, assumed to be 50% of the conventional pretreatment cost Capex: **0.4 US¢/m^3** of pretreated water Opex: **1.1 US¢/m^3** of pretreated water total: **1.5 US¢/m^3** of pretreated water	 **160 US$/d** **440 US$/d** **600 US$/d**	assumed costs for a single stage pressure DMF Capex: **0.7 US¢/m^3** Opex: **2.2 US¢/m^3** total: **2.9 US¢/m^3**	Conv. I / II in US$/d: **280 / 308** **880 / 968** **1160 / 1276**	assumed costs for a DOW SFP-2880 operated at 75 LMH Capex: **1.2 US¢/m^3** of filtrate Opex: **1.9 US¢/m^3** of filtrate total: **3.2 US¢/m^3** of filtrate	 **480 US$/d** **760 US$/d** **1280 US$/d**
Material use	⊕ binary	material is mostly recyclable, and no large material wastes are produced?	**yes**	material is mostly recyclable, and no large material wastes are produced?	**yes**	material is mostly recyclable, and no large material wastes are produced?	**no**
Potential ecological impact	⊖ binary	impingement/entrainment of organisms? discharge of backwash water/sludge?	**no** **yes**	impingement/entrainment of organisms? discharge of backwash water/sludge?	**yes** **yes for Conv. I** **no for Conv. II**	impingement/entrainment of organisms? discharge of backwash water/sludge?	**yes** **yes**
Water quality	⊕ binary	is an SDI<3 achieved?	**yes**	is an SDI<3 achieved?	**yes**	is an SDI<3 achieved?	**yes**

[c]Acid dosing is assumed to take place before the pretreatment and primarily for the purpose of coagulation.
[d]Antiscalant dosing takes place after the pretreatment system and concentrations were assumed to be the same for all four alternatives.

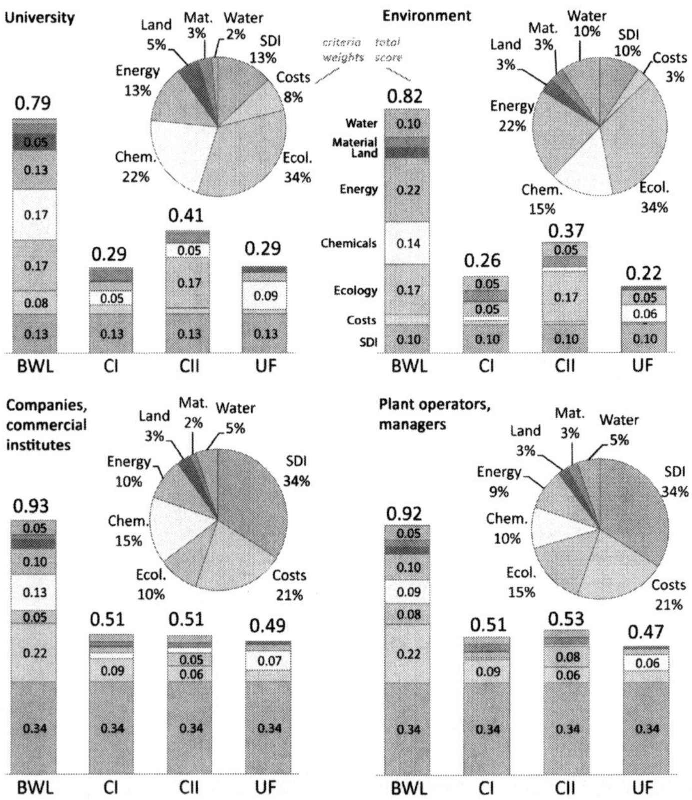

Figure 9.5 MCA ranking results of the four alternatives by four expert groups. BWL: beachwell, UF: ultrafiltration, CI: conventional pretreatment, CII: conventional pretreatment with waste (sludge) treatment (Lattemann, 2010). This figure is available in colour at http://www.iwapublishing.com/template.cfm?name=isbn9781843393214

The MCA ranking and sensitivity analysis support the conclusion that a beachwell is the preferred intake and pretreatment alternative for SWRO plants. Where a beachwell is not feasible, for example, due to an impermeable geologic substratum, the preferred option would most likely be a conventional pretreatment, either with or without sludge treatment, followed by UF.

The MCA decision problem was revised in two subsequent scenarios. In scenario 1, the specific operational conditions of selected full-scale plants were used as baseline for the MCA instead of selected literature values. Scenario1 resulted in a similar ranking of the alternatives, that is, natural intake systems such as horizontal offshore wells (i.e., Neodren) were also the preferred choice for larger SWRO plants, followed by conventional and UF pretreatment.

The scenario 2, in which different UF membranes and modes of operation were compared, showed that the successful operation of an integrated membrane system combining UF and SWRO is essentially an optimization problem, which has to be solved plant- and site-specifically, balancing energy demand, chemical use, filtration time, CEB and CIP intervals. Similarly to conventional pretreatment systems, which have diversified into various pretreatment options over the years ranging from minimalist to an extensive three stage design, not all UF systems are alike, let alone their modes of operation. The range of possible operation modes of UF systems shows that a more sustainable approach with a low energy and chemical demand is feasible, and that this approach could be altogether equal to or better than a conventional pretreatment.

The results of the MCA should be understood as an exercise, which can always be revised and refined in the light of better data and new information. The results are only valid for the given alternatives and criteria. The inclusion of new or modified alternatives and criteria may alter the ranking results. For example, the alternative UF pretreatment will score better on the ecology criterion if one assumes that the backwash water is treated instead of discharged, i.e., by modifying the present assumptions for the alternative. Although discharge seems to be the common practice of the few operational UF-SWRO plants today, future projects in Australia or California will probably require a treatment of the backwash. Also, UF might perform better if better indicators for water quality were available. Pilot studies often found UF pretreatment superior to conventional pretreatment in difficult feed waters, which may have secondary beneficial effects on plant operations, such as lower cleaning frequencies of SWRO membranes, lower energy demand or lower operating costs. However, quantifying these effects in an MCA would still be highly speculative at the moment, due to both, a lack of reliable indicators for water quality and data.

9.6 ENERGY AND EXERGY ANALYSIS OF INTEGRATED MEMBRANE SYSTEMS

Besides the MCA analysis, in the work package 9 of the MEDINA project, an energy and exergy analysis of the integrated membrane systems identified in the other work packages, was also carried out. The exergy is defined as the part of energy that, by reversible transformations, can be completely converted from one form to another (Molinari *et al.*, 1995). Since in any real process there is always an exergy loss by irreversibility, the convenience of an innovative process with respect to the traditional one can be evaluated in terms of exergetic efficiency of the process by calculating the entropic losses (or production of entropy).

In the work carried out, for the different flow-sheets studied the entropic losses were calculated and compared. The comparison was also made in terms of electrical energy/productivity ratio and productivity/size ratio.

9.6.1 Integrated flow-sheets

The flow sheet with membrane contactor (MC), microfiltration (MF) and reverse osmosis (RO) was selected as the "reference" scheme (FS1-Figure 9.6). Other flow sheets studied in WP8 were selected for comparing their performance in terms of exergy efficiency. The selected flowsheets were those having: membrane crystallizer working on the NF brine (FS2-Figure 9.7); membrane crystallizer working on the RO brine (FS3-Figure 9.8); membrane crystallizers working on the NF and RO brines (FS4-Figure 9.9).

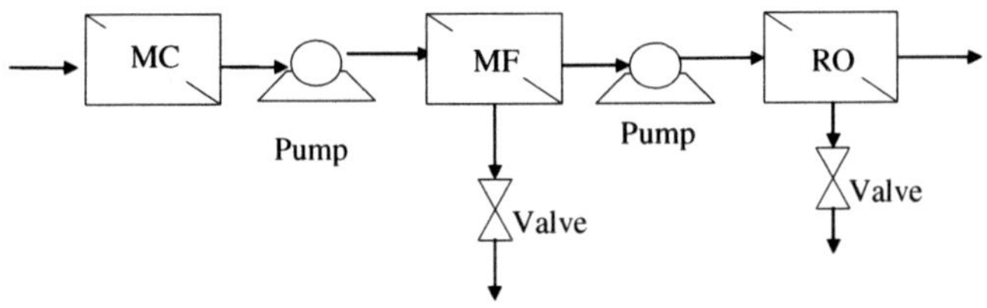

Figure 9.6 FS1

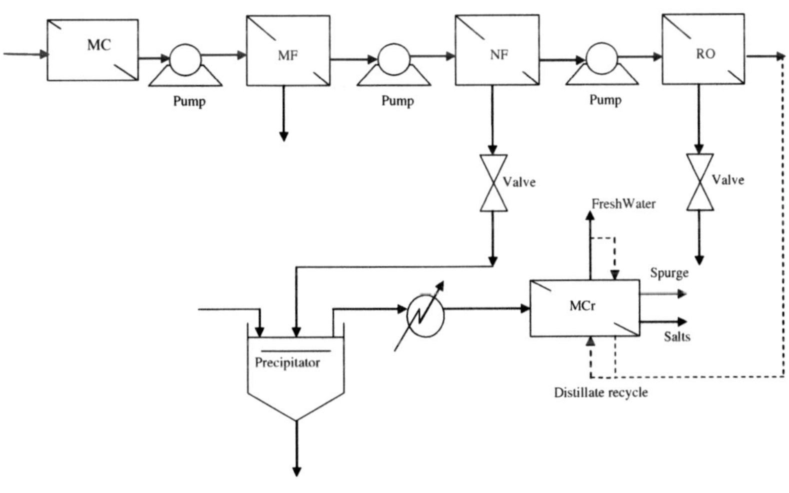

Figure 9.7 FS2

The three flow-sheets (FS2, FS3 and FS4) were chosen by considering their performance in terms of water recovery factor and, therefore, brine reduction, as resulted from WP8. In particular, FS4 was the flow-sheet with the highest recovery factor (and lowest brine production), whereas FS2 and FS3 behaved similarly.

9.6.2 Results and discussion

The information on the streams composition, temperature and pressure, obtained in WP8 for the different flow-sheets, were considered to carry out the exergy analysis, The entropy production was, therefore, calculated for each flow-sheet. Table 9.2 shows the entropy production (RsT_0), as well as the thermal (Wu') and electrical (Wu) exergy supplied to the different flow-sheets. As it can be observed, the highest exergy efficiency was obtained with the FS1 flow sheet, that does not need any thermal source and presents the lowest electrical demand (due to the fact that it does not include the NF unit, that works under pressure). Moreover, among the other flow-sheets, the highest entropy production was for FS4, because of the highest thermal demand related to the presence of membrane crystallizers on both brines (that implies the heating of two streams). However, if the thermal energy is available in the plant, the three flow-sheets behaved similarly and, if a energy recovery system is applied, their exergy efficiency was comparable to that of FS1.

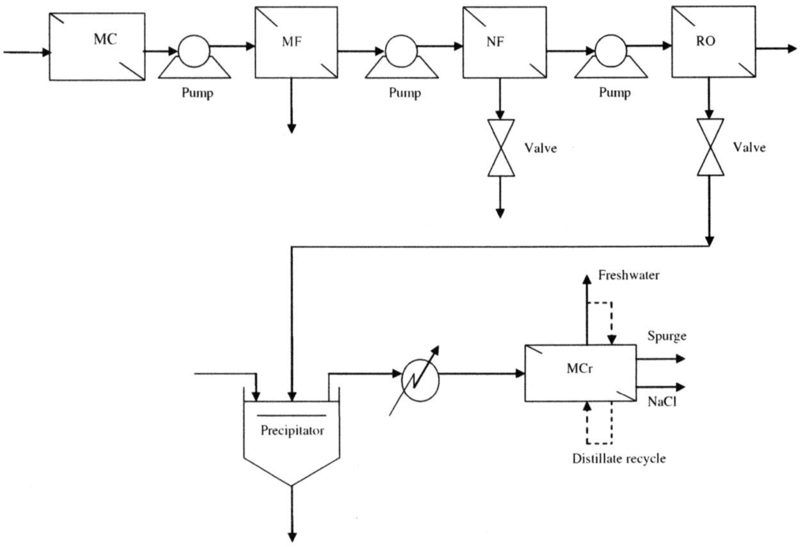

Figure 9.8 FS3

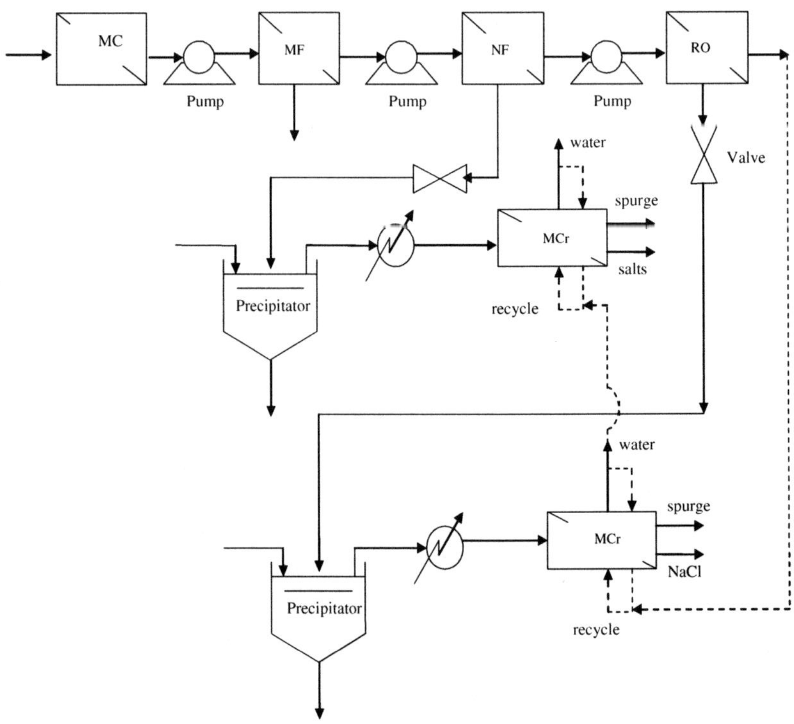

Figure 9.9 FS4

Table 9.2 Results of the exergy analysis of the different flow sheets

	FS1	FS2	FS3	FS4
W_u *(kJ/h)*	$5.84 \cdot 10^6$	$6.45 \cdot 10^6$	$6.45 \cdot 10^6$	$6.45 \cdot 10^6$
W'_u *(kJ/h)*	0	$9.60 \cdot 10^6$	$8.91 \cdot 10^6$	$18.5 \cdot 10^6$
RsT_0 *(kJ/h)*	$5.86 \cdot 10^6$	$1.67 \cdot 10^7$	$1.67 \cdot 10^7$	$2.67 \cdot 10^7$
RsT_0 *(kJ/h)* *[if thermal energy is already available]*	$5.86 \cdot 10^6$	$7.10 \cdot 10^6$	$7.77 \cdot 10^6$	$8.20 \cdot 10^6$
Electrical energy *(kWh/h)*	1623	1792	1792	1792
RsT_0 *(kJ/h)* *[without the electrical energy for pressurising the NF feed: energy recovery system]*	$5.86 \cdot 10^6$	$5.08 \cdot 10^6$	$5.76 \cdot 10^6$	$6.19 \cdot 10^6$

Besides the above analysis, the selected flow-sheets were also compared in terms of electrical energy/productivity ratio and productivity/size ratio. This last calculation was made by considering the performance and properties of commercial membrane modules. An overall comparison of the analysed flow-sheets is summarized in Table 9.3, where the best performance are underlined in bold.

Table 9.3 Overall comparison of the integrated flow-sheets

	Elect. En./Prod. (kWh/m^3)	Recovery factor (%)	Prod./Size ($m^3/h/m^3$)	Brine reduction (%)	RsT_0 (kJ/h)*
FS1	3.63	45.0	**18.2**	0.00	$5.86 \cdot 10^6$
FS2	2.39	71.5	8.38	48.49	$\mathbf{5.08 \cdot 10^6}$
FS3	2.45	69.9	8.49	45.56	$5.76 \cdot 10^6$
FS4	**1.85**	**92.4**	7.26	**86.41**	$6.19 \cdot 10^6$

**With thermal energy available and without the electrical energy for pressurising the NF feed: energy recovery system.*

Based on the analysis made, it can be concluded that the introduction of membrane crystallizers on both NF and RO brines could be an interesting option to consider for increasing the water recovery factor, reducing the brine production, and working with low electrical energy demand/productivity ratio. However, in this case, it is necessary to have thermal energy available in the plant, otherwise the thermal energy need will lead to a lower exergy efficiency with respect to the other flow-sheets. Even if the FS4 productivity/size ratio was lower than that of FS1, its value is comparable to those of the other integrated flow-sheets.

9.7 CONCLUSIONS

As stated in the introduction, desalination has been described as a water treatment process that is often "chemically, energetically and operationally intensive" (Shannon *et al.*, 2008). The logical solution to this problem must consequently be the development of greener, more sustainable technologies. This can be achieved by conducting EIA and environmental monitoring studies at the project level and introducing BAT standards at the technology level.

Public acceptance is a key issue of new desalination developments. Public concerns are often the impacts of desalination plants on the marine environment and energy consumption. A green image campaign to increase public acceptance is only credible if it is well-grounded. Some project developers have recognized this and started to work towards green, sustainable desalination. Sustainable desalination is not some distant utopia but can be achieved with existing technologies. In particular the Australian projects, including Sydney, Perth or the Gold Coast project, set a good example for incorporating environmental protection measures that will hopefully encourage others to follow in their footsteps. The industry, regulators and communities alike, however, have to pave the way by making a commitment to green, sustainable desalination projects which really live up to the expectations. It also requires a commitment to providing water at a price which does not only include the usual construction and operating costs, but also the costs that are necessary to reduce the environmental footprint through environmental studies, advanced technology, or compensation measures. Environmental protection measures will most certainly increase the cost of water production, but, as the Australian projects show, sustainable solutions are economically viable.

In the end, decisions about desalination developments revolve around local circumstances including the need for water and available alternatives to desalination, the costs of the project and financing options, the significance of environmental impacts, and the definition of significance. Desalination will not be the solution to all of the world's water problems, but modern SWRO projects seem to be a more sustainable alternative than many existing water supply schemes and can alleviate pressures on freshwater ecosystems. However, a project has to be designed and operated according to environmental criteria in order to not spread the problem from the freshwater to the marine ecosystems. Many good concepts have been implemented in the latest desalination projects in Australia to minimize the environmental footprint of desalination. What is furthermore needed is ongoing research and demonstration projects to gain experience, knowledge and trust in new, environmentally friendly technologies, as well as political incentives through policies or financial support to implement best available technologies.

9.8 ACKNOWLEDGEMENTS

The results reported in this chapter have been produced as part of work package 9 of the MEDINA project (project no. 036997), funded under the Sixth Framework Programme of the European Commission. The complete results have been published as deliverables on the project website (http://medina.unical.it/) and in various publications, including a guidance manual on environmental impact assessment for desalination plants published by the United Nations Environment Programme (UNEP, 2008) and as a dissertation published by CRC Press/Balkema (Lattemann, 2010).

REFERENCES

Abdel-Jawad M. and Al-Tabtabaei M. (1999, September). Impact of current power generation and water desalination activities on Kuwaiti marine environment. *Proceedings of IDA World Congress on Desalination and Water Reuse*, vol 3. San Diego, USA, pp. 231–240.

Beery M. and Repke J. U. (2010). Sustainability analysis of different SWRO pre-treatment alternatives. *Desal. Wat. Treat.*, **16**, 218–228.

Crisp G. (2006). Australia's first big plant thinks "green". *Desalination and Water Reuse*, **16**(3), 18–27.

Crisp G., Rhodes M. and Procter D. Perth Seawater Desalination Plant – Blazing a Sustainability Trail. *IDA World Congress on Desalination and Water Reuse*, Maspalomas, Gran Canaria.

EPA US. (2009). *Effluent Limitations Guidelines and Standards.*

Hoepner T. (1999). A procedure for environmental impact assessments (EIA) for seawater desalination plants. *Desalination*, **124**(1–3), 1–12.

Janssen R., Herwijnen M. V. and Beinat E. (2003). DEFINITE 3.0 manual. *Institute for Environmental Studies*, Vrije University Amsterdam, The Netherlands. http://www.definite-bosda.nl/.

Lattemann S. (2010). *Development of An Environmental Impact Assessment and Decision Support System for Seawater Desalination Plants.* CRC Press/Balkema, Leiden, The Netherlands, 276 p.

Lattemann S. and Höpner T. (2003). *Seawater Desalination – Impacts of Brine and Chemical Discharges on the Marine Environment*, Desalination Publications, L'Aquila, Italy, 142 p.

Molinari R., Gagliardi R. and Drioli E. (1995). Methodology for estimating saving of primary energy with membrane operations in industrial processes. *Desalination*, **100**, 125–137.

NRC. (2008). Desalination: A National Perspective. Committee on Advancing Desalination Technology, Water Science and Technology Board, Division on Earth and Life Studies, National Research Council (NRC) of the National Academies, United States.

Port C., Roddy S. and Trousdale S. (2009). Sydney's SWRO project – working towards sustainability. *The International Desalination and Water Reuse Quarterly*, **18**(4), 21–33.

Raluy G., Serra L. and Uche J. (2006). Life cycle assessment of MSF, MED and RO desalination technologies. *Energy*, **31**(13), 2025 2036.

Raluy R. G., Serra L. and Uche J. (2005). Life cycle assessment of water production technologies – Part 1: Life cycle assessment of different commercial desalination technologies (MSF, MED, RO). *International Journal of Life Cycle Assessment*, **10**(4), 285–293.

Raluy R. G., Serra L., Uche J. and Valero A. (2004). Life-cycle assessment of desalination technologies integrated with energy production systems. *Desalination*, **167**(1–3), 445–458.

Shannon M. A., Bohn P. W., Elimelech M., Georgiadis J. G., Marinas B. J. and Mayes A. M. (2008). Science and technology for water purification in the coming decades. *Nature*, **452**(20), 301–310.

UNEP. (2008). Desalination. In: *Resource and Guidance Manual for Environmental Impact Assessments.* S. Lattemann, K. H. Mancy, B. S. Damitz, H. K. Khordagui and G. Leslie (ed.), United Nations Environment Programme, Nairobi, Kenya. http://www.unep.org.bh/Publications/Type7.asp.

Voutchkov N. (2008). Carlsbad project plan for green SWRO. *The International Desalination and Water Reuse Quarterly*, **18**(3), 30–36.

Winters H., Isquith I. R. and Bakish R. (1979). Influence of desalination effluents on marine ecosystems. *Desalination*, **30**, 403–410.

Conclusions

In this volume the results reached in MEDINA project have been presented and discussed.

During the project, MEDINA consortium developed a programme which aimed at improving design and operation practices of the current membrane-based desalination plants. The proposed approach was based on the integration of different operations in reverse osmosis (RO) pre-treatment and post-treatment stages. The carried out research activities covered all the aspects related to membrane desalination. The most relevant results achieved in the thirty-nine months of the project are summarized as follow:

1. *various tools were developed/improved for the assessment of water quality*: laboratory protocols for organic matter characterization (LC-OCD and F-EEM) were improved; clone libraries for microorganisms found on membrane autopsies were constructed; a new set-up for measuring particulate and colloidal fouling (MFI-UF) was developed, tested and validated; a monitoring method for calcium carbonate scaling was developed and Pitzer activity coefficients were incorporated into the saturation index procedure; the effect of AOC on fouling was investigated in the laboratory and in full scale facilities using the Membrane Fouling Simulator (MFS) system; the ATP method has been adapted for seawater applications; the biofouling parameter has been developed and tested on real seawater in combination with LC-OCD.
2. In the *evaluation and comparison of pre-treatment processes* promising results were obtained on membrane bioreactor (MBR) as a means for reducing organic content from seawater and as a step towards reducing fouling potential (removal of TOC and EPS), particularly when membrane crystallisers (MCr) have been considered to operate on the NF (nanofiltration) and RO brine. The removal of particulate matter and organics from seawater by the use of biofiltration was also investigated. Granular activated carbon (GAC) and anthracite were tested as a filter medium at two different velocities.
3. *Various tools for RO fouling characterization and understanding were developed.* The filtrate quality was evaluated by measuring the silt density index (SDI), modified fouling index (MFI) and turbidity removal. Both biofilters demonstrated similar behavior in terms of SDI and MFI. In addition, fiber filter as pre-treatment of SWRO in terms of MFI, SDI_{10}, pressure drop, turbidity and MWD was investigated with and without precoagulation. Submerged microfiltration coupled with physic-chemical processes such as adsorption on powder activated carbon was investigated as well.

   Two handbooks on the methods and analytical tools used for performing an autopsy were prepared: one focuses on the basic methods whereas the second one describes the advanced analytical tools used by the different partners according to their expertise. In total, 23 analytical tools were used for autopsying 12 reverse osmosis modules, 10 from seawater pilot and full-scale plants and 2 from a brackish full-scale plant. The analyses aimed at enumerating bacteria, quantifying microbial activity, characterizing organic matter and identifying minerals involved in the fouling phenomenon. All the analyses showed that the phenomenon of biofouling was prevalent on all the sites but at different extent. Organic matter was dominant in all membrane samples even if inorganics, such as aluminosilicates and iron precipitates, could be put into in evidence on a few sites.

4. Progresses were made with *testing cleaning chemicals and procedures* in laboratory procedures and in a pilot plant and selected chemical agents were tested for biomass inactivation and removal.
5. Regarding the *optimization and modelling of RO processes*, Computerized Fluid Dynamics (CFD) was used to optimize characteristics of the spacer in spiral wound elements in terms of mass transport enhancement, fouling and scaling mitigation, and reduction of energy consumption. The performance of membrane contactors for removing the dissolved oxygen while controlling the pH in different parts of a desalination plant was optimized in terms of operating conditions and membrane properties. By selecting the appropriate stripping gas phase it was possible to increase or decrease the water pH up to 30% while removing oxygen. Results obtained on synthetic waters were reproduced also on real waters. It was, therefore, confirmed that by using membrane contactors it is possible to reduce the chemical consumption inside the plant, controlling the water pH and the oxygen content by simply using a gas stream.
6. Injection of RO concentrate in the subsurface, Vacuum Membrane Distillation (VMD), Membrane Crystallization (MCr) and Wind Intensified enhAnced eVaporation (WAIV) were studied as *technologies for improving the productivity of desalination systems*, reducing their environmental impact and recovering some of the valuable ions present in the highly concentrated streams of the desalination plants. In particular, the benefits and hurdles of using brackish groundwater as a drinking water source, as well as the practical guidelines for the design of BWRO plants were studied; microporous hydrophobic membranes were prepared and characterized for performing VMD tests on seawater and on reverse osmosis brine; the coupling of the VMD unit with solar energy was investigated through the modeling of four different configurations involving a salinity gradient solar pond and a solar thermal collector; a semi-industrial unit where the VMD units are coupled to solar collectors was built in Tunisia; a Vacuum Membrane Distillation Modelling Software was achieved; the economical evaluation of desalination systems with membrane crystallizers working on RO or on NF retentate was carried out; the use of wind to evaporate desalination brines and reduce their volume in an economic way with minimal energy inputs was realized using WAIV. Efforts were also made to combine WAIV with MCr. Particularly, tests were carried out on a brackish water integrated desalination system constituted by pre-treatment/RO/WAIV/MCr. The obtained results showed that water recoveries higher than 75% can be achieved from the integrated system while less than 0.75% of the raw water fed to the desalination system is discharged to the environment.
7. The results achieved in the RO pre-treatment and post-treatment stages were incorporated in the *integrated membrane-based desalination system*. Different combinations of membrane units were modelled. The different flow-sheets were compared in terms of quantitative indicators, such as water recovery factor, brine reduction, permeate quality improvements, energy consumption and costs. All flow-sheets were based on the presence of MF followed by NF and, finally, by RO. The difference among them was related to the type of concentration units (membrane crystallizers and/or membrane distillation) and the streams they treated (NF and/or RO brines). The highest water recovery factor (92.4%) and brine reduction (86.4%) were obtained in the desalination system with membrane crystallizers working both on NF and on RO brine. The cost analysis and energy consumption of the various integrated membranes systems have been also useful to identify the most interesting combination of the various membrane units. The achieved results showed that utilising available waste heat or solar energy, the integrated membrane desalination systems costs are quite competitive in energy consumption and overall costs than any other existing technology; moreover they offer the possibility to recover the salts present in the brine streams of the desalination plants as high quality crystals of significant industrial interest.
8. Finally, the *environmental concerns of desalination projects* were comprehensively evaluated in the project. In order to provide guidance to project designers, regulators and decision makers, a structured environmental impact assessment (EIA) approach was developed for desalination projects, including a 10-step methodological approach and a modular environmental monitoring framework. Furthermore, the concept of best available techniques (BAT) was applied to desalination processes and recommendations on which technologies and modes of operation can generally be considered as BAT were elaborated. A decision support system based on multi-criteria analysis (MCA) was used to compare different intake and pretreatment alternatives for seawater reverse osmosis (SWRO) plants in order to identify the best option.

The MEDINA activities have been summarised in a large number of Deliverables. In Appendix I all the produced Deliverables are listed. Most of them are today available to the large public and their access can be through the website http://medina.unical.it/.

In Appendix II, the list of published manuscripts, books, chapters of books and conference proceedings containing the results of MEDINA project can be found.

For the diffusion and discussion of the results of potential interest for industrial exploitation, a concrete dissemination policy was also started. One of the dissemination activity concerned the organization of the EU-CHINA Workshop on MEDINA project in China. The event was co-organized by the MEDINA Consortium, the Ocean University of China, the China Desalination Association and Qingdao International Desalination Center. It was held at the Academic Exchange Centre of Ocean University of China (Qingdao, China) on September 4th–6th, 2009.

Participants of the EU-China Workshop on MEDINA project – Academic Exchange Centre of Ocean University of China, Qingdao, China, September 04–06, 2009

It is interesting to notice as the strategy of integrated membrane operations is becoming today a reality. Most part of the RO feeds pre-treatments are now based on UF/MF/NF operations. Moreover, the potentialities of the MBR are today also considered, as suggested in the MEDINA project. The overall MEDINA results confirm also the potentialities of membrane distillation (MD), WAIV and membrane crystallisers in the RO post-treatments for increasing the water recovery factor and minimizing the brine disposal problem. It is interesting that various pilot plants and research proposal on MD for sea water and brackish water desalination are today in progress worldwide. The development of new MD polymeric membranes, and the design and optimization of MD modules for a better control of temperature polarization phenomena is however necessary for further progresses. Moreover, through MD/MCr, the various metals present in seawater can be recovered as high quality salts. It is well known today the necessity of realising metals recovery and reuse for avoiding the lack of various metals of strategic importance in modern industrial production . This is the case for example for Mg, As, Tallium, Iridium, Sr, etc. The use of MD/MCr on the brine of the NF and RO steps, after appropriate pre-treatment, might be an interesting solution to this problem.

Appendix I

Deliverables list of MEDINA project

Deliverable. no	Deliverable name	WP no.	Organization which provided the deliverable	Nature	Dissemination level
D 1.1.1	Report on raising public participation and awareness	1	UNICAL	R	Public
D.1.1.2	Protocol for Fluorescence Excitation-Emission Matrix (EEM) Test – Protocol for HPSEC-DOC/UVA/Fluorescence test – Protocol for LC-OCD (-OND) test	1	IHE	R	
D.1.1.3	Membrane desalination glossary	1	IHE	R	Public
D 1.1.4	Documentation of modified protocols of analytical tools applied to seawater/ brackish water sources: LC-OCD protocol modification	1	AR-VW	R	Confidential
D 1.2.1	*Protocol for MFI-UF test* Protocol for SDI test	1	IHE	R	
D.1.2.3	D.2.1.1: Documentation of new protocols: MFI-UF, CFS-MFIUF	1	UNESCO-IHE and UNSW	R	Public
D 1.3.1	*ScaleGuard® Description*	1	*IHE – KIWA*	R	
D.1.3.2 &	D.1.3.2: Incorporate Pitzer activity model in solubility calculations of different scalants mainly CaCO3 and CaSO4. &	1	UNESCO-IHE and UNSW	R	Public
D.1.3.3	D.1.3.3: Use the saturation Index incorporated with Pitzer model instead of normal solubility prediction methods namely LSI and S&DSI.				
D 1.4.1	Protocols for analysis of ATP and AOC	1	KWR	R	
	Protocol for Determination of Biodegradable Dissolved Organic Carbon (BDOC)		IWW	R	
	Documentation of protocols for biofouling indicators (AOC, BDOC) and supporting analytics (e.g. ATP)	1	KWR	R	Confidential
D1.4.2/1.4.3	Biodegradable organic compounds (AOC, BDOC) and adenosinetriphosphate (ATP) in seawater: adaptation of analysis and significance	1	KWR/IWW	R	Confidential
D.1.5.1	QA/QC protocol for sample preservation and shipment	1	IHE	R	Public
D 1.5.2	Report summarizing the ongoing water quality monitoring practices at full-scale desalination plants	1	AR-VW	R	Public
D 1.5.3	Workplan describing the implementation and validation of the water quality assessment tools at full scale desalination plants	1	AR-VW	R	Confidential
D 1.5.4	Report consolidating the first results on water quality collected at full-scale desalination facilities	1	AR-VW	R	Confidential
D 1.5.5	Final report consolidating the results collected at full-scale desalination facilities	1	AR-VW	R	Confidential
D.2.1.1	A comprehensive critical evaluation on conventional, membrane–based and novel pre-treatment procedures for SW and BW desalination	2	IHE, BGU	R	Public
D 2.1.2	A report on the current advantages and drawbacks of MBR as a pre-treatment step in the SW and BW desalination	2	UNICAL, BGU	R	Confidential
D 2.4	FINAL REPORT	2	BGU, UNICAL, AR-VW, INSA, UTS, UNSW	R	
D 3.1	A Report on the existing methodologies for performing autopsies as well as understanding fouling	3	AR-VW	R	Public

D 3.2	Handbook for best practice for basic membrane autopsies	3	IWW	R	Public
D 3.3	Handbook for best practice for advanced membrane autopsies	3	IWW	R	Public
D 3.4	Detailed report on autopsies results of RO membranes from a full scale desalination	3	AR-VW	R	Confidential
D 3.5	Final report on all autopsies summarizing main conclusions	3	AR-VW	R	Confidential
D 3.6	Detailed report on the biodegradability of antiscalants and their effects on biofouling	3	AR-VW, IWW, BGU, KWR	R	Confidential
D 3.7	D 3.7_Contribution BGU and UNSW Final report on the understanding of fouling mechanism and the factors influencing its formation D 3.7_Contribution KWR Selected biomass parameters to determine the degree of biofouling of spiral-wound membranes	3	UNSW, BGU, KWR	R	Confidential
D 4.1.1	Documentation on the state of the art of membrane cleaning, in combination with an evaluation of data (including autopsies) collected from full scale plants and pilot plants	4	KWR	R	Confidential
D 4.2.1	Documentation of a protocol for a laboratory test method to assess the effect of a cleaning agent on biomass attached to membranes	4	KWR	R	Confidential
D4.2.2/D4.2.4	Documentation on the effect of cleaning strategies on biofilms	4	LCME & IWW	R	Confidential
D 4.2.3	Generation of defined fouling layers for testing the efficacy of cleaners towards different foulants	4	IWW	R	Public
D 4.2.5	Documentation on the effects of cleaning procedures on mixed fouling layers	4	IWW	R	Public
D4.2.6	Documentation on the effects of cleaning on membrane performance	4	AR-VW	R	Confidential
D4.2.7	Cleaning of spiral-wound membranes	4	KWR	R	Confidential
D 5.1.1A	Benefits and hurdles of using brackish groundwater as a drinking water source in the Netherlands	5	KWR	R	Public
D 5.1.1B	Guidelines for the design of brackish water reverse osmosis plants: selection of source and disposal aquifers and design of RO installation, injection wells and monitoring programs	5	KWR	R	Confidential
D5.1.2	VMD Modelling Software version 2 User's Guide	5	INSA	R	Confidential
D 5.2.1	Protocols for salts crystallization	5	UNICAL	R	Confidential
D 5.2.2	Estimate of the extent that brine superconcentrates can be evaporated by WAIV without causing plugging or deposits	5	BGU	R	Confidential
D 5.2.3	Defined methods for removing cleanable deposits	5	UNICAL	R	Confidential
D 5.3	Report on the economic analysis	5	UNICAL	R	Confidential
D6.1.3	Lab-scale Prototypes of membrane distillation modules	6	GVS	R	Public
D.6.1.4–7	Synthetic Report including deliverables D.6.1.4, D.6.1.5, D.6.1.6, D.6.1.7	6	INSA	R	Confidential
D.6.1.8	Intermediate report on the chosen configuration and design of a semiindustrial scale pilot plant	6	INSA, ENIG	R	Confidential
D 7.1.1	A state of art report on the use of nanofiltration (NF) as a pre-treatment for RO in sea and brackish water desalination applications	7	BGU	R	Public

(*Continued*)

Deliverable. no	Deliverable name	WP no.	Organization which provided the deliverable	Nature	Dissemination level
D7.2.2	Report on optimizing membrane contactors	7	UNICAL	R	Public
D7.3.1	A Critical Literature Review about Models to Optimize Hydraulics and to Predict Permeability in SW and BW Desalination	7	IWW/UNSW	R	Public
D7.3.2	Laboratory Test Unit for small spiral wound elements	7	IWW	R	Confidential
D7.3.4	Preliminary results of ANN-modelling and simulation	7	IWW	R	Public
D7.3.5	A practical software tool to predict the performance of the RO process in terms of permeability, permeate and brine quality	7	IWW/UNSW	R	Public
D7.3.6	Report and manual on the software tool	7	IWW/UNSW	R	Public
D 8.1	Report Of Critical Analysis On The Desalination Technologies	8	All partners	R	Public
D 8.2	Lab-scale unit of an integrated membrane system for carrying out experimental tests	8	UNICAL	P	Public
D 8.3.1, D 8.3.2 and D 8.3.3	Common report for D 8.3.1 Definition of the best membrane integrated system for carrying out both SW and BW desalination – D 8.3.2 Definition of the best operating conditions for the optimal membrane system identified – D 8.3.3 Simulation code the modelling the integrated membrane system	8	UNICAL	R	Confidential
D 8.4.1, D 8.4.2 and D 8.5	Common report for D 8.4.1 Economic evaluation of the cost associated to the integrated membrane system – D 8.4.2 Report on the comparison of performance and cost of the integrated membrane system with those related with conventional technologies – D 8.5 Report on the evaluation and application of quantitative indicators	8	UNICAL	R	Confidential
D 9.1	EIA checklist/contents list	9	ICBM	R	Public
D 9.2.1	Overview and description of the relevant impacts of membrane desalination plants, with emphasis on concentrate and chemical discharges	9	ICBM	R	Public
D 9.2.2	Matrix of cause-effect relationships	9	ICBM	R	Public
D 9.3	Development of a methodological EIA approach for membrane desalination plants	9	ICBM	R	Public
D 9.3.1 and D 9.3.2	Terms of Reference for the collection of relevant data and information, and for evaluation and assessment of this data	9	ICBM	R	Public
D 9.4	Recommendations for process-design and site-selection for impact mitigation	9	ICBM	R	Public
D 9.5.1 & D 9.5.3	Development of a decision support system (DSS) for environmental impact assessment studies (EIA) using multi-criteria analysis (MCA)	9	ICBM	R	Public
D 9.5.2	Energy and exergy analysis	9	UNICAL	R	Public
D 9.6	Development of a decision support system (DSS) for environmental impact assessment studies (EIA) using multi-criteria analysis (MCA)	9	ICBM	R	Public

R = REPORT
P = PROTOTYPE.

Appendix II

List of publications

JOURNAL ARTICLES

1. Baghoth S. A., Maeng S. K., Salinas Rodriguez S. G., Ronteltap M., Sharma S., Kennedy M. D. and Amy G. (2008). An urban water cycle perspective of Natural Organic Matter (NOM): NOM in drinking water, wastewater effluent, storm water, and seawater. *Water Science & Technology: Water Supply*, **8**, 701–707.
2. Chinu K. J., Johir A. H., Vigneswaran S., Shon H. K. and Kandasamy J. (2009). Biofilter as pretreatment to membrane based desalination: Evaluation in terms of fouling index. *Desalination*, **247**(1–3), 77–84.
3. Chinu K., Johir A. H., Vigneswaran, S., Shon H. K. and Kandasamy J. (2010). Assessment of pretreatment to microfiltration for desalination in terms of fouling index and molecular weight distribution. *Desalination*, **250**(2), 644–647.
4. Criscuoli A., Carnevale M. C., Mahmoudi H., Gaeta S., Lentini F. and Drioli E. (2010). Membrane contactors for the oxygen and pH control in desalination. *Journal of Membrane Science.*
5. Curcio E., Ji X., Di Profio G., Sulaiman A. O., Fontananova E. and Drioli E. (2010). Membrane distillation operated at high seawater concentration factors: Role of the membrane on CaCO3 scaling in presence of humic acid. *Journal of Membrane Science*, **346**, 263–269.
6. de O. Manes C.-L., West N. J., Rapenne S. and Lebaron P. (2011). Dynamic bacterial communities on reverse-osmosis membranes in a full-scale desalination plant. *Biofouling*, **27**, 47–58.
7. Di Profio G., Ji X., Curcio E. and Drioli E. (2011). Submerged hollow fiber ultrafiltration as seawater pretreatment in the logic of integrated membrane desalination systems. *Desalination*, **269**, 128–135.
8. Hijnen W. A. M., Biraud D., Cornelissen E. R. and van der Kooij D. (2009). Threshold concentration of easily assimilable organic carbon in feedwater for biofouling of spiral-wound membranes. *Environ. Sci. Techn.*, **43**, 4890–4895.
9. Hijnen W. A., Cornelissen E. R. and van der Kooij D. (2011). Threshold concentrations of biomass and iron for pressure drop increase in spiral-wound membrane elements. *Water Res*, **45**, 1607–1616.
10. Hijnen W. A. M., Biraud D., Cornelissen E. R. and van der Kooij D. (2011). Threshold concentration of easily assimilable organic carbon in feedwater for biofouling of spiral-wound membranes (additions/corrections). *Environ Sci Technol*, **45**, 1744–1745.
11. Ji X., Curcio E., Al Obaidani S., Di Profio G., Fontananova E. and Drioli E. (2010). Membrane distillation-crystallization of seawater reverse osmosis brines. *Separation and Purification Technology*, **71**, 76–82.
12. Johir A. H., Khorshed C., Vigneswaran S. and Shon H. K. (2009). In-line-flocculation- Filtration as Pre-treatment to Reverse Osmosis Desalination. *Desalination*, **247**(1–3), 85–93.
13. Katzir L., Volkmann Y., Daltrophe N., Korngold E., Mesalem R., Oren Y. and Gilron J. WAIV- Wind Aided Intensified eVaporation for brine volume reduction and generating mineral byproducts.
14. Lattemann S. and Höpner T. (2008). Environmental Impact and Impact Assessment of Seawater Desalination. *Desalination*, **220**, 1–15.
15. Lattemann S. and El Habr H. (2009). UNEP resource and guidance manual for environmental impact assessment of desalination projects. *Desalination and Water Treatment*, **3**, 217–228.
16. Lee J. J., Johir M. A. H., Chinu K. H., Shon H. K., Vigneswaran S., Kandasamy J., Kim C.W. and Shaw K. (2009). Hybrid Filtration Method for Pre-treatment of Seawater Reverse Osmosis (SWRO). *Desalination*, **247**(1–3), 15–24.
17. Lee J. J., Johir M. A. H., Chinu K. H., Shon H. K., Vigneswaran S., Kandasamy J., Kim C.W. and Shaw K. (2010). Novel Pre-Treatment Method for Seawater Reverse Osmosis: Fibre Media Filtration. *Desalination*, **250**(1–32), 557–561.

18. Macedonio F. and Drioli E. (2010). An exergetic analysis of a Membrane Desalination System. *Desalination – special issue honoring Sidney Loeb*, **261**, 293–299.
19. Macedonio F. and Drioli E. (2010). Hydrophobic membranes for salts recovery from desalination plants. *Desalination and Water Treatment*, **18**, 224–234.
20. Macedonio F., Drioli E., Curcio E. and Di Profio G. (2009). Experimental and economical evaluation of a membrane crystallizer plant. *Desalination and Water Treatment*, **9**, 49–53.
21. Macedonio F., Katzir L., Geisma N., Simone S., Drioli E. and Gilron J. (2011). Wind-Aided Intensified eVaporation (WAIV) and Membrane Crystallizer (MCr) Integrated Brackish Water Desalination Process: Advantages and Drawbacks. *Desalination*, doi:10.1016/j.desal.2010.12.002. *In press*.
22. Mericq J. P., Laborie S. and Cabassud C. (2009). Vacuum membrane distillation for an integrated seawater desalination process. *Desalination and Water Treatment*, **9**, 287–296.
23. Mericq J. P., Laborie S. and Cabassud C. (2010), Vacuum membrane distillation of seawater reverse osmosis brines. *Water Research*, **44**, 5260–5273.
24. Mericq J. P., Laborie S. and Cabassud C. (2011). Evaluation of systems coupling vacuum membrane distillation and solar energy for seawater desalination. *Chemical Engineering Journal*, **166**, 596–606.
25. Mondamert L., Labanowski J., Berjeaud (Poitiers) J. M., Rapenne (Maisons-Laffitte) S. and Croué J. P. (2009). Autopsy of RO desalination membrane: Part 2. Chemical characterisation of the foulant. *Desalination and Water treatment*, **9**, 73–81.
26. Salinas Rodríguez S. G., Al-Rabaani B., Kennedy M. D., Schippers J. C. and Amy G. L. (2009). MFI-UF constant pressure at high ionic strength conditions. *Desalination and Water Treatment*, **10**, 64–72.
27. Salinas Rodríguez S. G., Kennedy M. D., Schippers J. C. and Amy G. L. (2009). Organic foulants in estuarine and bay sources for seawater reverse osmosis – Comparing pre-treatment processes with respect to foulant reductions. *Desalination and Water Treatment*, **9**, 155–164.
28. Schaule G., Rapenne S., Strathmann M., Grobe S., Robert C. and Jacquemet V. (2009). Autopsy of RO desalination membranes. Part 1. Microbial characterization of foulants. *Desalination and Water treatment*, **9**, 66–73.
29. Sim L. N., Ye Y., Chen V. and Fane A. G. (2010). Crossflow Sampler Modified Fouling Index Ultrafiltration (CFS-MFIUF) – an alternative fouling index. *Journal of Membrane Science*, **360**(1–2), 174–184.
30. Sim L. N., Ye Y., Chen V. and Fane A. G. (2011a). Comparison of MFI-UF constant pressure, MFI-UF constant flux and Crossflow Sampler – Modified Fouling Index Ultrafiltration (CFS-MFIUF). *Water Research*, **45**(4), 1639–1650.
31. Sim L. N., Ye Y., Chen V. and Fane A. G. (2011b). Investigations of the coupld effect of cake-enhanced osmotic pressure and colloidal fouling in RO using crossflow sampler-modified fouling index ultrafiltration. *Desalination*, doi:10.1016/j. desal.2011.01.059.
32. Simone S., Figoli A., Criscuoli A., Carnevale M. C., Rosselli A. and Drioli E. (2010). Hollow fibre membranes prepared from PVDF/PVP blends: VMD tests. *Journal of Membrane Science*, **364**, 219–232.
33. Stuyfzand P. J. and Raat K. J. (2010). Benefits and hurdles of using brackish groundwater as a drinking water source in the Netherlands. *Geohydrology Journal*, DOI:10.1007/s10040-009-0527-y.
34. Tansakul C., Laborie S. and Cabassud C. (2009). Membrane hybrid processes for pretreatment before seawater reverse osmosis desalination. *Desalination and Water Treatment*, **9**, 279–286.
35. Tansakul C., Laborie S. and Cabassud C. (2010). Study on performance of ultrafiltration membrane-based pretreatment for application to seawater reverse osmosis desalination. *Water Science & Technology*, **62**(9), 1984–1990.
36. Ye Y., Sim L. N., Herulah B., Chen V. and Fane A. G. (2010). Effects of operating conditions on submerged hollow fibre membrane systems used as pre-treatment for seawater reverse osmosis. *Journal of Membrane Science*, **365**(1–2), 78–88.

MAGAZINE ARTICLES

1. Lattemann S. (2008). Le dessalement, est-il écologique? *La Recherche*, **421**(July/August), p. 62–65.
2. Lattemann S. and Höpner T. (2007). Why we must have impact studies and mitigation. *Desalination and Water Reuse*, **17**(2), pp. 36–44.

BOOKS, BOOK CHAPTERS, REPORTS

1. Lattemann S., Kennedy M. D., Amy G. and J. Schippers (in review): Global Desalination Situation. In: I. Escobar & A.I. Schäfer (Eds.), Sustainable Water for the Future – Water Recycling versus Desalination. Elsevier Science.
2. Lattemann S. (2009). Protecting the Marine Environment, In: Micale G., Cipollina A., Rizzuti L. (Eds.), Seawater Desalination. Springer-Verlag GmbH Berlin Heidelberg.
3. Lattemann S., Mancy K. H., Khordagui H., Damitz B. and G. Leslie (2008). Desalination Resource and Guidance Manual for Environmental Impact Assessments. United Nations Environment Programme, Nairobi, Kenya.

CONFERENCE PRESENTATIONS AND PROCEEDINGS

1. Baghoth S. A., Maeng S. K., Salinas Rodriguez S. G., Ronteltap M., Sharma S., Kennedy M. D. and Amy G. (2008). An urban water cycle perspective of Natural Organic Matter (NOM): NOM in drinking water, wastewater effluent, storm water, and seawater, In: IWA (Ed.), Natural Organic Matter: From Source to Tap. IWA – Cranfield University, Bath, UK.
2. Bleninger T., Lattemann S., Purnama A., Al-Barwani H. H., Niepelt A., Doneker R. L. and Jirka G. H. (2009). Environmental planning, prediction, and management of brine discharges from desalination plants, International Desalination Association (IDA) Congress, The Palm, Dubai, November 2009.
3. Criscuoli A., Carnevale M. C., Mahmoudi H., Gaeta S., Lentini F., Reggiani S. and Drioli E. "Oxygen and carbon dioxide control by membrane contactors in desalination", *International Conference on Membranes, ICOM 2008*, July 12–18 2008, Honolulu, Hawaii (USA).

4. Criscuoli A., Figoli A., Simone S., Carnevale M. C. and Drioli E. Preparation and characterization of PVDF hollow fibre membranes for vacuum membrane distillation, oral presentation at Euromembrane 2009, September 6–10 2009, Montpellier (France).
5. Drioli E. and Macedonio F. New Integrated Water Treatments and Production Modes for City Planning. New Technology for Infrastructure – The World of Tomorrow. 2006 Invitation Symposium of the Australian Academy of Technological Sciences and Engineering (ATSE), Sidney, Australia, 20–21, November 2006.
6. Drioli E. and Macedonio F. Progresses in Integrated Membrane Operations for Desalination and Wastewater Treatment. Asian Conference on Desalination & Water Reuse 2007, Qingdao, China, July 4–6, 2007.
7. Drioli E. and Macedonio F. Progresses on Seawater Desalination and Wastewater Treatment in the Logic of Process Intensification Strategy. European Congress of Chemical Engineering (ECCE-6), Copenhagen, Denmark, 16–20, September, 2007. Keynote lecture.
8. Drioli E. and Macedonio F. Advanced Membrane Research in Desalination and Water Recycle. JDA Forum 2007, Tokyo, Japan, November 12th, 2007.
9. Drioli E. and Macedonio F. Progresses in Membrane Processes for the Rationalization of Industrial Productions and Waste Water Treatment. All India Seminar on "Zero Effluent Discharge-Latest Development in Recycling", Kolkata, India, December 22–23, 2007.
10. Drioli E., Macedonio F., Di Profio G. and Curcio E. Membrane crystallizers for reducing brine disposal and increasing recovery factor in seawater and brackish water desalination. AMS5 in Kobe – The Fifth Conference of Aseanian Membrane Society, July 12–14, 2009, Kobe, Japan.
11. Drioli E. and Macedonio F. Hydrophobic Membranes For Salts Recovery From Desalination Plants. 5th IWA Specialised Membrane Technology Conference for Water and Wastewater Treatment (IWA-MTC2009), September 1–3, 2009, Beijing, China.
12. Drioli E. and Macedonio F. Membrane Engineering for a Process Intensification Strategy: Potentialities and Realizations. EMCC-6, 6th Chemical Engineering Conference for Collaborative research in Eastern Mediterranean Countries. March 7–12, 2010. Antalya (Turkey).
13. Drioli E. and Macedonio F. Membrane systems for possible solutions to water stress. Advances in science and Engineering for Brackish Water and seawater Desalination. May 8–12, 2010. Cetraro (ITALY).
14. Drioli E., Macedonio F., Katzir L., Geisma N. and Gilron J., Integrated Brackish Water Desalination Process optimized for approching the "zero-liquid discharge". AMS6/IMSTEC10, November 22–26, 2010. Sidney, Australia.
15. Drioli E. and Macedonio F. Membrane Engineering for the Treatment of High Concentrated Solutions in Desalination. CHISA 2010 – ECCE 7, 28 August – 1 September, 2010. Prague, Czech Republic.
16. Lattemann S. (2007). WHO Guidance "Desalination for Safe Water Supply" – Results of the Work Group on Environmental Impacts, International Desalination Association (IDA) Congress on Desalination and Water Reuse, Maspalomas, Gran Canaria, October 2007.
17. Lattemann S. Environmental impact and impact assessment of seawater desalination, 4th International Water Association (IWA) Young Water Professionals Conference, University of California, Berkeley, USA, July 2008.
18. Lattemann S. Environmental impact and impact assessment of seawater desalination, European Water Research Day, European Commission RTD Framework Programmes, Zaragoza, Spain, September 2008.
19. Leparc J., Schrotter J.-C., Rapenne S., Croué J.-P., Lebaron P., Gaid A. and Lafon D. Use of advanced analytical tools for monitoring performance of seawater pre-treatment processes, International Desalination Association World Congress on Desalination and Water Reuse, October 21–26, 2007 in Gran Canaria (Spain).
20. Macedonio F., Curcio E., Di Profio G. and Drioli E. Integrated Membrane Systems for Water Treatment. IMSTEC 07, The 6th International Membrane Science and Technology Conference, The University of New South Wales, Sydney, Australia, 5–9, November, 2007.
21. Macedonio F., Xiaosheng J., Drioli E., Curcio E. and Di Profio G. Crystals recovery from NF retentate streams through membrane crystallizer devices. 5th Chemical Engineering Conference for Collaborative Research in Eastern Mediterranean Countries – EMCC5, Cetraro (Italy), 24–29, May 2008.
22. Macedonio F., Curcio E., Di Profio G. and Drioli E. MF/NF/RO and Membrane Crystallizers for seawater desalination. "IFMSTED 2008 Asian-Pacific Conference on Desalination & Water Reuse", Qingdao, China, May 28–30, 2008.
23. Macedonio F., Drioli E., Curcio E. and Di Profio G. Experimental and economical evaluation of a membrane crystallizer plant. Membranes in Drinking Water Production and Wastewater Treatment, Toulouse, France, October 20–22, 2008.
24. Macedonio F., Drioli E., Curcio E. and Di Profio G. Membrane Engineering for High Recovery Factors in Seawater Desalination. IDA World Congress 2009 on Desalination and Water Reuse, Atlantis-The Palm Jumeirah, Dubai, UAE, 7–12 November, 2009.
25. Macedonio F., Drioli E., Curcio E. and Di Profio G. "New Metrics" for Evaluating the Performance of Desalination Processes. IDA World Congress 2009 on Desalination and Water Reuse, Atlantis-The Palm Jumeirah, Dubai, UAE, 7–12 November, 2009.
26. Macedonio A. F. and Drioli E. Water cost of a desalination system with membrane crystallizer operation. Advances in science and Engineering for Brackish Water and seawater Desalination. May 8–12, 2010. Cetraro (ITALY).
27. Macedonio F. and Drioli E. Integrated Membrane-Based Operations for Seawater and Brackish Water Desalination. EuroMed 2010 – Desalination for Clean Water and Energy Cooperation among Mediterranean Countries of Europe and the MENA Region, October 3–7, 2010, Tel Aviv, Israel.
28. Rapenne S., Barbé C., Mondamert L., Manes C., Schaule G., Lebaron P. and Croué J.-P. (2009). Seawater Characterization and Membrane Autopsy Tools for Better Operation of SWRO Plants, Proc. IDA World Congress on Desalination and Water Reuse, Dubaï.
29. Rapenne S., Port C. L., Roddy S. J. and Croue J.-P. Pretreatment prior to RO for seawater desalination – The Sydney pilot-scale study, International Desalination Association World Congress on Desalination and Water Reuse, October 21–26, 2007 in Gran Canaria (Spain).
30. Salinas Rodríguez S. G., Al-Rabaani B., Kennedy M. D., Amy G. and Schippers J. C. (2008). MFI-UF at constant pressure for seawater applications. In: EDS (ed.) Euromed 2008. European Desalination Society, Dead Sea, Jordan.
31. Salinas Rodriguez S. G., Gonzales T. A., Kennedy M. and Amy G. (2008). Fluorescence of selected organic matter compounds: looking at the effect of concentration, ionic strength and pH (Poster). In: Birmingham U.O. (Ed.), AGU Chapman Conference on Organic Matter Fluorescence. Edgbaston, Birmingham, UK.

32. Salinas Rodríguez S. G., Kennedy M. D., Amy G. and Schippers J. C. (2010). Flux dependency of particulate fouling by MFI-UF measurements in seawater reverse osmosis systems. In: EDS (Ed.), EuroMed 2010: Desalination for Clean Water and Energy. EDS, Tel-Aviv, Israel.
33. Salinas Rodríguez S. G., Kennedy M. D., Amy G. and Schippers J. C. (2010). Flux dependency of particulate fouling in seawater reverse osmosis systems. In: EDS (Ed.), Membranes in Drinking Water Production and Waste Water Treatment. EDS/IWA, Trondheim, Norway.
34. Salinas Rodríguez S. G., Kennedy M. D., Schippers J. C. and Amy G. (2008). Identification of organic foulants in estuarine and seawater reverse osmosis systems – Comparison for different pre-treatments. In: EDS-INSA (Ed.), Membranes in Drinking Water Production and Waste Water Treatment. European desalination society, Toulouse, France.
35. Salinas Rodríguez S. G., Kennedy M. D., Schippers J. C. and Amy, G. (2009). Identification of organic foulants in seawater reverse osmosis (SWRO). In: Association A.W.W. (Ed.), Membrane Technology Conference. AWWA and AMTA, Memphis, Tennessee.
36. Salinas Rodríguez S. G., Kennedy M. D., Schippers J. C. and Amy G. L. (2009). Tracking organic matter in four SWRO plants – Comparing pre-treatment processes with respect to organic foulant reductions. In: DUT (Ed.), High Quality Drinking Water Conference. Delft.
37. Salinas Rodríguez S. G., Mamoun A., Schurer R., Kennedy M. D., Amy G. L. and Schippers J. C. (2009). Modified Fouling Index (MFI-UF) at constant flux for seawater RO applications. In: EDS (Ed.), Desalination for the Environment: Clean water and Energy. European desalination society, Baden-Baden, Germany.

WORKSHOPS AND TRAINING

1. Abdallah S. B., Frikha N., Chaouachi B. and Gabsi S. Study Of The Performances Of Different Configurations For Seawater Desalination With A Solar Membrane Distillation. EU-CHINA Workshop on "Membrane-Based Desalination: An Integrated Approach (acronym MEDINA)", Qingdao (China), 4–6, September 2009.
2. Arbel R., Koren O., Oren Y. and Gilron J. Interaction Of Biofouling And Mineral Scaling In Reverse Osmosis Operation. EU-CHINA Workshop on "Membrane-Based Desalination: An Integrated Approach (acronym MEDINA)", Qingdao (China), 4–6, September 2009.
3. Chinu K. J., Johir A. H., Vigneswaran S. and Kandasamy J. Comparison Of Pre-Treatment Methods For RO Desalination. EU-CHINA Workshop on "Membrane-Based Desalination: An Integrated Approach (acronym MEDINA)", Qingdao (China), 4–6, September 2009.
4. Cornelissen E. R., Hijnen W. A. M. and Van Der Kooij D. Assessment Of The Efficiency Of Hydraulic And Chemical Cleaning For The Removal Of Biomass From Surfaces In A Laboratory Test. EU-CHINA Workshop on "Membrane-Based Desalination: An Integrated Approach (acronym MEDINA)", Qingdao (China), 4–6, September 2009.
5. Curcio E., Di Profio G., Ji X., Sulaiman A.O., Macedonio F. and Drioli E. Pre-treatment for membrane distillation operated at high seawater concentration factors. EU-CHINA Workshop on "Membrane-Based Desalination: An Integrated Approach (acronym MEDINA)", Qingdao (China), 4–6, September 2009.
6. Drioli E. and Macedonio F. New Trends and Technologies for Membrane Desalination. Inaugural KAUST Economic Development International Symposium at the China Water Show in Shanghai. May 6, 2010.
7. Drioli E. and Macedonio F. Membrane Engineering for Process Intensification Strategy. 7th Italy-Korea Workshop "Membranes for a clean environment" June 4–6, 2010. Ravello (ITALY).
8. Drioli E. and Macedonio F. Recent Progresses and Perspectives in desalination with integrated membrane systems. The 3rd International Desalination Workshop organized by the Center for Seawater Desalination Plant and European Desalination Society, 3–6 November 2010 at Hyatt Regency Jeju on Jeju Island, Korea.
9. Frikha N., Chaouachi B. and Gabsi S. Design Of A Semi Industrial Scale Pilot Plant For Seawater Desalination With A Solar Membrane Distillation. EU-CHINA Workshop on "Membrane-Based Desalination: An Integrated Approach (acronym MEDINA)", Qingdao (China), 4–6, September 2009.
10. Hijnen W. A. M., Cornelissen E. R. and Van Der Kooij D. Biofilm And Pressure Drop Formation In Spiral-Wound Membranes Induced By Low Concentrations Of Easily Assimilable Organic Carbon In The Feed Water. EU-CHINA Workshop on "Membrane-Based Desalination: An Integrated Approach (acronym MEDINA)", Qingdao (China), 4–6, September 2009.
11. Katzir L., Volkmann Y., Daltrophe N., Korngold E., Mesalem R., Oren Y. and Gilron J. Waiv- Wind Aided Intensified Evaporation For Brine Volume Reduction And Generating Mineral Byproducts. EU-CHINA Workshop on "Membrane-Based Desalination: An Integrated Approach (acronym MEDINA)", Qingdao (China), 4–6, September 2009.
12. Keller M. and Panglisch S. Using Artificial Intelligence For Analyzing Ro Performance. EU-CHINA Workshop on "Membrane-Based Desalination: An Integrated Approach (acronym MEDINA)", Qingdao (China), 4–6, September 2009.
13. Lattemann S. and Bleninger T., et al., Desalination and the Environment (two day workshop), organized by the German Desalination Society (EDS) at the Water Berlin Exhibition and Congress, April 2009.
14. Lattemann S., Höpner T. and Bleninger T. Seawater Desalination and the Environment. Impact of Brine and Chemical Discharges on the Marine Environment (three day workshop), organized by the Middle East Desalination Research Center (MEDRC) and the Government of Germany through its Embassy in Muscat, Sultanate of Oman, Amman, Jordan, December 2008.
15. Lattemann S. Seawater Desalination and the Marine Environment (half-day seminar), organized by the Water Desalination Conference in Arab Countries (ARWADEX 6), Riyadh, Saudi Arabia, April 2008.
16. Lattemann S. and Höpner T. Seawater Desalination and the Marine Environment (two day seminar), organized by the European Desalination Society and University of L'Aquila, Faculty of Engineering, Italy, February 2008.
17. Lattemann S. Results From The Medina Work Package On Environmental Impact Assessment And Life Cycle Analysis Of Desalination Plants. EU-CHINA Workshop on "Membrane-Based Desalination: An Integrated Approach (acronym MEDINA)", Qingdao (China), 4–6, September 2009.

18. Leparc J., Remize P. J., Laroche J. F., Rapenne S., Barbé C., Schrotter J.-C. and Bréant P. Towards A Better Characterisation Of Pretreatment Performance Through The Use Of Modified Fouling Index And Membrane Autopsies. EU-CHINA Workshop on “Membrane-Based Desalination: An Integrated Approach (acronym MEDINA)”, Qingdao (China), 4–6, September 2009.
19. Macedonio F., Drioli E. and Curcio E. Membrane crystallizer as tool for water recovery factor increase and salts recovery from RO/NF brine: simulation of the system and experimental results. EU-CHINA Workshop on “Membrane-Based Desalination: An Integrated Approach (acronym MEDINA)”, Qingdao (China), 4–6, September 2009.
20. Mericq J.-P., Laborie S. and Cabassud C. Vacuum Membrane Distillation For An Integrated Seawater Desalination Process. EU-CHINA Workshop on “Membrane-Based Desalination: An Integrated Approach (acronym MEDINA)”, Qingdao (China), 4–6, September 2009.
21. Rapenne S., Barbe C., Leparc J., Schrotter J. C., Schaule G., de O. Manes C.-L., Lebaron P., Mondamert L., Croue J.-P., Salinas Rodriguez S. and Amy G. Medina Project: Seawater Characterisation and Membrane Autopsies Tools for Better Design and Operation of SWRO plants 5 mars 2009. Thessaloniki, Greece.
22. Salinas Rodríguez S. G., Althuluth M., Schurer R., Kennedy M.D., Amy G.L. and Schippers J.C. Modified Fouling Index (MFI-UF) At Constant Flux For Seawater RO Applications. EU-CHINA Workshop on “Membrane-Based Desalination: An Integrated Approach (acronym MEDINA)”, Qingdao (China), 4–6, September 2009.
23. Salinas Rodríguez S. G., Kennedy M. D., Schippers J. C. and Amy G. L. Organic Foulants In Estuarine And Bay Sources For Seawater Reverse Osmosis – Comparing Pre-Treatment Processes With Respect To Foulant Reductions. EU-CHINA Workshop on “Membrane-Based Desalination: An Integrated Approach (acronym MEDINA)”, Qingdao (China), 4–6, September 2009.
24. Schaule G., Petrowski K., Rapenne S. and Huber S. Assessment Of Biofouling Potential Of Salt And Brackish Water For Ro Desalination Plants. EU-CHINA Workshop on “Membrane-Based Desalination: An Integrated Approach (acronym MEDINA)”, Qingdao (China), 4–6, September 2009.
25. Sim L.-N., Ye Y., Chen V. and Fane A. G. Development Of Crossflow Sampler Modified Fouling Index Ultrafiltration (CFS-MFI_{UF}) For Water Treatment Application. EU-CHINA Workshop on “Membrane-Based Desalination: An Integrated Approach (acronym MEDINA)”, Qingdao (China), 4–6, September 2009.
26. Profio G. D., Ji X., Curcio E. and Drioli E. Exploring MBR technology as pre-treatment in reverse osmosis seawater desalination, WORKSHOP “Membrane technologies for alternative water resources” 5 March 2009. Thessaloniki, Greece.
27. Tansakul C., Laborie S. and Cabassud C. Membrane Hybrid Processes For Pre-Treatment Before Seawater Reverse Osmosis Desalination. EU-CHINA Workshop on “Membrane-Based Desalination: An Integrated Approach (acronym MEDINA)”, Qingdao (China), 4–6, September 2009.
28. Van Der Kooij D. and Veenendaal H. Assessment Of The Concentration Of Easily Biodegradable Compounds In Seawater With Growth Tests Using Halomonas Cupida. EU-CHINA Workshop on “Membrane-Based Desalination: An Integrated Approach (acronym MEDINA)”, Qingdao (China), 4–6, September 2009.
29. Ye Y., Sim L., Chen V. and Fane A. G. Submerged Hollow Fiber Systems As Potential Pretreatment For Reverse Osmosis: Effect Of Operating Modes And Fouling Characterization. EU-CHINA Workshop on “Membrane-Based Desalination: An Integrated Approach (acronym MEDINA)”, Qingdao (China), 4–6, September 2009.

FURTHER PRESENTATIONS

1. Drioli E. Membrane Engineering Contribution for the Water Stress Problem Worldwide, International Conference and Exhibition on Water & Wastewater Treatment (Asia Pro Eco Project Innowa Results), 01–04 April, 2007, Sylhet, Bangladesh.
2. Lattemann S. Environmental aspects, Symposium of the German Desalination Society (DME) and Outokumpu on Thermal Desalination, Essen/Germany, January 2008.
3. Lattemann S. UNEP guidance manual for environmental impact assessment of desalination projects, EUROMED 2008: Desalination Cooperation among Mediterranean Countries of Europe and the MENA Region, Dead Sea/Jordan, November 2008.
4. Lattemann S. Desalination plant impacts on the marine environment, AQUA 2008 Conference, Valencia, October 2008.
5. Lattemann S. Environmental issues in Desalination, EXPO Zaragoza 2008, Thematic week 10: New water resources: reuse and desalination, Zaragoza/Spain, September 2008.
6. Lattemann S. Seawater desalination in the Mediterranean– solution to the region’s water woes? Seminaire international sur le dessalement des aux – percees technologiques et Maitrise des couts, SONEDE, Tunis, July 1st, 2008.
7. Lattemann S. Water Environmental Aspects, Symposium of the German Desalination Society (DME) and Solar Institute Jülich (SIJ) on Desalination and Renewable Energies, Jülich/Germany, June 2008.
8. Lattemann S. Meerwasserentsalzung: Umweltbeeinträchtigungen und Umweltverträglichkeitsprüfung, Kolloquium Series at the Institute for Chemistry and Biology of the Marine Environment (ICBM) of Oldenburg University, Oldenburg/Germany, May 2008.
9. Lattemann S. Desalination’s environmental footprint, International Desalination Association (IDA) and Global Water Intelligence (GWI) Conference on Water, Finance & Sustainability, New Directions for a Thirsty Planet, London/UK, April 2008.
10. Lattemann S. WHO project on Desalination for Safe Water Supply: Guidance for the Health and Environmental aspects applicable to Desalination Projects, IDA World Congress on Desalination and Water Reuse, Maspalomas/Gran Canaria, November 2007.
11. Lattemann S. Desalination Projects – a Need for Environmental Impact Assessment and Impact Mitigation. Inaugural congress of the Wetsus Technological Top Institute for Water Technology, Leeuwarden/The Netherlands, October 2007.
12. Lattemann S. Environmental impact and impact assessment of seawater desalination plants, European Desalination Society (EDS) conference on desalination and the environment, Halkidiki, Greece, April 2007 (keynote).
13. Lattemann S. Environmental Impacts of Seawater Desalination, Symposium of the German Desalination Society (DME) “Introduction to Seawater Desalination”, Halkidiki/Greece, April 2007.

Lightning Source UK Ltd.
Milton Keynes UK
13 April 2011

170858UK00001B/4/P

9 781843 393214